T0181345

Design of Experiments

An Introduction
Based on Linear Models

CHAPMAN & HALL/CRC
Texts in Statistical Science Series

Series Editors
Bradley P. Carlin, *University of Minnesota, USA*
Julian J. Faraway, *University of Bath, UK*
Martin Tanner, *Northwestern University, USA*
Jim Zidek, *University of British Columbia, Canada*

Texts in Statistical Science

Design of Experiments

An Introduction
Based on Linear Models

Max D. Morris

Iowa State University

Ames, U.S.A.

CRC Press
Taylor & Francis Group
Boca Raton London New York

CRC Press is an imprint of the
Taylor & Francis Group an **informa** business

A CHAPMAN & HALL BOOK

Chapman & Hall/CRC
Taylor & Francis Group
6000 Broken Sound Parkway NW, Suite 300
Boca Raton, FL 33487-2742

First issued in paperback 2017

© 2011 by Taylor and Francis Group, LLC
Chapman & Hall/CRC is an imprint of Taylor & Francis Group, an Informa business

No claim to original U.S. Government works

ISBN-13: 978-1-58488-923-6 (hbk)
ISBN-13: 978-1-138-11178-3 (pbk)

This book contains information obtained from authentic and highly regarded sources. Reasonable efforts have been made to publish reliable data and information, but the author and publisher cannot assume responsibility for the validity of all materials or the consequences of their use. The authors and publishers have attempted to trace the copyright holders of all material reproduced in this publication and apologize to copyright holders if permission to publish in this form has not been obtained. If any copyright material has not been acknowledged please write and let us know so we may rectify in any future reprint.

Except as permitted under U.S. Copyright Law, no part of this book may be reprinted, reproduced, transmitted, or utilized in any form by any electronic, mechanical, or other means, now known or hereafter invented, including photocopying, microfilming, and recording, or in any information storage or retrieval system, without written permission from the publishers.

For permission to photocopy or use material electronically from this work, please access www.copyright.com (http://www.copyright.com/) or contact the Copyright Clearance Center, Inc. (CCC), 222 Rosewood Drive, Danvers, MA 01923, 978-750-8400. CCC is a not-for-profit organization that provides licenses and registration for a variety of users. For organizations that have been granted a photocopy license by the CCC, a separate system of payment has been arranged.

Trademark Notice: Product or corporate names may be trademarks or registered trademarks, and are used only for identification and explanation without intent to infringe.

Library of Congress Cataloging-in-Publication Data

Morris, Max, 1950-
    Design of experiments : an introduction based on linear models / Max Morris.
        p. cm. -- (Chapman & Hall/CRC texts in statistical science series)
    Includes bibliographical references and index.
    ISBN 978-1-58488-923-6 (hardcover : alk. paper)
    1. Experimental design. 2. Linear models (Statistics) I. Title. II. Series.

QA279.M67 2010
519.5'7--dc22   2010022635

Visit the Taylor & Francis Web site at
http://www.taylorandfrancis.com

and the CRC Press Web site at
http://www.crcpress.com

Dedication

*With love and thanks to my two favorite teachers,
Alice Jane and Cecil.*

Contents

Contents

Preface

Experimental design is a subject that can be successfully taught at many levels, and good textbooks have been written for many different audiences. A few of the excellent and frequently cited texts on the subject are by Cobb (1999), which assumes little or no mathematical or statistical background; Hinkelmann and Kempthorne (2005), written primarily for advanced students of statistics; Box, Hunter, and Hunter (2005), with specific appeal to engineers and physical scientists with some knowledge of statistics; and Dean and Voss (1999) for more general audiences at the advanced undergraduate or beginning graduate levels. Some texts such as the book by Wu and Hamada (2009) describe experimental designs for several specific settings, while others such as Cox and Reid (2000) focus more on general principles.

In graduate programs in statistics, courses in experimental design are often electives to be taken after students have completed a few "core courses" in theory and methods. This is the audience for which this book is primarily intended. In a course taught at this level, connections between the structure of an experimental design and the performance of a data analysis can be made specific through connection of, for example, the balance and blocking properties of the design, to the model matrices, to the form of a noncentrality parameter, to the power of an F-test. The fact that orthogonal blocks can be "ignored" in the estimation of treatment contrasts can be explicitly discussed through the matrix form of the reduced normal equations. Previous exposure to the basic general linear model allows a more unified presentation of design ideas using, for example, design information matrices, than is practical at a more elementary level. A brief review of some of the more relevant elements of general linear models is given in Chapter 2.

But while the intent is to offer a presentation that takes advantage of about one year of graduate-level "statistical maturity," the goal is certainly *not* to present a "theoretical" course. The ideas and methods presented in a first year of graduate study, including introduction to general linear models, equip students to understand the basic concepts and techniques of experimental design at a deeper level than would have been possible one year earlier. The intent of this book is to use that background knowledge to help students more fully appreciate the fundamental concepts and techniques of experimental design, but not to reduce it to a theoretical exercise. To emphasize the "real world" value of design, most chapters contain a short reference example to a relevant experiment described in the scientific or technical literature.

Since many of the analysis methods used in experimental statistics are fairly elementary, second-year graduate students will have already been introduced to the most commonly used data analysis techniques associated with basic designs. As a result, data analysis is not a central focus of this book. Chapter 6 does review some fundamental techniques concerning residuals and a few methods of particular value in experimental statistics, and later chapters develop specific forms of ANOVA decompositions and standard errors, as appropriate. But the emphasis is on how design influences the *quality* of basic data analysis that should already be familiar, through the resulting degrees of freedom, noncentrality parameters, and the form of standard errors.

Statistical computing is an important component of almost any aspect of applied statistics, but it is a tool to be knowledgeably used, rather than a substitute for knowledge of methodology. Furthermore, statistical computing can be accomplished in many ways, and none of the currently available options is necessarily best or most relevant to a course in experimental design. Still, students who have minimal experience with statistical computing tools may need some hints regarding how to get started. While not a focus of the book, each demonstration calculation was performed using R, with details provided in Appendix A along with a brief overview of the commands most often used here. These are referenced throughout the text with parenthetical pointers "(R#.#)". A few end-of-chapter exercises ask students to write programs to perform small simulation studies or solve simple optimization problems; these can be avoided in classes where students have not had programming experience.

The material in Chapters 1–13 forms the core of a second-year M.S.-level course. Material from Chapters 15 and 16 can be added, as time permits, as an introduction to design for regression models. Chapter 14 (Factorial Group Screening Experiments) is an optional topic that can be included if it matches the interests of the class, and Chapter 17 (Introduction to Optimal Design) may be of particular interest to students planning further study of experimental design.

Finally, I want to acknowledge the contributions made to this work by the 300 or so graduate students who have taken Statistics 512 from me at Iowa State University over the last decade. As all educators know very well, teaching is a two-way process. Through their feedback, questions, and discussion, my students have taught me much about how to think and talk about experimental design, and I owe them thanks for their resulting contributions to this effort. That having been said, I take full responsibility for anything that could have been said better, and of course, for any errors in this presentation.

Max Morris

Ames, Iowa

2010

CHAPTER 1

Introduction

Statistical analysis is an approach to answering questions using data. In many contexts, this is done through techniques designed to detect or understand interesting patterns or *signals* hidden in, or masked by, uninteresting *noise*. Empirical studies based on "observational" or "available" data are sometimes inconclusive or misleading, because the noise component of such data may be very substantial and its sources not well understood. In order to avoid these shortcomings, *experiments* are sometimes performed with the intent of facilitating the separation of signal from noise, clarifying the first and minimizing the ambiguity caused by the second, to an extent that cannot be achieved in observational studies. As a result, experiments are usually carried out under circumstances that are at least somewhat artificial, and must be interpreted carefully when inferences are to be drawn about the "real world." But where they are feasible and can be reasonably interpreted, carefully conducted experiments often provide the best opportunity to understand complex phenomena of interest.

In experimental contexts, data usually result from measurement processes of one sort or another. The questions of interest take various forms, and depending on their context may be addressed by statistical "answers" in the form of estimates, hypothesis tests, decisions, or less formal analyses. The quality of the statistical answers obtained through experiments, characterized by quantities such as the bias and standard errors of estimators and the power of tests, depends on many things. Some of these, e.g., the relative magnitudes of signal and noise in the system under study, usually lie outside of the influence of the investigator. But choices *are* often available concerning sample sizes, organization of experimental material, specific experimental conditions included in the study, and other aspects of the planning and control of the experiment. In experimental studies, "planning and control" includes the specification of operational details that are usually unknown or ignored in observational studies. *Experimental design* is an approach to arranging these operational details so that the quality of the statistical answers to be derived from the data is as high as possible.

1.1 Example: rainfall and grassland

Fay et al. (2000) described an experiment performed to investigate how rainfall patterns associated with predicted global climate changes might affect prairie tallgrass. The study involved the use of fifteen 9×14 m "rainfall manipulation shelters" – open-air structures with translucent roofs, rainfall collection systems, and sprinkler systems allowing for simulated rainfall on the grass and soil beneath the roof. *"Four rainfall manipulation treatments (three replicates) then were assigned to 12 of the plots ... The three remaining plots [shelters with the roofs removed] serve as unsheltered controls for effects caused by the shelters and irrigation systems."* The four experimental *treatments* (not counting the unsheltered controls, which would sometimes also be called a treatment) are the artificial environments created within the structures, described as follows:

1. *"Natural quantity and interval. This treatment replicates the naturally occurring rainfall regime. Each time a natural rainfall event occurs, the quantity of rain that fell is immediately applied to the plots.*

2. *Increased interval. Rather than immediately applying rainfall as it occurs, rainfall is stored and accumulated to lengthen the dry intervals by 50%. The accumulated rainfall is applied as a single large event at the end of the dry interval. Over the season, the naturally occurring quantity of rainfall is applied, but the timing of events is altered, repackaging the rain into fewer, larger events.*

3. *Reduced quantity. In this treatment, only 70% of the naturally occurring rainfall is immediately applied. This imposes reduced amounts of rainfall without altering the timing of rainfall events.*

4. *Reduced quantity and increased interval. Rainfall intervals are lengthened by 50%, and only 70% of the accumulated rainfall is applied, imposing both reduced quantity and altered timing of events."*

The authors report that 14 different kinds of measurements were made on a variety of timing schedules to assess the productivity and other indicators of health of the grass within the area of each shelter. Aboveground net primary productivity (ANPP) and soil CO_2 flux were two of these measurements or responses: *"Soil CO_2 flux was measured with a closed-flow gas exchange system ... ANPP was estimated from dry weights of early November harvests of all aboveground vegetation."*

1.2 Basic elements of an experiment

In this or any experiment, and with any form of inference, some assumptions are necessary to enable the drawing of general conclusions from specific observations. In the best case, assumptions may reflect known physical facts about a particular experiment. In other settings, they reflect general knowledge of what has often occurred in previous related experiments. The least desirable assumptions are statements that are accepted primarily for reasons of convenience or practical necessity. A primary goal of experimental design is to arrange the operational details of the experiment to be consistent with these assumptions, and to the degree possible, to enhance the measurable quality of the analysis. The following sections briefly describe a few of the operational issues that are most important in this regard, and are common in some form to almost all experiments.

1.2.1 Treatments and material

The classical ideas in the study of statistical experimental design are formed around the management of *experimental treatments* and *experimental material*. Treatments generally represent the entities of scientific interest in the study; the four experimental treatments included in the rainfall experiment are specified by the quantity of rainfall delivered and time between rainfall events. However, these four treatments really are *just* rainfall specifications (relative to natural precipitation during the experiment), or "recipes." They don't represent anything physical, and so cannot be "evaluated" through data analysis until they are carried out using experimental material – in this case the tallgrass and soil beneath each of the shelters. The language of experimental design has been developed using a general concept that treatments are *applied to* experimental material. Alternatively, it is sometimes convenient to say that experimental material is *allocated* or *assigned to* a treatment. Each application of a treatment to a quantity of experimental material is called an *experimental trial*, or sometimes a *run*. The effects of the specific treatments on the responses are of interest; the effects of the specific quantities of material used in the experiment are not, but it is understood that the experimental material can also introduce variation in the experimental process. This material-induced variability is a major component of the "noise" in many experiments and so it must be managed carefully.

For purposes of statistical modeling, treatments can be represented in a number of different ways. In this book, the treatment representations, or treatment *structures*, discussed are described as *unstructured, factorial,* and *functional*. By *unstructured*, we mean a collection of treatments that are generally represented by a single nominal variable, or by a collection of indicator variables, one associated with each treatment. This is the most general treatment structure we discuss, but it is also the least physically meaningful because the "coding" implies nothing about relationships among treatments. In other instances, treatments are characterized by a set of *factors*, with any particular treatment defined by selecting a *level* for each factor. In the rainfall experiment, the four treatments of primary interest are defined by setting the "quantity" factor to its "natural" or "reduced" level, and the "interval" factor to its "natural" or "increased" level, where each of these terms is given precise definition in the referenced paper. A factorial representation implies specific relationships among treatments that have physical meaning, and leads to statistical designs and models that naturally address the most important aspects of the experimenter's questions. By *functional* treatments, we mean treatments that are identified by a set of (usually) continuous controlled variables. We might use this designation for the rainfall experiment had the "quantity" characteristic been specified by inches of rainfall per event, and the "frequency" characteristic been specified by days between events. Functional relationships potentially represent substantial information about treatment structure; with appropriate selection of models, they can often provide the basis for understanding treatments not included in the experiment. It is important to understand that there is often a choice of how the treatments of a particular experiment will be represented in statistical modeling. However, these three categories offer a convenient framework for introducing and discussing the material in Chapters 3 to 8, 9 to 14, and 15 and 16, respectively.

When a treatment is physically applied to material, the specific quantity of material used in one application is called an *experimental unit*, or usually simply a *unit*. In the rainfall experiment, the soil and biomass beneath one shelter form an experimental unit; application of one of the treatments is the physical process by which one of the rainfall recipes is simulated. This terminology fits the conceptually simple rainfall experiment well, but it may not be so clear in other contexts. Sometimes there is also physical material actually associated with each treatment, such as the addition of a fertilizer to a test plot in an agricultural study. In other cases, the experimental material may not seem to be physical at all, such as in an experiment to determine which of two computer programs executes more quickly in a specified computing environment. There are also situations in which the most important component in the definition of an experimental unit is the period of time in which the treatment is applied. These considerations can introduce complications in the selection of experimental material, but in most experiments the basic idea of the experimental unit is applicable in some form.

1.2.2 Control and comparison

The rainfall experiment described by Fay et al. is a *controlled* study in which data are collected through a planned sequence of activities. It is also a *comparative* study focused on whether there are *differences* between the properties of the tallgrass plots that can be associated with the different treatments. (In contrast, a study in which only one of the simulated rainfall recipes is used could not be comparative.) These two properties are closely related. Experiments are controlled as carefully as possible so as to isolate the differences between the treatments of interest, and to minimize extraneous variability so as to enable the sharpest possible statistical analyses (e.g., narrow confidence intervals or powerful tests). In many instances, this high degree of control means that the data collected are actually representative of only a very special situation, reflecting the particular laboratory procedures, batch of experimental material, et cetera, used in the performance of the experiment. As a result, meaningful inferences usually need to be based on comparisons *within an experiment*, with the idea that anything unusual, but common, to all trials in the experiment will "cancel out" in the analysis. This emphasis on comparative structure and its implication for experimental design and analysis is a recurring theme throughout this book. (See Youden's (1972) classic paper for a historical account of difficulties encountered in the experimental estimation of "absolute," rather than comparative, quantities.)

This emphasis on comparison often leads to the inclusion of one or more *experimental controls*, or simply *controls*, in the experimental plan. For example, in addition to the four carefully defined "experimental treatments" in the rainfall experiment, three "sham shelters" were left open to natural rainfall so as to provide a comparison to what might have happened without experimental manipulation. In this context, the actual level of ANPP (say) in a given control area is not especially interesting; it is simply a reflection of what happens to the tallgrass in the corresponding unit, during the time period of the study, if there is no experimental interference. But the control is included in the experiment so that there is an internal "benchmark" against which the more interesting experimental treatments can be compared. In this case, the most important comparison to the control might involve the "natural quantity, natural interval" treatment. A large difference between responses for these two conditions could indicate unanticipated influences of the experimental procedure *per se*; a small or negligible difference might be viewed as support for the investigators' intent that the "natural/natural" treatment be a reasonable representation of the current environment. In an experiment involving the growth of cell cultures in response to hormone exposure, the control might consist of cultures grown without the addition of any hormones, while the more interesting cultures would be grown in the presence of one or more hormones at specified concentrations. *On their own*, experimental controls usually represent conditions that are of little real scientific interest, but including them in the experiment gives us the ability to directly compare treatments that represent both well-understood and novel conditions, to see how the differences affect responses.

1.2.3 Responses and measurement processes

In order to determine the effect of treatments applied to units, we need some means of evaluating the result of that application. This is accomplished by obtaining values of one or more *responses* – the data to be analyzed – for each experimental unit. If an experiment is well designed, the resulting response values can be statistically analyzed to answer the questions that led to the study. Selection of appropriate response variables is as critical to the success of the experimental program as any other decision made. Units may be selected with the greatest care, and treatments defined and applied with precise control, but unless the responses are relevant to the questions at hand the experiment will be of little or no practical value. The 14 biological variables monitored by Fay et al. in the rainfall study reflect the specific aspects of tallgrass growth and health of interest to them.

Having said this, we must also understand that in all but the simplest studies, response values don't just appear, but must be determined by what we shall call a *measurement process*. These processes may be relatively straightforward, such as the weighing of biomass to determine ANPP, or more complex, such as the use of the closed-flow gas exchange system to determine CO_2 flux. However, the measurement process, like the collection of experimental units used in the experiment, is generally of no real interest to the experimenter; it is only an operational means to obtain response values that *are* of interest. Hence, just as uninteresting units must be available if interesting treatments are to be applied, an uninteresting measurement process will be required to produce interesting response data. And, just as selection of units may have some influence on the experimental results, so the particular way in which the measurement process is employed can have an effect on the data. (Of course, the collection or production of good experimental units and good measurement techniques certainly *are* interesting in other contexts, but these are different from the context of an experiment carried out to study the effects of treatments.)

1.2.4 Replication, blocking, and randomization

Control and comparison are two experimental devices used to reduce sources of variability that are uninteresting or irrelevant in the context of the questions being asked. However, these strategies almost never *completely* eliminate the undesirable noise in the data to be collected. *Replication, blocking*, and *randomization* are aspects of statistical design that are used to further reduce extraneous variation in responses. Experiments make use of replication when they contain multiple trials that are executed under circumstances that are nominally identical. The word "nominally" refers to the degree of experimental control that can actually be exerted in the study. For example, the rainfall experiment contains replication because each specified rainfall condition (treatment) was simulated in three different rainfall manipulation shelters. We know that the three areas of sod receiving any one treatment are *physically different* and so are not likely to result in responses that are identical, but in a well-controlled experiment they will be prepared in a manner and with material that is as homogeneous as possible. Furthermore, collecting data from these three units will require three separate uses of a measurement process; some degree of random *measurement error* is a source of variability in most real experiments, the effects of which may be minimized through good experimental design. Within this degree of attainable control, replication effectively reduces the random variation or noise in the comparisons examined in the analysis, and provides an opportunity to estimate the typical size of this random component in individual measurements.

Some other sources of uncontrolled variation might not be reasonably characterized as random. If a potential source of systematic variation is known, the experiment can sometimes be designed in *blocks* to minimize its effect. For example, the rainfall experiment was actually conducted in three complete blocks of five shelters/plots. Within a block, the five shelters were physically located in close proximity, and each of the treatments (four simulated conditions and one control) was applied to one unit in each block. Distances between blocks were large relative to the distances between units within a block. The blocking strategy was used to facilitate "correction" for any systematic differences associated with block locations.

The decision to organize an experiment in blocks often follows from a recognition that experimental units may be available in recognizable "batches," such that units from the same batch tend to be more similar than units from different batches. For example, in the cellular biology experiment mentioned above it might be necessary to incubate the cell cultures in several blocks due to equipment limitations on the number of cultures that can be processed together. The practical concern here is that when the same experimental treatment is applied to two units from the same batch, the corresponding responses are likely to be at least somewhat more similar than when the same experimental treatment is applied to two units from different batches. Ignoring such differences in the analysis of an experiment can call into question any assumption that random effects are "independently and identically distributed." An analysis that accounts for potential systematic differences associated with batches/blocks – following an experimental design arranged to make such an analysis efficient – can ensure the validity of the conclusions. In still other cases, all the physical material needed to perform an experiment may be relatively homogeneous, but each experimental trial may be time-consuming; it may be feasible to complete (say) four trials in a day, leading to a 32-run experiment that requires 8 days of laboratory time. If the measurement instruments used are calibrated at the beginning of each day, the laboratory personnel are different each day, the temperature and/or humidity in the laboratory are different each day, et cetera, it may be prudent to regard the four units processed in one day as a block so as to account for any effects such systematic differences may have on the measured responses.

In still other cases, the sources of undesirable variation may not be associated with patterns that can be identified beforehand. Fay et al. make use of randomization to avoid any uncontrolled systematic differences that may exist between units within a block, by randomly pairing the five treatments to the five shelters in each block. (This is an example of the Randomized Complete Block Design discussed in Chapter 4.) Here, the physical identity of each unit is not necessarily the important point; protection is sought from *any* unforeseen systematic differences among the shelters and their contents other than the differences intentionally introduced by applying the rainfall recipes. The assignments are made *randomly* so that any of the 5! possible pairings of treatments to units within a block is equally likely. Most experimental designs we discuss will incorporate both blocking and randomization to control possible patterns of variation that can or cannot, respectively, be anticipated when the experiment is designed.

1.2.5 Validity and optimality

In a sense, one goal of experimental design is always to determine a specific plan for dealing with the treatments of interest and the material that can be used, that is best or *optimal* in some sense that relates to the study objectives. However, the *first* goal of experimental design – the one most often stressed in introductory discussions about unit and block selection – is that the anticipated analysis be *valid*, even if not entirely optimal. Randomization and blocking are employed *first* to assure that the analysis to be used accurately reflects the physical sources of variation, whether of interest or not, that may be present in the data.

Given that the general structure of the design satisfies the goals of validity and is consistent with the relevant operational constraints, the *second* goal of experimental design is to lead to a statistical analysis that is optimal or near-optimal – for which tests of relevant hypotheses are as powerful as possible, or confidence intervals of interest have the smallest possible expected lengths. In the simplest cases, design optimality is addressed through controlling sample sizes, that is, determining the number of units *allocated* to each treatment so as to result in estimates of desired precision or tests of desired power. Historically, much of the theory of optimal experimental design was developed after the fundamental work on blocking and randomization was carried out. Still, the two objectives are not mutually exclusive. For example, an experiment may be set up in blocks of homogeneous units to ensure validity, while the number of units allocated to each treatment within a block might be chosen to minimize variability in the estimation of the expected differences between a response under various treatments.

1.3 Experiments and experiment-like studies

In contrast to the carefully controlled rainfall experiment described in Section 1.1, suppose a health research organization was interested in conducting a study to assess the effect of having experienced a heart attack on certain indices of health, such as blood pressure, in adults. This might be done by recruiting some people (who would take the role of "units" in experimental terms) who have had heart attacks, and others who have not (the two "treatments" of interest) for comparison. In such studies, a kind of blocking can sometimes be achieved by recruiting pairs of siblings, one of each category, to reduce variability in the comparisons of interest.

Such a study might, if carefully conducted, yield valuable information on the connection between heart attacks and subsequent health, but it would not be an experiment of the kind we are discussing because it cannot allow for the randomized application of treatments to units. The individuals come to the study already labeled as "previous heart attack" or "no previous heart attack." This means that there is a chance that unrecognized bias can be present in the comparisons of interest. For example, suppose that individuals with a certain genetic trait were actually more likely to suffer heart attacks than those who do not have the trait. If this were not known by the study designers, it is likely that the genetic trait would be more prevalent in the members of the "previous heart attack" group, and *this* might be responsible for any observed differences we mistakenly interpret as being due to having had a heart attack. Were it possible to randomly assign people to the two treatments in a true experiment, systematic biases of this kind could be avoided.

Epidemiologists who undertake nonexperimental studies are extremely careful to account for such known or suspected patterns in the selection of individuals. Researchers in the social sciences who rely on surveys often face similar challenges. In many cases, the kind of data analysis performed is very similar to what would be used in the analysis of experimental data, and these analyses can often lead to valuable information. Nevertheless, there is always more risk involved in the interpretation of data from these "pseudo-experiments" than in the case of true experiments, because we can never be completely sure that all potential *confounders* (a term used in epidemiology) have been accounted for or eliminated.

1.4 Models and data analysis

Our approach to experimental design is motivated largely by its relationship to the quality of data analysis that can be performed. In order to be specific about performance, we develop and examine designs in the context of the general linear model. Forms of the model vary with different designs and treatment structures, and depend on the nature of potential sources of variability. For most of the designs we cover, a fixed effects model (i.e., only one random term) will be adequate, but split-plot designs (Chapter 10) will require a mixed effects model. All designs are formulated with the idea that the motivating questions can be answered through inferences about the fixed model effects. In many chapters, we describe graphical summaries of experimental data that are primarily motivated by an assumed linear structure in the data. Chapter 2 is a brief review of the results we will use from the theory of linear models.

Having said this, it should also be noted that there certainly are many important questions for which experimental information about variances, quantiles of distributions, and other statistical indices are informative. There are also important experimental settings in which the most appropriate data models are not linear. In fact, the designs we discuss here, and some variants of them, are often useful in these situations also. We won't attempt to offer every statistical setting that might serve to motivate any one design. But the general linear model provides a rich framework for studying how basic experimental designs "work."

1.5 Conclusion

The chapters of this book describe specific forms or *classes* of experimental designs. These differ in the details of how units, blocks, replication, and randomization are managed, and how the effects of treatments are represented. The scientific and operational characteristics of an experimental context usually determine which class or classes of designs can best assure the validity of the intended analysis. Within a class of designs, a particular design can often be selected by balancing statistical performance (optimality) with operational demands (e.g., cost). While the details are always specific to the application, the fundamental issues associated with these details are common to nearly all experiments. The intent of this book is to present an organized framework within which the statistical aspects of experimental design can be understood as a "whole" within the structure provided by general linear models, rather than as a collection of seemingly unrelated solutions to unique problems.

It should be admitted from the outset that many aspects of practical experimental design are *not* covered in this book. Coleman and Montgomery (1993) describe a useful approach to making many of the decisions that must be made *before* a specific experimental design can be formatted, such as the selection of variables and measurements to be considered. Barton (1999) also addresses some of these issues, and describes ways that graphical representations can be helpful throughout the process of planning an experiment. The designs and ideas discussed in this book have proven helpful in a wide variety of experimental contexts, but successful experimentation in any *specific* case requires careful communication involving all those involved. The statistical goal must always be to ensure that a design has been formulated that fits the problem at hand, not the reverse.

1.6 Exercises

1. A classic and famous example of a hypothetical and simple, but illuminating, experiment was offered by R.A. Fisher (1971):

   "A lady declares that by tasting a cup of tea made with milk she can discriminate whether the milk or the tea infusion was first added to the cup. We will consider the problem of designing an experiment by means of which this assertion can be tested. For this purpose let us first lay down a simple form of experiment with a view to studying its limitations and its characteristics, both those that appear to be essential to the experimental method, when well developed, and those that are not essential but auxiliary.

   Our experiment consists in mixing eight cups of tea, four in one way and four in the other, and presenting them to the subject for judgment in a random order. The subject has been told in advance of what the test will consist, namely that she will be asked to taste eight cups, that these shall be four of each kind, and that they shall be presented to her in a random order, that is in an order not determined arbitrarily by human choice, but by the actual manipulation of the physical apparatus used in games of chance, cards, dice, roulettes, etc., or, more expeditiously, from a published collection of random sampling numbers purporting to give the actual results of such manipulation. Her task is to divide the 8 cups into two sets of 4, agreeing, if possible, with the treatments received."

   (a) Carefully identify the units in this experiment.

   (b) Carefully define the treatments in this experiment.

   (c) Fisher mentioned several physical devices that might be used to determine a random temporal order of treatments to the available units. Carefully describe exactly how this might be done using any of these devices, while honoring the sample size constraints of the problem.

(d) Suppose eight (physical cups) are available for the execution of this experiment, but they are from two sets. Four are made from heavy, thick porcelain, while the other four are made from much lighter china. If operational restrictions are such that each cup can only be used once, how might this fact be incorporated into the experimental design?

2. Suppose you are interested in learning which of three "recipes" for making a paper airplane results in (actual) paper airplanes with the greatest average length-of-flight when launched from a height of five feet above the floor in a draft-free room. Design a simple, unblocked experiment, involving the construction and testing of five airplanes of each kind, to address this question. Write one concise paragraph (including rough sketches where this is helpful) to describe each of the following:

   (a) The three experimental treatments you decide to compare.

   (b) The experimental units you will need, and what you will do to see that these units are as homogeneous as possible.

   (c) A randomization process for pairing units with treatments, and for determining the time-order of the experimental trials, assuming you can only test one airplane at a time.

   (d) A procedure for applying a given treatment to a given unit (remembering that paper airplanes don't "just happen," and that you want to introduce as little airplane-to-airplane "noise" as possible).

   (e) A measurement process, selected to be relevant to your experimental goal, offering as much control (e.g., introducing as little measurement error) as reasonably possible.

3. Suppose you wish to conduct a study of three brands of gasoline to see which leads to the best mileage in your car. You contemplate three different studies to accomplish this:

   • Design A: Poll 100 persons drawn randomly from a phone book, asking each of them what brand of gasoline they generally buy and what mileage they generally experience in their cars.

   • Design B: Over a period of three months, use only one of the three brands each month and record the mileage you experience in your car.

   • Design C: Over a period of three months, rotate among Brand 1, Brand 2, and Brand 3 with each fill-up and carefully read the mileage you experience in your car. (Assume you must buy gasoline about once per week.)

   (a) Compare the strengths and weaknesses of each of these study plans relative to the others.

   (b) Which (if any) of the proposed studies is a "true experiment," and why?

4. Suppose that a certain baseball league contains eight teams, and that each pair of teams plays the same number of games in a season. The season can be thought of as an experiment carried out to determine which team is best. We might make this idea somewhat more precise (and artificial!) by defining the "best" team to be the one that would have the greatest expected number of victories in a single round-robin tournament. In this setting, identify the treatments and units, and describe what it means for a treatment to be "applied to" a unit. (Remember that only *one* treatment can be applied to a unit.) How might randomization and replication be used? Does blocking seem applicable in this context, and if so, how?

5. As a home gardener, you would like to perform an experiment to determine which combination of two kinds of fertilizer and three varieties of tomato plants yields the largest crop of tomatoes (per plant, in weight of fruit over the entire growing season). Describe how you might conduct this experiment. How might you use replication and randomization to make the experiment more efficient? If you continued your experiment for a second season, what additional design principle(s) discussed in this chapter would be relevant?

## CHAPTER 2

Linear statistical models

In this book, much of the discussion about the properties of experimental designs is presented in the context of the general linear statistical model, usually in partitioned form. It is assumed that the reader has been introduced to the fundamental results and techniques associated with the matrix form of linear statistical models. This chapter presents an overview of the main results and some notation used in this text. Textbooks on linear models by Christensen (2002) and Ravishanker and Dey (2002) are good references for readers who need a more thorough treatment of this subject. In addition, Graybill (1983) and Harville (2008) are good resources for matrix algebra used in linear models.

2.1 Linear vector spaces

Much of the structure and mathematics associated with experimental design and linear models is most easily described and discussed in terms of linear vector spaces, usually of dimension N, the number of response values obtained in the experiment. A specific N-dimensional linear vector space (or, in this book, just "vector space") is generally defined by a *spanning set* of p N-element vectors

$$\{\mathbf{v}_1, \mathbf{v}_2, \mathbf{v}_3, \ldots, \mathbf{v}_p\}.$$

The vector space itself is the collection of all N-dimensional vectors that can be formed as a linear combination of the vectors of the spanning set, that is, all vectors $\mathbf{v}$ that can be expressed as

$$\mathbf{v} = c_1\mathbf{v}_1 + c_2\mathbf{v}_2 + c_3\mathbf{v}_3 + \cdots + c_p\mathbf{v}_p$$

for any collection of real-valued constants $c_1, c_2, c_3, \ldots, c_p$. When the vectors in a spanning set are *linearly independent*—that is, when none of them can be formed as a linear combination of the others—the spanning set is also called a *basis*. So, for example, the vector space *spanned by* the basis

$$\left\{ \begin{pmatrix} 1 \\ 0 \\ 1 \end{pmatrix}, \begin{pmatrix} 1 \\ 1 \\ 0 \end{pmatrix} \right\}$$

is the set of all 3-dimensional vectors in which the first element is the sum of the second and third elements. When there are $p = N$ vectors in the basis the vector space contains all N-dimensional real-valued vectors, and is often denoted by $\mathcal{R}^N$. When there are fewer than N linearly independent N-dimensional vectors in the spanning set or basis, the vector space does not contain all vectors in $\mathcal{R}^N$.

Two vectors $\mathbf{w}$ and $\mathbf{v}$ of the same dimension are said to be *orthogonal* if $\mathbf{w}'\mathbf{v} = 0$. Let S be a vector space in $\mathcal{R}^N$, and let $\mathbf{w}$ be an N-dimensional vector such that $\mathbf{w}'\mathbf{v} = 0$ for every $\mathbf{v} \in S$. The set of all such vectors $\mathbf{w}$ is a second vector space called the *orthogonal complement* of S in $\mathcal{R}^N$, or sometimes just the complement of S.

In most of our discussion, the vector spaces of interest will be those defined by spanning sets consisting of the columns of model matrices, and the complements of these vector spaces. We will sometimes say "the column space of $\mathbf{X}$" to mean the vector space associated with the spanning set made up of columns from that matrix.

2.2 Basic linear model

The basic form of a statistical linear model of N related measurements or statistical observations is:

$$\mathbf{y} = \mathbf{X}\boldsymbol{\theta} + \boldsymbol{\epsilon} \tag{2.1}$$

where $\mathbf{y}$ is the N-element column vector of observable responses, $\mathbf{X}$ is an N-by-k model matrix of (in our context) controlled quantities representing the details of the experimental design, $\boldsymbol{\theta}$ is a k-element column vector of unknown parameters, and $\boldsymbol{\epsilon}$ is an N-element column vector of random variables, each with mean zero. Except as noted, we assume that $N \geq k$. The questions motivating the execution of an experiment are generally answered, at least partially, by estimation of, or hypothesis tests about, the elements of $\boldsymbol{\theta}$. The elements of $\boldsymbol{\epsilon}$ represent the statistical "noise" associated with their counterparts in $\mathbf{y}$. The simplest and most common assumption about the second moments of $\boldsymbol{\epsilon}$ is that the elements are uncorrelated and have equal, generally unknown variance, i.e., $Var(\boldsymbol{\epsilon}) = \sigma^2\mathbf{I}$; we take this assumption as given in all that follows except where noted. Sometimes it is additionally assumed that the elements of $\boldsymbol{\epsilon}$ follow a normal distribution. However, when N is large, and under fairly general conditions, Central Limit Theory can be used to asymptotically justify most finite-sample distributional results stemming from the normal-errors assumption. In practice, equality of variance and independence between observations are usually more critical than normality to the validity of the analysis.

2.3 The hat matrix, least-squares estimates, and design information matrix

Given any experimental design and model, and the resulting model matrix $\mathbf{X}$, the associated "hat" matrix is defined as

$$\mathbf{H} = \mathbf{X}(\mathbf{X}'\mathbf{X})^-\mathbf{X}'$$

where the superscript "$-$" denotes a generalized inverse of its matrix argument. In turn, a generalized inverse of a square, symmetric matrix $\mathbf{A}$ is any matrix $\mathbf{A}^-$ that satisfies $\mathbf{A}\mathbf{A}^-\mathbf{A} = \mathbf{A}$. The largest number of linearly independent columns of $\mathbf{A}$ is called the *rank* of $\mathbf{A}$, denoted rank($\mathbf{A}$). When $\mathbf{A}$ is a matrix of full rank, i.e., when rank($\mathbf{A}$) is the dimension of $\mathbf{A}$, the only generalized inverse is the unique matrix inverse $\mathbf{A}^{-1}$; when $\mathbf{A}$ is of less than full rank, $\mathbf{A}^-$ is not unique. In our context, $\mathbf{X}'\mathbf{X}$ is of less than full rank when one or more of the columns of $\mathbf{X}$ can be expressed as a linear combination of the remaining columns. The largest number of linearly independent columns of $\mathbf{X}$, rank($\mathbf{X}$), is also rank($\mathbf{X}'\mathbf{X}$). So, when rank($\mathbf{X}$) = rank($\mathbf{X}'\mathbf{X}$) < k, $(\mathbf{X}'\mathbf{X})^-$ is not unique. However, $\mathbf{H}$ *is* unique even in this case; that is, $\mathbf{X}(\mathbf{X}'\mathbf{X})^-\mathbf{X}'$ is the same matrix regardless of the choice of generalized inverse $(\mathbf{X}'\mathbf{X})^-$ used in evaluating it. The *trace* of a square matrix such as $\mathbf{X}'\mathbf{X}$ or $\mathbf{H}$, denoted trace($-$), is the sum of its diagonal elements.

$\mathbf{H}$ is an important factor in many statistical formulae, and it is a matrix that has a number of special properties, including the following:

- $\mathbf{H}$ is symmetric, that is $\{\mathbf{H}\}_{ij} = \{\mathbf{H}\}_{ji}$.
- $\mathbf{H}$ is idempotent, that is $\mathbf{H}^2 = \mathbf{H}$.
- rank($\mathbf{H}$) = rank($\mathbf{X}$) = trace($\mathbf{H}$).

In thinking about the role of the hat matrix in statistical expressions, it is especially helpful to understand that $\mathbf{H}$ is the *projection operator* associated with the column space of $\mathbf{X}$. This means that for any N-element vector $\mathbf{v}$, $\mathbf{w} = \mathbf{H}\mathbf{v}$ is as close as possible to $\mathbf{v}$ in the least-squares sense (that is, it minimizes $(\mathbf{w}-\mathbf{v})'(\mathbf{w}-\mathbf{v})$) subject to the constraint that $\mathbf{w}$ must be expressible as a linear combination of the columns of $\mathbf{X}$ (that is, $\mathbf{w} = \mathbf{X}\mathbf{z}$ for some k-element vector $\mathbf{z}$, or $\mathbf{w}$ is in the column space of $\mathbf{X}$). This, in fact, is where the "hat" designation has its origin. Letting the data vector $\mathbf{y}$ play the role of $\mathbf{v}$, the least-squares estimate of $E(\mathbf{y})$ is $\hat{\mathbf{y}} = \mathbf{H}\mathbf{y}$, and any vector $\mathbf{w}$ for which $\hat{\mathbf{y}} = \mathbf{X}\mathbf{w}$ is a least-squares estimate of θ. This implies that the vector of least-squares coefficient estimates, $\hat{\theta}$, must satisfy $\mathbf{X}\hat{\theta} = \mathbf{H}\mathbf{y}$. Pre-multiplying this equation by $\mathbf{X}'$, and realizing that $\mathbf{X}'\mathbf{H}$ must be $\mathbf{X}'$ (since multiplying by $\mathbf{H}$ does not change vectors already in the column space of $\mathbf{X}$) yields the usual form of the *normal equations* for $\hat{\theta}$,

$$\mathbf{X}'\mathbf{X}\hat{\theta} = \mathbf{X}'\mathbf{y}. \tag{2.2}$$

The matrix complement of $\mathbf{H}$, $\mathbf{I} - \mathbf{H}$, also possesses the special properties listed above, and is the projection operator associated with the complement, within $\mathcal{R}^N$, of the column space of $\mathbf{X}$. This means that for any vector $\mathbf{v}$, $\mathbf{w} = (\mathbf{I} - \mathbf{H})\mathbf{v}$ is as close as possible to $\mathbf{v}$ in the least-squares sense, subject to the constraint that $\mathbf{w}$ must be *orthogonal* to every column in $\mathbf{X}$ (that is, $\mathbf{X}'\mathbf{w} = \mathbf{0}$). In linear statistical modeling, $(\mathbf{I} - \mathbf{H})\mathbf{y} = \mathbf{y} - \hat{\mathbf{y}} = \mathbf{r}$ is the vector of residuals from the least-squares fit.

The normal equations (2.2) can be further rewritten by substituting the model expression (2.1) for $\mathbf{y}$:

$$\mathbf{X}'\mathbf{X}\hat{\theta} = \mathbf{X}'(\mathbf{X}\theta + \epsilon) = \mathbf{X}'\mathbf{X}\theta + \mathbf{X}'\epsilon$$

or, letting $\delta = \mathbf{X}'\epsilon$,

$$\mathbf{X}'\mathbf{X}\hat{\theta} = \mathbf{X}'\mathbf{X}\theta + \delta$$

$$E(\delta) = \mathbf{0}, \quad Var(\delta) = \sigma^2\mathbf{X}'\mathbf{X}.$$

Hence the entire character of the *information* relevant to inference about θ, at least through statistical moments of second order, is characterized by the (usually) unknown scalar σ^2 and the known matrix

$$\mathcal{I} = \mathbf{X}'\mathbf{X}.$$

We shall refer to $\mathcal{I}$ as the *design information matrix* for θ, to distinguish it from Fisher's information matrix $\mathcal{I}/\sigma^2$. $\mathcal{I}$ appears in formulae for the power of hypothesis tests and variances of estimates, and so is a critical link between the experimental design and the quality of statistical inference that can be drawn from the experimental data collected. Among its special properties are the following:

- $\mathcal{I}$ is symmetric.
- $\mathcal{I}$ is positive semi-definite, that is, $\mathbf{v}'\mathcal{I}\mathbf{v} \geq 0$ for all conformable $\mathbf{v}$.
- $\text{rank}(\mathcal{I}) = \text{rank}(\mathbf{X})$.

$\mathcal{I}$ can be regarded as a kind of matrix-valued generalization of a sample size. In single-sample inferences about means, hypothesis tests are generally more powerful, and estimates more precise, when σ^2/N is relatively small. Our context is complicated by the fact that multiple parameters are of interest, but hypothesis tests are generally more powerful when the elements of $\mathcal{I}$ are large and the value of σ^2 is small, and parameter estimates are generally more precise when the elements of $\mathcal{I}^-$ or $\mathcal{I}^{-1}$ are small (roughly coinciding with "large" $\mathcal{I}$) and the value of σ^2 is small. Information matrices also generalize an intuitive additivity property of sample sizes. If two independent random samples of size N_1 and N_2 are drawn from a common population for which the variance is σ^2, the power and precision associated with pooled-sample inferences about the mean are characterized by $\sigma^2/(N_1 + N_2)$. Similarly, if two independent experiments are conducted for which the values of θ and σ^2 are the same, and for which the design information matrices are $\mathcal{I}^1$ and $\mathcal{I}^2$, respectively, inference about θ based on the combined data is characterized by the information matrix $\mathcal{I}^1 + \mathcal{I}^2$.

2.3.1 Example

A small study is designed to compare means associated with conditions that can be regarded as cells in a two-way table:

As indicated in the figure, three cells contain two observations each, and the remaining cells contain only one each. Denote by y_{ijk} the kth observation from row i and column j. If it is assumed that rows and columns have *additive effects* on the response, i.e., that there is no row-column interaction, we might consider a model of form:

$$
\begin{pmatrix} y_{111} \\ y_{112} \\ y_{121} \\ y_{122} \\ y_{131} \\ y_{211} \\ y_{212} \\ y_{221} \\ y_{231} \end{pmatrix}
=
\begin{pmatrix}
1 & 0 & 1 & 0 & 0 \\
1 & 0 & 1 & 0 & 0 \\
1 & 0 & 0 & 1 & 0 \\
1 & 0 & 0 & 1 & 0 \\
1 & 0 & 0 & 0 & 1 \\
0 & 1 & 1 & 0 & 0 \\
0 & 1 & 1 & 0 & 0 \\
0 & 1 & 0 & 1 & 0 \\
0 & 1 & 0 & 0 & 1
\end{pmatrix}
\begin{pmatrix} \alpha_1 \\ \alpha_2 \\ \beta_1 \\ \beta_2 \\ \beta_3 \end{pmatrix}
+
\begin{pmatrix} \epsilon_{111} \\ \epsilon_{112} \\ \epsilon_{121} \\ \epsilon_{122} \\ \epsilon_{131} \\ \epsilon_{211} \\ \epsilon_{212} \\ \epsilon_{221} \\ \epsilon_{231} \end{pmatrix}.
$$

In this case,

$$
\mathcal{I} = \mathbf{X}'\mathbf{X} =
\begin{pmatrix}
5 & 0 & 2 & 2 & 1 \\
0 & 4 & 2 & 1 & 1 \\
2 & 2 & 4 & 0 & 0 \\
2 & 1 & 0 & 3 & 0 \\
1 & 1 & 0 & 0 & 2
\end{pmatrix}.
$$

It is easy to see that neither $\mathbf{X}$ nor $\mathbf{X}'\mathbf{X}$ is of full rank because, for either matrix, the sum of the first two columns equals the sum of the last three. Since any four columns of $\mathbf{X}$ (or of $\mathbf{X}'\mathbf{X}$) are linearly independent, but all five columns do not have this property, $\text{rank}(\mathbf{X}) = \text{rank}(\mathbf{X}'\mathbf{X}) = 4$.

There are infinitely many generalized inverses of $\mathbf{X}'\mathbf{X}$ in this case. Here is one numerical technique for computing a generalized inverse of a square symmetric matrix $\mathbf{M}$:

1. Identify a maximal set of linearly independent columns in $\mathbf{M}$. "Maximal" means that this set should contain as many columns as possible; when more than one such set exists, any one of them may be selected. Let L be the set of column numbers of these columns. Note that the number of elements of L is $\text{rank}(\mathbf{M})$.

2. Form a new matrix $\mathbf{M}^*$ starting with $\mathbf{M}$, and removing columns *and* rows with numbers not in L. Note that $\mathbf{M}^*$ will also be square and symmetric, and will be of dimension $\text{rank}(\mathbf{M})$.

3. $\mathbf{M}^*$ is of full rank, and so $\mathbf{M}^{*-1}$ exists and is unique. Compute $\mathbf{M}^{*-1}$. Now add rows and columns of zeros to $\mathbf{M}^{*-1}$, corresponding to the rows and columns removed from $\mathbf{M}$ in step (2). For example, if the second and third rows/columns had been deleted from $\mathbf{M}$ to form $\mathbf{M}^*$, insert two rows and columns of zeros between the first and remaining rows and columns of $\mathbf{M}^{*-1}$. The resulting symmetric matrix is a generalized inverse of $\mathbf{M}$.

For our example, any four rows/columns of $\mathbf{X}'\mathbf{X}$ are linearly independent; suppose we select $L = \{2, 3, 4, 5\}$. Then (R2.1):

$$(\mathbf{X}'\mathbf{X})^* = \begin{pmatrix} 4 & 2 & 1 & 1 \\ 2 & 4 & 0 & 0 \\ 1 & 0 & 3 & 0 \\ 1 & 0 & 0 & 2 \end{pmatrix}$$

$$(\mathbf{X}'\mathbf{X})^{*-1} = \begin{pmatrix} 0.4615 & -0.2308 & -0.1538 & -0.2308 \\ -0.2308 & 0.3654 & 0.0769 & 0.1154 \\ -0.1538 & 0.0769 & 0.3846 & 0.0769 \\ -0.2308 & 0.1154 & 0.0769 & 0.6154 \end{pmatrix}$$

$$(\mathbf{X}'\mathbf{X})^- = \begin{pmatrix} 0.0000 & 0.0000 & 0.0000 & 0.0000 & 0.0000 \\ 0.0000 & 0.4615 & -0.2308 & -0.1538 & -0.2308 \\ 0.0000 & -0.2308 & 0.3654 & 0.0769 & 0.1154 \\ 0.0000 & -0.1538 & 0.0769 & 0.3846 & 0.0769 \\ 0.0000 & -0.2308 & 0.1154 & 0.0760 & 0.6154 \end{pmatrix}.$$

Pre-multiplying $(\mathbf{X}'\mathbf{X})^-$ by $\mathbf{X}$, and post-multiplying by $\mathbf{X}'$, yields the unique hat matrix $\mathbf{H}$, in this case:

$$\begin{pmatrix} 0.3654 & 0.3654 & 0.0769 & 0.0769 & 0.1154 & 0.1346 & 0.1346 & -0.1538 & -0.1154 \\ 0.3654 & 0.3654 & 0.0769 & 0.0769 & 0.1154 & 0.1346 & 0.1346 & -0.1538 & -0.1154 \\ 0.0769 & 0.0769 & 0.3846 & 0.3856 & 0.0769 & -0.0769 & -0.0769 & 0.2308 & -0.0769 \\ 0.0769 & 0.0769 & 0.3846 & 0.3856 & 0.0769 & -0.0769 & -0.0769 & 0.2308 & -0.0769 \\ 0.1154 & 0.1154 & 0.0769 & 0.0769 & 0.6154 & -0.1154 & -0.1154 & -0.1538 & 0.3846 \\ 0.1346 & 0.1346 & -0.0769 & -0.0769 & -0.1154 & 0.3654 & 0.3654 & 0.1538 & 0.1154 \\ 0.1346 & 0.1346 & -0.0769 & -0.0769 & -0.1154 & 0.3654 & 0.3654 & 0.1538 & 0.1154 \\ -0.1538 & -0.1538 & 0.2308 & 0.2308 & -0.1538 & 0.1538 & 0.1538 & 0.5385 & 0.1538 \\ -0.1154 & -0.1154 & -0.0769 & -0.0769 & 0.3846 & 0.1154 & 0.1154 & 0.1538 & 0.6154 \end{pmatrix}.$$

2.4 The partitioned linear model

In experimental contexts, it is common that some of the elements of $\boldsymbol{\theta}$ are not related to any of the experimental questions of interest, but are necessary nuisance parameters such as those associated with the blocks in the design. In such cases, it is useful to think of the model structure as *partitioned*. We partition $\boldsymbol{\theta}$ into two sub-vectors, $\boldsymbol{\theta}_1$ (k_1 elements) and $\boldsymbol{\theta}_2$ (k_2 elements, $k_1 + k_2 = k$) containing the nuisance parameters and those of interest, respectively, and partition the columns of $\mathbf{X}$ into two corresponding sets of columns, $\mathbf{X} = (\mathbf{X}_1 | \mathbf{X}_2)$. The basic linear model can then be rewritten as a partitioned linear model:

$$\mathbf{y} = \mathbf{X}_1 \boldsymbol{\theta}_1 + \mathbf{X}_2 \boldsymbol{\theta}_2 + \boldsymbol{\epsilon}. \tag{2.3}$$

The partitioned linear model is perhaps most often used in discussing the *analysis of covariance* (e.g., Searle, 1971; Wildt and Ahtola, 1978; Milliken and Johnson, 2002), where the elements of $\mathbf{X}_1$ are *covariates*, collected as observations along with the elements of $\mathbf{y}$. These are included in the analysis to account for some of the variability in the response, in order to improve the quality of inferences about the parameters of interest. The data analysis is sometimes said to focus on inferences about $\boldsymbol{\theta}_2$ *after correcting for* the effects represented by $\boldsymbol{\theta}_1$. In our context, the motivation is somewhat different. $\mathbf{X}_1$ is known in advance because it represents part of the structure of the selected experimental design, and $\mathbf{X}_1\boldsymbol{\theta}_1$ is generally viewed as an acknowledged component of the mean of the observed data which must be accommodated in the model to assure that the analysis concerning the parameters of interest $(\boldsymbol{\theta}_2)$ is valid. However, the development of the analysis generally parallels that of analysis of covariance, and leads to inferences about the effect of experimental treatments *after correcting for* the effect of blocks.

2.5 The reduced normal equations

Recalling that the basic and partitioned models are related by $\mathbf{X} = (\mathbf{X}_1|\mathbf{X}_2)$ and $\boldsymbol{\theta}' = (\boldsymbol{\theta}_1'|\boldsymbol{\theta}_2')$, and using the same pattern to express the vector of least-squares estimates $\hat{\boldsymbol{\theta}}' = (\hat{\boldsymbol{\theta}}_1'|\hat{\boldsymbol{\theta}}_2')$, the normal equations

$$\mathbf{X}'\mathbf{X}\hat{\boldsymbol{\theta}} = \mathbf{X}'\mathbf{y}$$

can be rewritten as a system of two matrix equations:

$$\mathbf{X}_1'\mathbf{X}_1\hat{\boldsymbol{\theta}}_1 + \mathbf{X}_1'\mathbf{X}_2\hat{\boldsymbol{\theta}}_2 = \mathbf{X}_1'\mathbf{y}$$
$$\mathbf{X}_2'\mathbf{X}_1\hat{\boldsymbol{\theta}}_1 + \mathbf{X}_2'\mathbf{X}_2\hat{\boldsymbol{\theta}}_2 = \mathbf{X}_2'\mathbf{y}. \tag{2.4}$$

By pre-multiplying the first of these equations by $\mathbf{X}_2'\mathbf{X}_1(\mathbf{X}_1'\mathbf{X}_1)^-$, and then subtracting it from the second equation, we have:

$$\mathbf{X}_2'[\mathbf{I} - \mathbf{X}_1(\mathbf{X}_1'\mathbf{X}_1)^-\mathbf{X}_1']\mathbf{X}_1\hat{\boldsymbol{\theta}}_1 + \mathbf{X}_2'[\mathbf{I} - \mathbf{X}_1(\mathbf{X}_1'\mathbf{X}_1)^-\mathbf{X}_1']\mathbf{X}_2\hat{\boldsymbol{\theta}}_2$$
$$= \mathbf{X}_2'[\mathbf{I} - \mathbf{X}_1(\mathbf{X}_1'\mathbf{X}_1)^-\mathbf{X}_1']\mathbf{y},$$

or:

$$\mathbf{X}_2'[\mathbf{I} - \mathbf{H}_1]\mathbf{X}_1\hat{\boldsymbol{\theta}}_1 + \mathbf{X}_2'[\mathbf{I} - \mathbf{H}_1]\mathbf{X}_2\hat{\boldsymbol{\theta}}_2 = \mathbf{X}_2'[\mathbf{I} - \mathbf{H}_1]\mathbf{y}, \tag{2.5}$$

where $\mathbf{H}_1 = \mathbf{X}_1(\mathbf{X}_1'\mathbf{X}_1)^-\mathbf{X}_1'$ is the hat matrix for the model containing only terms associated with $\boldsymbol{\theta}_1$. Here it is convenient to recall the interpretation of $\mathbf{H}_1$ as the projection operator associated with the column space of $\mathbf{X}_1$. Because $\mathbf{H}_1$ projects the columns of $\mathbf{X}_1$ "into themselves," i.e., $\mathbf{H}_1\mathbf{X}_1 = \mathbf{X}_1$, the first term in equation (2.5) is zero regardless of the value of $\hat{\boldsymbol{\theta}}_1$, leading to the *reduced normal equations* in only $\hat{\boldsymbol{\theta}}_2$:

$$\mathbf{X}_2'[\mathbf{I} - \mathbf{H}_1]\mathbf{X}_2\hat{\boldsymbol{\theta}}_2 = \mathbf{X}_2'[\mathbf{I} - \mathbf{H}_1]\mathbf{y}. \tag{2.6}$$

It is important to understand that equation (2.6) is completely consistent with equation (2.2). For any data vector $\mathbf{y}$, the estimate or estimates of $\boldsymbol{\theta}_2$ derived from solving the reduced normal equations are precisely the same as the second segment of the estimate or estimates of $\boldsymbol{\theta}$ derived from solving the full normal equations. The reduced form is of value to us primarily because it eliminates the estimates of nuisance parameters, providing a more focused expression of the information available concerning the parameters of interest. The reduced normal equations can be written in the same form as equation (2.2) if we define a "corrected" version of $\mathbf{X}_2$:

$$\mathbf{X}_{2|1} = (\mathbf{I} - \mathbf{H}_1)\mathbf{X}_2. \tag{2.7}$$

Given our earlier discussion of complementary hat matrices, each column of $\mathbf{X}_{2|1}$ could be constructed as the set of residuals that would result from fitting the corresponding column of $\mathbf{X}_2$ as "data" to the linear model containing only mean structure $\mathbf{X}_1\boldsymbol{\theta}_1$. $\mathbf{X}_{2|1}$ can be thought of intuitively as what remains of $\mathbf{X}_2$ after everything that could be explained by $\mathbf{X}_1$ has been removed or accounted for. Using this notation, equation (2.6) can be rewritten as:

$$\mathbf{X}'_{2|1}\mathbf{X}_{2|1}\hat{\boldsymbol{\theta}}_2 = \mathbf{X}'_{2|1}\mathbf{y}. \tag{2.8}$$

Following our introduction of a design information matrix in Section 2.3, substituting equation (2.3) into equation (2.8) yields:

$$\mathbf{X}'_{2|1}\mathbf{X}_{2|1}\hat{\boldsymbol{\theta}}_2 = \mathbf{X}'_{2|1}(\mathbf{X}_1\boldsymbol{\theta}_1 + \mathbf{X}_2\boldsymbol{\theta}_2 + \boldsymbol{\epsilon}) = \mathbf{X}'_{2|1}\mathbf{X}_2\boldsymbol{\theta}_2 + \mathbf{X}'_{2|1}\boldsymbol{\epsilon}$$

with the simplification due to the fact that the columns of $\mathbf{X}_{2|1}$ are orthogonal to those of $\mathbf{X}_1$. Furthermore, since

$$\mathbf{X}'_{2|1}\mathbf{X}_2 = \mathbf{X}'_2(\mathbf{I} - \mathbf{H}_1)\mathbf{X}_2 = \mathbf{X}'_2(\mathbf{I} - \mathbf{H}_1)(\mathbf{I} - \mathbf{H}_1)\mathbf{X}_2 = \mathbf{X}'_{2|1}\mathbf{X}_{2|1}$$

we have

$$\mathbf{X}'_{2|1}\mathbf{X}_{2|1}\hat{\boldsymbol{\theta}}_2 = \mathbf{X}'_{2|1}\mathbf{X}_{2|1}\boldsymbol{\theta}_2 + \mathbf{X}'_{2|1}\boldsymbol{\epsilon}$$

or letting $\boldsymbol{\delta} = \mathbf{X}'_{2|1}\boldsymbol{\epsilon}$,

$$\mathbf{X}'_{2|1}\mathbf{X}_{2|1}\hat{\boldsymbol{\theta}}_2 = \mathbf{X}'_{2|1}\mathbf{X}_{2|1}\boldsymbol{\theta}_2 + \boldsymbol{\delta},$$

$$E(\boldsymbol{\delta}) = \mathbf{0}, \quad Var(\boldsymbol{\delta}) = \sigma^2\mathbf{X}'_{2|1}\mathbf{X}_{2|1}.$$

So in the context of a model including both $\boldsymbol{\theta}_1$ and $\boldsymbol{\theta}_2$, the information about $\boldsymbol{\theta}_2$ is characterized by σ^2 and the design information matrix:

$$\mathcal{I}_{2|1} = \mathbf{X}'_{2|1}\mathbf{X}_{2|1}. \tag{2.9}$$

As with information matrices for an unpartitioned model, the quality of the inferences we can draw about $\boldsymbol{\theta}_2$, when $\boldsymbol{\theta}_1$ is also included in the model, is determined by the amount of random "noise" in the data characterized by σ^2 and by the experimental design characterized by $\mathcal{I}_{2|1}$. The complete analogy to sample size described in Section 2.3 does not always hold because design information matrices for partitioned models are not always additive for combined experiments. However, the general role of $\mathcal{I}_{2|1}$ in statistical formulae derived from the partitioned linear model is basically the same as that of $\mathbf{X}'\mathbf{X}$ in the context of the basic linear model; other things begin equal, hypothesis tests are generally more powerful when the elements of $\mathcal{I}_{2|1}$ are large and the value of σ^2 is small, and parameter estimates are generally more precise when the elements of $\mathcal{I}_{2|1}^{-}$ or $\mathcal{I}_{2|1}^{-1}$ are small and the value of σ^2 is small.

When $\mathbf{X}_{2|1}$ is of full rank, the least-squares solution of the reduced normal equations is unique and can be written in terms of a unique matrix inverse as $\hat{\boldsymbol{\theta}}_2 = (\mathbf{X}'_{2|1}\mathbf{X}_{2|1})^{-1}\mathbf{X}'_{2|1}\mathbf{y}$. When $\mathbf{X}_{2|1}$ is not of full rank, a similar expression characterizes the nonunique solutions, $\hat{\boldsymbol{\theta}}_2 = (\mathbf{X}'_{2|1}\mathbf{X}_{2|1})^{-}\mathbf{X}'_{2|1}\mathbf{y}$; this can be demonstrated by substituting this expression back into the reduced normal equations form:

$$\mathbf{X}'_{2|1}\mathbf{X}_{2|1}(\mathbf{X}'_{2|1}\mathbf{X}_{2|1})^{-}\mathbf{X}'_{2|1}\mathbf{y} = \mathbf{X}'_{2|1}\mathbf{y}$$

or

$$\mathbf{X}'_{2|1}\mathbf{H}_{2|1}\mathbf{y} = \mathbf{X}'_{2|1}\mathbf{y}$$

where $\mathbf{H}_{2|1} = \mathbf{X}_{2|1}(\mathbf{X}'_{2|1}\mathbf{X}_{2|1})^{-}\mathbf{X}'_{2|1}$ is the projection operator associated with the column space of $\mathbf{X}_{2|1}$. It is clear that this holds for any vector $\mathbf{y}$ because $\mathbf{X}'_{2|1}\mathbf{H}_{2|1} = \mathbf{X}'_{2|1}$.

When $\mathbf{X}_{2|1}$ is of less than full rank, $\hat{\boldsymbol{\theta}}_2$ is of relatively little practical value in its own right because estimates specified by different generalized inverses have different expectations – that is, they don't even unbiasedly estimate the same vector quantity. On the other hand, linear functions of parameters $\mathbf{c}'\boldsymbol{\theta}_2$ have unique least-squares estimates $\widehat{\mathbf{c}'\boldsymbol{\theta}_2} = \mathbf{c}'\hat{\boldsymbol{\theta}}_2$ so long as $\mathbf{c}'$ can be represented as a linear combination of the rows of $\mathbf{X}_{2|1}$, i.e., $\mathbf{c}' = \mathbf{l}'\mathbf{X}_{2|1}$ for some N-element vector $\mathbf{l}$, because in this case

$$\mathbf{c}'\hat{\boldsymbol{\theta}}_2 = \mathbf{l}'\mathbf{X}_{2|1}(\mathbf{X}'_{2|1}\mathbf{X}_{2|1})^{-}\mathbf{X}'_{2|1}\mathbf{y} = \mathbf{l}'\mathbf{H}_{2|1}\mathbf{y}. \tag{2.10}$$

Equation (2.10) shows that such estimates are unique because $\mathbf{H}_{2|1}$ is invariant to the generalized inverse chosen. $\mathbf{c}'\boldsymbol{\theta}_2$ is said to be *estimable* in this case. An analogous result holds in the simpler case of unpartitioned models; $\mathbf{c}'\boldsymbol{\theta}$ is estimable if and only if $\mathbf{c}'$ can be written as $\mathbf{l}'\mathbf{X}$ for some N-element vector $\mathbf{l}$.

2.5.1 Example

Continuing the example of subsection 2.3.1, suppose that the rows of the table represent two batches of material used in the experiment, and that the three columns represent the treatment conditions of interest to the experimenter.

We would partition the previous matrix $\mathbf{X}$ as:

$$\mathbf{X}_1 = \begin{pmatrix} 1 & 0 \\ 1 & 0 \\ 1 & 0 \\ 1 & 0 \\ 1 & 0 \\ 0 & 1 \\ 0 & 1 \\ 0 & 1 \\ 0 & 1 \end{pmatrix} \qquad \mathbf{X}_2 = \begin{pmatrix} 1 & 0 & 0 \\ 1 & 0 & 0 \\ 0 & 1 & 0 \\ 0 & 1 & 0 \\ 0 & 0 & 1 \\ 1 & 0 & 0 \\ 1 & 0 & 0 \\ 0 & 1 & 0 \\ 0 & 0 & 1 \end{pmatrix}.$$

In this case,

$$(\mathbf{X}_1'\mathbf{X}_1)^{-1} = \begin{pmatrix} \frac{1}{5} & 0 \\ 0 & \frac{1}{4} \end{pmatrix} \qquad \mathbf{H}_1 = \mathbf{X}_1(\mathbf{X}_1'\mathbf{X}_1)^{-1}\mathbf{X}_1' = \begin{pmatrix} \frac{1}{5}\mathbf{J}_{5\times5} & \mathbf{0}_{5\times4} \\ \mathbf{0}_{4\times5} & \frac{1}{4}\mathbf{J}_{4\times4} \end{pmatrix}$$

where $\mathbf{J}$ and $\mathbf{0}$ are matrices of the indicated dimension in which each element is 1 or 0, respectively. Further,

$$\mathbf{X}_{2|1} = (\mathbf{I} - \mathbf{H}_1)\mathbf{X}_2 = \mathbf{X}_2 - \mathbf{H}_1\mathbf{X}_2 = \begin{pmatrix} 0.60 & -0.40 & -0.20 \\ 0.60 & -0.40 & -0.20 \\ -0.40 & 0.60 & -0.20 \\ -0.40 & 0.60 & -0.20 \\ -0.40 & -0.40 & 0.80 \\ 0.50 & -0.25 & -0.25 \\ 0.50 & -0.25 & -0.25 \\ -0.50 & 0.75 & -0.25 \\ -0.50 & -0.25 & 0.75 \end{pmatrix},$$

and the design information matrix for $\boldsymbol{\theta}_2$, recognizing that $\boldsymbol{\theta}_1$ must also be included in the model, is

$$\mathcal{I}_{2|1} = \mathbf{X}_{2|1}'\mathbf{X}_{2|1} = \begin{pmatrix} 2.20 & -1.30 & -0.90 \\ -1.30 & 1.95 & -0.65 \\ -0.90 & -0.65 & 1.55 \end{pmatrix}.$$

$\mathcal{I}_{2|1}$ is not of full rank because the sum of the three rows or columns is a vector of zeros; this should not be surprising, since we have already "accommodated" $\boldsymbol{\theta}_1$, and know from the discussion in subsection 2.3.1 that the sum of the columns in $\mathbf{X}_1$ equals the sum of columns in $\mathbf{X}_2$. Hence, the reduced normal equations:

$$\begin{pmatrix} 2.20 & -1.30 & -0.90 \\ -1.30 & 1.95 & -0.65 \\ -0.90 & -0.65 & 1.55 \end{pmatrix} \hat{\boldsymbol{\theta}}_2 = \mathbf{X}'_{2|1}\mathbf{y}$$

have infinitely many solutions. However $\{\boldsymbol{\theta}_2\}_1 - \{\boldsymbol{\theta}_2\}_2$ (for example) is uniquely estimable because the coefficients of this linear combination can be written as a linear combination of the rows of $\mathbf{X}_{2|1}$; one of the infinitely many such linear combinations is:

$$(+1, -1, 0) = (0, 0, 0, 0, 0, +1, 0, -1, 0) \begin{pmatrix} 0.60 & -0.40 & -0.20 \\ 0.60 & -0.40 & -0.20 \\ -0.40 & 0.60 & -0.20 \\ -0.40 & 0.60 & -0.20 \\ -0.40 & -0.40 & 0.80 \\ 0.50 & -0.25 & -0.25 \\ 0.50 & -0.25 & -0.25 \\ -0.50 & 0.75 & -0.25 \\ -0.50 & -0.25 & 0.75 \end{pmatrix}.$$

2.6 Linear and quadratic forms

Many of the statistics with which we deal in experimental design and analysis of variance are linear and quadratic forms in the N-element vector of data, $\mathbf{y}$, i.e., functions of form:

$$\mathbf{Ly} \quad \text{and} \quad \mathbf{y}'\mathbf{Qy}$$

respectively, where $\mathbf{L}$ $(m \times N)$ and $\mathbf{Q}$ $(N \times N)$ are specified constant matrices. Here we limit attention to symmetric $\mathbf{Q}$. The properties of statistics of these forms are used repeatedly, and we review a few of these here. In particular, if $E(\mathbf{y}) = \mathbf{m}$ and $Var(\mathbf{y}) = \boldsymbol{\Sigma}$, then

- $E(\mathbf{Ly}) = \mathbf{Lm}$
- $Var(\mathbf{Ly}) = \mathbf{L\Sigma L}'$
- $E(\mathbf{y}'\mathbf{Qy}) = \mathbf{m}'\mathbf{Qm} + \text{trace}(\mathbf{Q\Sigma})$.

If it is also the case that $\mathbf{y}$ has a multivariate normal distribution (we use notation $\mathbf{y} \sim N(\mathbf{m}, \boldsymbol{\Sigma})$ to indicate this), additional statements can be made about the properties of these statistics. In particular,

- $Var(\mathbf{y}'\mathbf{Qy}) = 2\,\text{trace}(\mathbf{Q\Sigma})^2 + 4\mathbf{m}'\mathbf{Q\Sigma Qm}$,
- $\mathbf{Ly} \sim N(\mathbf{Lm}, \mathbf{L\Sigma L}')$,
- $\mathbf{Ly}$ and $\mathbf{y}'\mathbf{Qy}$ are statistically independent if and only if $\mathbf{L\Sigma Q} = \mathbf{0}$.

If in addition to multivariate normality, the elements of $\mathbf{y}$ are independent with the same variance, $\boldsymbol{\Sigma} = \sigma^2\mathbf{I}$, then for any two positive semi-definite symmetric $N \times N$ matrices $\mathbf{Q}_1$ and $\mathbf{Q}_2$, if:

- $\mathbf{Q}_1$ and $\mathbf{Q}_2$ are both idempotent, i.e., $\mathbf{Q}_1\mathbf{Q}_1 = \mathbf{Q}_1$ and $\mathbf{Q}_2\mathbf{Q}_2 = \mathbf{Q}_2$, and
- $\mathbf{Q}_1\mathbf{Q}_2 = \mathbf{0}$,

then:

- $\mathbf{y}'\mathbf{Q}_1\mathbf{y}/\sigma^2 \sim \chi^{2'}(\text{rank}(\mathbf{Q}_1), \mathbf{m}'\mathbf{Q}_1\mathbf{m}/\sigma^2)$,
- $\mathbf{y}'\mathbf{Q}_2\mathbf{y}/\sigma^2 \sim \chi^{2'}(\text{rank}(\mathbf{Q}_2), \mathbf{m}'\mathbf{Q}_2\mathbf{m}/\sigma^2)$, and
- $\mathbf{y}'\mathbf{Q}_1\mathbf{y}$ and $\mathbf{y}'\mathbf{Q}_2\mathbf{y}$ are independent statistics,

where $\chi^{2'}(-,-)$ denotes the noncentral chi-square distribution with degrees of freedom specified by the first argument and noncentrality parameter specified by the second. Finally, if $\mathbf{m}'\mathbf{Q}_2\mathbf{m} = 0$, an additional result is:

- $[\mathbf{y}'\mathbf{Q}_1\mathbf{y}/\text{rank}(\mathbf{Q}_1)]/[\mathbf{y}'\mathbf{Q}_2\mathbf{y}/\text{rank}(\mathbf{Q}_2)]$
  $\sim F'(\text{rank}(\mathbf{Q}_1), \text{rank}(\mathbf{Q}_2), \mathbf{m}'\mathbf{Q}_1\mathbf{m}/\sigma^2)$

where $F'(-,-,-)$ denotes the noncentral F distribution with degrees of freedom specified by the first two arguments and noncentrality parameter specified by the third. In subsequent chapters, we shall often refer to the noncentrality parameter as λ; when this parameter is zero, the distribution is said to be a central F distribution (or just "F distribution"). In particular, this last result is the basis for the ANOVA test for equality of treatments, where $\mathbf{y}'\mathbf{Q}_1\mathbf{y}$ and $\mathbf{y}'\mathbf{Q}_2\mathbf{y}$ are the sums of squares for treatments and residuals (error), respectively.

2.7 Estimation and information

For our purposes, the linear forms of greatest interest are the least-squares estimates of the treatment-related model coefficients, and estimable linear combinations of them. For the partitioned linear model in which $\boldsymbol{\theta}_2$ is associated with treatments and the elements of $\boldsymbol{\theta}_1$ are nuisance parameters, these can be written as:

$$\hat{\boldsymbol{\theta}}_2 = (\mathbf{X}'_{2|1}\mathbf{X}_{2|1})^{-}\mathbf{X}'_{2|1}\mathbf{y},$$

and

$$\widehat{\mathbf{C}\boldsymbol{\theta}}_2 = \mathbf{C}(\mathbf{X}'_{2|1}\mathbf{X}_{2|1})^{-}\mathbf{X}'_{2|1}\mathbf{y} = \mathbf{L}\mathbf{H}_{2|1}\mathbf{y} \quad \text{where} \quad \mathbf{C} = \mathbf{L}\mathbf{X}_{2|1},$$

respectively. Since estimable functions have unique least-squares estimators, these estimators have unique sample variances. If $Var(\mathbf{y}) = \sigma^2\mathbf{I}$, then

$$Var(\widehat{\mathbf{C}\boldsymbol{\theta}}_2) = Var(\mathbf{L}\mathbf{X}_{2|1}(\mathbf{X}'_{2|1}\mathbf{X}_{2|1})^{-}\mathbf{X}'_{2|1}\mathbf{y})$$

$$= \sigma^2\mathbf{L}\mathbf{X}_{2|1}(\mathbf{X}'_{2|1}\mathbf{X}_{2|1})^{-}\mathbf{X}'_{2|1}\mathbf{X}_{2|1}(\mathbf{X}'_{2|1}\mathbf{X}_{2|1})^{-}\mathbf{X}'_{2|1}\mathbf{L}'$$

$$= \sigma^2\mathbf{L}\mathbf{X}_{2|1}(\mathbf{X}'_{2|1}\mathbf{X}_{2|1})^{-}\mathbf{X}'_{2|1}\mathbf{L}'$$

$$= \sigma^2\mathbf{C}(\mathbf{X}'_{2|1}\mathbf{X}_{2|1})^{-}\mathbf{C}'$$

$$= \sigma^2\mathbf{C}\mathcal{I}_{2|1}^{-}\mathbf{C}'$$

where the generalized inverse is replaced by a unique matrix inverse if $\mathbf{X}_{2|1}$ is of full rank. The functional form of this expression clearly separates the influence of the *noise* characterized by σ^2, the *parametric functions of interest* characterized by $\mathbf{C}$, and the *design* characterized by $\mathcal{I}_{2|1}$ on the precision of estimation resulting from an experiment.

The quadratic form of greatest interest in estimation is the residual, or error, mean square from the fit of the full model:

$$MSE = \frac{1}{N - \text{rank}(\mathbf{X})} \, \mathbf{y}'(\mathbf{I} - \mathbf{H})\mathbf{y}.$$

If $Var(\epsilon) = \sigma^2\mathbf{I}$ and the form of the linear model is correct, then from the general results for quadratic forms,

$$E(MSE) = \frac{1}{N - \text{rank}(\mathbf{X})} \, \boldsymbol{\theta}'\mathbf{X}'(\mathbf{I} - \mathbf{H})\mathbf{X}\boldsymbol{\theta} + \frac{\sigma^2}{N - \text{rank}(\mathbf{X})} \, \text{trace}(\mathbf{I} - \mathbf{H}).$$

The first term is zero, regardless of the value of $\boldsymbol{\theta}$, because $\mathbf{H}$ projects the columns of $\mathbf{X}$ "into themselves," that is, $\mathbf{HX} = \mathbf{X}$. The second term can be written as

$$\frac{\sigma^2}{N - \text{rank}(\mathbf{X})} \, [\text{trace}(\mathbf{I}) - \text{trace}(\mathbf{H})].$$

Because $\mathbf{H}$ is idempotent, $\text{trace}(\mathbf{H}) = \text{rank}(\mathbf{H}) = \text{rank}(\mathbf{X})$, and it immediately follows that MSE is an unbiased estimate of σ^2:

$$E(MSE) = \sigma^2.$$

When ϵ is normally distributed, $\widehat{\mathbf{C}\boldsymbol{\theta}}_2$ and MSE are independent statistics because

$$[\mathbf{LH}_{2|1}][\sigma^2\mathbf{I}] \left[\frac{1}{N - \text{rank}(\mathbf{X})} \, (\mathbf{I} - \mathbf{H}) \right]$$

$$= \frac{\sigma^2}{N - \text{rank}(\mathbf{X})} \, [\mathbf{LX}_{2|1}(\mathbf{X}'_{2|1}\mathbf{X}_{2|1})^-\mathbf{X}'_2(\mathbf{I} - \mathbf{H}_1)(\mathbf{I} - \mathbf{H})],$$

and the last three matrix factors on the right side of this equation can be written as:

$$\mathbf{X}'_2[\mathbf{I} - \mathbf{H}_1 - \mathbf{H} + \mathbf{H}_1\mathbf{H}] = \mathbf{X}'_2[\mathbf{I} - \mathbf{H}] = \mathbf{0},$$

because $\mathbf{H}_1\mathbf{H} = \mathbf{H}_1$.

Under the assumption of normality for ϵ, the quality of MSE as an estimator of σ^2 is affected by the design only through the value of $N - \text{rank}(\mathbf{X})$, the associated degrees of freedom. The estimator is more precise and, other things being equal, leads to narrower confidence intervals (on average) and more powerful hypothesis tests when $N - \text{rank}(\mathbf{X})$ is large. Once the model has been established, this value is determined by the size of the data set collected in the experiment; as usual, larger N is better. In most experimental settings, estimation of σ^2 is (in its own right) not of primary interest, since it is often more related to the homogeneity of experimental material and precision of measurement processes than to the experimental treatments. But the

precision of MSE also influences the quality of inference that can be made
about estimable functions of $\boldsymbol{\theta}_2$. For example, the expected squared length of
a $(1 - \alpha)100\%$ confidence interval for estimable $\mathbf{c}'\boldsymbol{\theta}_2$ is:

$$4\, t_{1-\frac{\alpha}{2}}^2 (N - \text{rank}(\mathbf{X}))\, \sigma^2 \mathbf{c}' \mathcal{I}_{2|1}^- \mathbf{c}.$$

(We square the length of the confidence interval to avoid dealing with the expectation of $\sqrt{MSE}$.) As noted above, precision is enhanced (or, the expected
squared interval length is made small) for designs leading to relatively small
values of $\mathbf{c}'\mathcal{I}_{2|1}^- \mathbf{c}$. However $N - \text{rank}(\mathbf{X})$ also plays a role since for given level
(or type I error probability) α, the t-quantile decreases with increasing degrees
of freedom. When competing experimental designs all lead to relatively large
but different values of $N - \text{rank}(\mathbf{X})$, this influence is relatively small. But for
smaller experiments, changing the design to increase the residual degrees of
freedom from, say, 2 to 5, can make a substantial difference in the quality of
formal inferences that can be drawn.

2.7.1 Pure error and lack of fit

Validity of MSE as an estimate of σ^2 relies primarily on two assumptions: (1)
the homogeneity of variance and independence of the elements of $\boldsymbol{\epsilon}$, and (2)
the assumed functional form of $E(\mathbf{y}) = \mathbf{X}\boldsymbol{\theta}$. However, there are situations,
such as in diagnostic analysis, where it is desirable to have an estimate for
which validity requires fewer assumptions. For designs that contain replicate
runs – groups of trials that are "coded" with identical rows in $\mathbf{X}$ – such an
estimate can be formed based on what is sometimes called a *pure error* sum
of squares.

Formally, suppose the N rows of $\mathbf{X}$ are actually each copies of one of the
N^* *unique* rows of $\mathbf{X}^*$, $N^* < N$. Now propose a *more general* model for $\mathbf{y}$:

$$\mathbf{y} = \mathbf{Z}\boldsymbol{\phi} + \boldsymbol{\epsilon},$$
$$E(\boldsymbol{\epsilon}) = \mathbf{0}, \quad Var(\boldsymbol{\epsilon}) = \sigma^2 \mathbf{I}$$

where each column of $\mathbf{Z}$ contains values of an indicator variable associated with
one of the unique rows of $\mathbf{X}$. For example, consider again the unbalanced,
partially replicated two-way ANOVA example of subsection 2.3.1, with no
interaction term in the model:

$$\mathbf{X} = \begin{pmatrix} 1 & 0 & 1 & 0 & 0 \\ 1 & 0 & 1 & 0 & 0 \\ 1 & 0 & 0 & 1 & 0 \\ 1 & 0 & 0 & 1 & 0 \\ 1 & 0 & 0 & 0 & 1 \\ 0 & 1 & 1 & 0 & 0 \\ 0 & 1 & 1 & 0 & 0 \\ 0 & 1 & 0 & 1 & 0 \\ 0 & 1 & 0 & 0 & 1 \end{pmatrix}, \quad \mathbf{Z} = \begin{pmatrix} 1 & 0 & 0 & 0 & 0 & 0 \\ 1 & 0 & 0 & 0 & 0 & 0 \\ 0 & 1 & 0 & 0 & 0 & 0 \\ 0 & 1 & 0 & 0 & 0 & 0 \\ 0 & 0 & 1 & 0 & 0 & 0 \\ 0 & 0 & 0 & 1 & 0 & 0 \\ 0 & 0 & 0 & 1 & 0 & 0 \\ 0 & 0 & 0 & 0 & 1 & 0 \\ 0 & 0 & 0 & 0 & 0 & 1 \end{pmatrix},$$

for which $N = 9$ and $N^* = 6$.

Define $\mathbf{H_Z} = \mathbf{Z}(\mathbf{Z'Z})^{-1}\mathbf{Z'}$, noting that by its definition $\mathbf{Z}$ must be of full rank (N^*), and so $\mathbf{Z'Z}$ must have a unique inverse. It should be clear that each column of $\mathbf{X}$ can be expressed as a linear combination of those of $\mathbf{Z}$; in fact $\mathbf{X} = \mathbf{ZX^*}$. This immediately implies that the columns of $\mathbf{X}$ lie in the space spanned by those of $\mathbf{Z}$:

$$\mathbf{H_Z X} = \mathbf{H_Z Z X^*} = \mathbf{Z X^*} = \mathbf{X}.$$

This, in turn, leads to a decomposition of the error sum of squares, SSE, as:

$$\mathbf{y'}(\mathbf{I} - \mathbf{H})\mathbf{y} = \mathbf{y'}(\mathbf{H_Z} - \mathbf{H})\mathbf{y} + \mathbf{y'}(\mathbf{I} - \mathbf{H_Z})\mathbf{y}$$

$$SSE = SSLOF + SSPE$$

where "LOF" and "PE" stand for *Lack of Fit* and *Pure Error*, respectively. The corresponding mean squares are

$$MSLOF = \frac{1}{N^* - \text{rank}(\mathbf{X})}SSLOF \quad \text{and} \quad MSPE = \frac{1}{N - N^*}SSPE.$$

Using the results described in Section 2.6, it can be quickly verified that when ϵ is normally distributed, $SSLOF$ and $SSPE$ are independent, each is independent of the treatment sum of squares, $SST = \mathbf{y'}(\mathbf{H} - \mathbf{H_1})\mathbf{y}$, and each, if divided by σ^2, would have a chi-squared distribution. If the model assumptions are correct, both $SSLOF/\sigma^2$ and $SSPE/\sigma^2$ follow *central* chi-squared distributions because $(\mathbf{H_Z} - \mathbf{H})\mathbf{X} = (\mathbf{I} - \mathbf{H_Z})\mathbf{X} = \mathbf{0}$. However, $MSPE$ is an unbiased estimate of σ^2 even if $E(\mathbf{y})$ is actually different from the form specified in the model, so long as the expectation of the response is the same for all trials within a "replication group." In fact, the functional form of $SSPE$ is simply a within-group sum of squares for these groups of runs which, according to our model, have common expectation.

Continuing our example, suppose that $E(\mathbf{y})$ actually includes a two-factor interaction, so that a "true" model could be written using:

$$\mathbf{X}_{true} = \begin{pmatrix} 1 & 0 & 1 & 0 & 0 & 1 & 0 & 0 & 0 & 0 & 0 \\ 1 & 0 & 1 & 0 & 0 & 1 & 0 & 0 & 0 & 0 & 0 \\ 1 & 0 & 0 & 1 & 0 & 0 & 1 & 0 & 0 & 0 & 0 \\ 1 & 0 & 0 & 1 & 0 & 0 & 1 & 0 & 0 & 0 & 0 \\ 1 & 0 & 0 & 0 & 1 & 0 & 0 & 1 & 0 & 0 & 0 \\ 0 & 1 & 1 & 0 & 0 & 0 & 0 & 0 & 1 & 0 & 0 \\ 0 & 1 & 1 & 0 & 0 & 0 & 0 & 0 & 1 & 0 & 0 \\ 0 & 1 & 0 & 1 & 0 & 0 & 0 & 0 & 0 & 1 & 0 \\ 0 & 1 & 0 & 0 & 1 & 0 & 0 & 0 & 0 & 0 & 1 \end{pmatrix}.$$

We have already seen that $\mathbf{H_Z X} = \mathbf{X}$, and since in this case (but not always) the new columns of $\mathbf{X}_{true}$ are the same as those in $\mathbf{Z}$, it should be clear that

$\mathbf{H_Z X}_{true} = \mathbf{X}_{true}$ and $E(MSPE) = \sigma^2$. In this example,

$$SSPE = \sum_{i=1}^{2}(y_i - \bar{y}_{1:2})^2 + \sum_{i=3}^{4}(y_i - \bar{y}_{3:4})^2 + \sum_{i=6}^{7}(y_i - \bar{y}_{6:7})^2$$

where indexing refers to row number in $\mathbf{X}$, and averages are over the indicated ranges of response values. However $(\mathbf{H_Z} - \mathbf{H})\mathbf{X}_{true} = \mathbf{X}_{true} - \mathbf{H}(\mathbf{X}|\mathbf{Z}) = (\mathbf{X}|\mathbf{Z}) - (\mathbf{X}|\mathbf{HZ}) = (\mathbf{0}|(\mathbf{I} - \mathbf{H})\mathbf{Z})$. But $\mathbf{HZ}$, the projection of $\mathbf{Z}$ into the space spanned by the columns of $\mathbf{X}$, is *not* the same as $\mathbf{Z}$, so $E(MSLOF)$ can be greater than σ^2.

Intuitively, if rank$(\mathbf{X}) <$ rank$(\mathbf{Z}) = N^*$ the experimental design has more "estimation capacity" than is minimally required to fit the assumed model, and if $N^* < N$ the design provides information about σ^2 that does not depend upon the assumed form of $E(\mathbf{y})$. Under these circumstances, $MSPE$ is an alternative estimator of σ^2 which is generally less precise than MSE when the assumed model is correct (since the former is associated with fewer degrees of freedom than the latter), but is not biased when the assumed model form is incorrect. A diagnostic test for adequacy of the model (or conversely, for model "lack of fit"):

$$\text{Hyp}_0 : E(\mathbf{y}) = \mathbf{X}\boldsymbol{\theta}$$
$$\text{Hyp}_A : E(\mathbf{y}) \neq \mathbf{X}\boldsymbol{\theta}$$

can be carried out by comparing $MSLOF/MSPE$ to $F_{1-\alpha}(N^* - \text{rank}(\mathbf{X}), N - N^*)$.

2.8 Hypothesis testing and information

In most experimental design settings where $\mathbf{X}_1$ represents an intercept or constant and/or block effects, the overall test for differences among treatments corresponds to:

$$\text{Hyp}_0 : \mathbf{y} = \mathbf{X}_1\boldsymbol{\theta}_1 + \boldsymbol{\epsilon}$$
$$\text{Hyp}_A : \mathbf{y} = \mathbf{X}\boldsymbol{\theta} + \boldsymbol{\epsilon} = \mathbf{X}_1\boldsymbol{\theta}_1 + \mathbf{X}_2\boldsymbol{\theta}_2 + \boldsymbol{\epsilon}.$$

When the elements of $\boldsymbol{\epsilon}$ are independent and normally distributed, the test is based on a statistic comprised of the ratio of the treatment mean square (given all effects represented by $\mathbf{X}_1$):

$$MST = \frac{1}{\text{rank}(\mathbf{X}) - \text{rank}(\mathbf{X}_1)} \, \mathbf{y}'(\mathbf{H} - \mathbf{H}_1)\mathbf{y}$$

and the residual mean square:

$$MSE = \frac{1}{N - \text{rank}(\mathbf{X})} \, \mathbf{y}'(\mathbf{I} - \mathbf{H})\mathbf{y}.$$

The central matrices of these quadratic forms are both idempotent, and aside from scalar factors, their product is:

$$(\mathbf{H} - \mathbf{H}_1)(\mathbf{I} - \mathbf{H}) = \mathbf{H} - \mathbf{H}_1 - \mathbf{H}^2 + \mathbf{H}_1\mathbf{H} = \mathbf{H} - \mathbf{H}_1 - \mathbf{H} + \mathbf{H}_1 = \mathbf{0},$$

proving that the two quadratic forms are independent as required for construction of the F-statistic. Under Hyp_0, $F = MST/MSE$ has a central F distribution with $\text{rank}(\mathbf{X}) - \text{rank}(\mathbf{X}_1)$ and $N - \text{rank}(\mathbf{X})$ degrees of freedom, so the critical value of the test at level α is $F_{1-\alpha}(\text{rank}(\mathbf{X}) - \text{rank}(\mathbf{X}_1), N - \text{rank}(\mathbf{X}))$. Under Hyp_A, F has a noncentral F distribution with noncentrality parameter

$$\lambda = E(\mathbf{y})'(\mathbf{H} - \mathbf{H}_1)E(\mathbf{y})/\sigma^2$$

$$= (\boldsymbol{\theta}_1'\mathbf{X}_1' + \boldsymbol{\theta}_2'\mathbf{X}_2')(\mathbf{H} - \mathbf{H}_1)(\mathbf{X}_1\boldsymbol{\theta}_1 + \mathbf{X}_2\boldsymbol{\theta}_2)/\sigma^2 = \boldsymbol{\theta}_2'\mathbf{X}_2'(\mathbf{H} - \mathbf{H}_1)\mathbf{X}_2\boldsymbol{\theta}_2/\sigma^2$$

because $\mathbf{H}\mathbf{X}_1 = \mathbf{H}_1\mathbf{X}_1 = \mathbf{X}_1$. Furthermore, since $\mathbf{H}\mathbf{X}_2 = \mathbf{X}_2$, λ can be written more simply as:

$$\lambda = \boldsymbol{\theta}_2'\mathbf{X}_2'(\mathbf{I} - \mathbf{H}_1)\mathbf{X}_2\boldsymbol{\theta}_2/\sigma^2 = \boldsymbol{\theta}_2'\mathcal{I}_{2|1}\boldsymbol{\theta}_2/\sigma^2.$$

For an α level test, the power is

$$\text{Prob}\{W > F_{1-\alpha}(\text{rank}(\mathbf{X}) - \text{rank}(\mathbf{X}_1), N - \text{rank}(\mathbf{X}))\}$$

where

$$W \sim F'(\text{rank}(\mathbf{X}) - \text{rank}(\mathbf{X}_1), N - \text{rank}(\mathbf{X}), \lambda).$$

The noncentrality parameter plays an important role in experimental design because, other quantities being equal, the power of an F-test increases with λ. As in estimation, the design information matrix $\mathcal{I}_{2|1}$ is critical to the performance of the hypothesis test. Intuitively, we want $\mathcal{I}_{2|1}$ to be as "large" as possible (in some sense), because this leads to relatively large λ when Hyp_0 is false and hence relatively large power, with the understanding that this cannot be said more precisely until the potential values of $\boldsymbol{\theta}_2/\sigma$ have been characterized.

2.8.1 Example

Recall that for the numerical example discussed in subsections 2.3.1 and 2.5.1, the design information matrix for the column parameters, adjusting for the row parameters, in the two-by-three table is (R2.2):

$$\mathcal{I}_{2|1} = \mathbf{X}_{2|1}'\mathbf{X}_{2|1} = \begin{pmatrix} 2.20 & -1.30 & -0.90 \\ -1.30 & 1.95 & -0.65 \\ -0.90 & -0.65 & 1.55 \end{pmatrix}.$$

Should it be the case that $\boldsymbol{\theta}_2'$ is actually $(-2, 1, 1)$, and σ is actually 1.25, the noncentrality parameter associated with the F-test for equality of column effects is:

$$\lambda = (-2 \ 1 \ 1) \begin{pmatrix} 2.20 & -1.30 & -0.90 \\ -1.30 & 1.95 & -0.65 \\ -0.90 & -0.65 & 1.55 \end{pmatrix} \begin{pmatrix} -2 \\ 1 \\ 1 \end{pmatrix} /1.25^2 = 12.672.$$

If the test is performed at level 0.05, the critical value of the test will be $F_{0.95}(2, 5) = 5.786$ and the probability with which the null hypothesis would be rejected is $\text{Prob}(W > 5.786)$, where W has a $F'(2, 5, 12.672)$ distribution, or 0.6348 (R2.3).

$\mathcal{I}_{2|1}$ is not of full rank, but since any two rows/columns of this matrix are linearly independent, a generalized inverse can be constructed as described in Section 2.3, by removing the last row and column, inverting the resulting 2×2 submatrix, and "padding" the result with a row and column of zeros:

$$\mathcal{I}_{2|1}^{-} = \begin{pmatrix} 0.7500 & 0.5000 & 0 \\ 0.5000 & 0.8462 & 0 \\ 0 & 0 & 0 \end{pmatrix}.$$

We demonstrated in subsection 2.5.1 that $(+1, -1, 0)\boldsymbol{\theta}_2$ is estimable under this design; the least-squares estimate of this quantity would have variance:

$$1.25^2 \times (+1, -1, 0) \begin{pmatrix} 0.7500 & 0.5000 & 0 \\ 0.5000 & 0.8462 & 0 \\ 0 & 0 & 0 \end{pmatrix} \begin{pmatrix} +1 \\ -1 \\ 0 \end{pmatrix} = 0.8584.$$

2.9 Blocking and information

There are at least two different but closely related ways to think about blocking in the context of the linear model. One is to regard blocking as a representation of unavoidable systematic differences between "sections" of an experiment. From this perspective, we account for blocking in design and data analysis so as to *ensure validity* of our results, e.g., to eliminate estimation bias that would damage our inference if these differences were ignored. Here a primary goal of blocking is to protect the unbiased nature of $\widehat{C\boldsymbol{\theta}}_2$, although blocking can also have an effect on the form of $\mathcal{I}_{2|1}$ and so also on the formulae for standard errors of these estimates.

Another view of blocking is that it represents an opportunity to *reduce uncontrolled variation* (represented by σ^2) through experimental control. For example, suppose a biologist plans a study in which mice are the experimental units. The biologist can choose one of three designs:

1. An unblocked design in $N = 30$ mice,

2. A blocked design in 3 "segments," each containing 10 mice, or

3. A blocked design in 10 "segments," each containing 3 mice.

If plan 1 is adopted, units will be randomly selected from the colony of all mice (of a certain strain) available to the investigator. If plan 2 is used, each group of 10 units corresponding to a single block can be selected from a set of animals of the same age, raised in the same cage. If plan 3 is implemented, the three mice in each block can be selected from the same litter of mice (i.e., born at the same time to the same animal). Because the units can be expected to be more *alike* in the experiments with smaller blocks, it is reasonable to expect that uncontrolled variation caused by unit-to-unit differences *within blocks* (often a large portion of what is represented by ϵ) is smaller. In this sense, selection of an experimental design may also influence the value of σ^2, the divisor in Fisher's information.

2.10 Conclusion

The influence of the experimental design on the performance of formal inference concerning the parameters representing treatments is summarized by the design information matrix, $\mathcal{I}_{2|1} = \mathbf{X}_2'(\mathbf{I} - \mathbf{H}_1)\mathbf{X}_2 = \mathbf{X}_{2|1}'\mathbf{X}_{2|1}$. Good designs are structured so that, to the extent possible and with respect to a statistical model that realistically describes the system under study,

$$\lambda = \boldsymbol{\theta}_2'\mathcal{I}_{2|1}\boldsymbol{\theta}_2/\sigma^2$$

is as large as possible for realistic nonzero values of $\boldsymbol{\theta}_2$, and

$$Var(\widehat{\mathbf{c}'\boldsymbol{\theta}_2}) = \sigma^2\mathbf{c}'\mathcal{I}_{2|1}^-\mathbf{c}$$

is as small as possible for interesting estimable functions $\mathbf{c}'\boldsymbol{\theta}_2$.

Given a model, $\mathcal{I}_{2|1}$ is determined entirely by the experimental design selected. A goal of experimental design is that $\mathcal{I}_{2|1}$ should be "large" and $\mathcal{I}_{2|1}^-$ should be "small" in such a way as to provide powerful tests and precise estimates concerning treatment effects. This use of the words "large" and "small" is vague, because one design may have larger λ than another design for one value of $\boldsymbol{\theta}_2$, but not for different values of the parameters. The situation is a little clearer for estimation since we often have some idea of the contrasts of interest ($\mathbf{c}$) before the study begins. The practical problem is often identification of a class of designs that are "good" for many possible values of $\boldsymbol{\theta}_2$ and contrasts $\mathbf{c}$, rather than a single design that is "best" for only one test or estimate.

2.11 Exercises

1. Consider a small experiment and partitioned model for which:

$$\mathbf{X}_1 = \begin{pmatrix} 1 & 1 & 0 \\ 1 & 1 & 0 \\ 1 & 1 & 0 \\ 1 & 0 & 1 \\ 1 & 0 & 1 \\ 1 & 0 & 1 \end{pmatrix} \qquad \mathbf{X}_2 = \begin{pmatrix} 1 & 0 & 0 \\ 0 & 1 & 0 \\ 0 & 0 & 1 \\ 1 & 0 & 0 \\ 0 & 1 & 0 \\ 0 & 0 & 1 \end{pmatrix}$$

(a) Compute the hat matrix for a model containing only the mean structure described in the first partition, $\mathbf{H}_1$.

(b) Compute the "corrected" model matrix associated with the second partition of the model, $\mathbf{X}_{2|1}$.

(c) Using your answer to part (b), characterize the estimable functions of $\boldsymbol{\theta}_2$ given that $\boldsymbol{\theta}_1$ must also be modeled.

2. Continue using $\mathbf{X}_1$ and $\mathbf{X}_2$ from exercise 1.

   (a) Compute the design information matrix for $\boldsymbol{\theta}_2$ given that $\boldsymbol{\theta}_1$ is to be simultaneously considered for this experiment of $N = 6$ observations.

   (b) Show that if this entire experiment is repeated r times (i.e., $\mathbf{X}_1$ and $\mathbf{X}_2$ are as described above, but enlarged to contain r "copies" of each row), that the resulting information matrix would be r times the matrix you computed in part (a).

   (c) Show that the result you proved in part (b) is true for *any* design and partitioned linear model. That is, if $\mathcal{I}_{2|1}^1$ is the information matrix for a given design and model, the information matrix for r replicates of that design and model is $\mathcal{I}_{2|1}^r = r\mathcal{I}_{2|1}^1$.

3. Suppose $\mathbf{y} \sim N(\mu\mathbf{1}_N, \boldsymbol{\Sigma}_{N \times N})$.

   (a) Suppose $\boldsymbol{\Sigma} = \sigma^2\mathbf{I}$. Using properties of linear and quadratic forms, prove that the sample mean and sample variance of the elements of $\mathbf{y}$ are independent statistics.

   (b) Suppose now that

   $$\boldsymbol{\Sigma} = \sigma^2 \begin{pmatrix} 1 & \rho & \rho & \cdots & \rho \\ \rho & 1 & \rho & \cdots & \rho \\ \rho & \rho & 1 & \cdots & \rho \\ \cdots & \cdots & \cdots & \cdots & \cdots \\ \rho & \rho & \rho & \cdots & 1 \end{pmatrix},$$

   where $\rho \in (\frac{-1}{N-1}, 1]$. Prove or disprove that the sample mean and sample variance of the elements of $\mathbf{y}$ are independent statistics.

4. Consider two competing designs, denoted A and B, of the same size (number of runs) which could be used in collecting data to fit a given partitioned model. The design information matrices for these two designs, for the set of parameters of interest $\boldsymbol{\theta}_2$, correcting for the nuisance parameters $\boldsymbol{\theta}_1$, are:

   $$\mathcal{I}_{2|1}^A = \begin{pmatrix} 8 & 8 & 0 \\ 8 & 8 & 0 \\ 0 & 0 & 4 \end{pmatrix} \qquad \mathcal{I}_{2|1}^B = \begin{pmatrix} 6 & 0 & 0 \\ 0 & 6 & 0 \\ 0 & 0 & 6 \end{pmatrix}$$

   Assume that the value of σ^2 would be the same for an experiment conducted using either design.

   (a) Give a vector value for $\boldsymbol{\theta}_2$ for which the noncentrality parameter associated with the test for $\text{Hyp}_0 : \mathbf{y} = \mathbf{X}_1\boldsymbol{\theta}_1 + \boldsymbol{\epsilon}$, $\text{Hyp}_A : \mathbf{y} = \mathbf{X}\boldsymbol{\theta} + \boldsymbol{\epsilon}$ is larger under design A than under design B.

(b) Give a value for $\boldsymbol{\theta}_2$ for which the reverse is true, i.e., the non-centrality parameter is larger under design B.

(c) Note that in parts (a) and (b), you were *not* asked for values of $\boldsymbol{\theta}_2$ that would lead to greater power for one of the designs. What additional information would you need if the question had been posed in this way?

5. Continue to use the two design information matrices from exercise 4:

(a) Identify a linear combination of parameters that is estimable under either design, $\mathbf{c}'\boldsymbol{\theta}_2$, that can be more precisely estimated under design A than under design B.

(b) Identify a linear combination of parameters that is estimable under either design for which the opposite is true, i.e., that can be more precisely estimated under design B.

(Hint: In Section 2.5 we noted that estimable functions $\mathbf{c}'\boldsymbol{\theta}_2$ are those for which $\mathbf{c}'$ can be written as a linear combination of the rows of $\mathbf{X}_{2|1}$. Equivalently, $\mathbf{c}'\boldsymbol{\theta}_2$ is estimable if $\mathbf{c}'$ can be written as a linear combination of the rows of $\mathcal{I}$.)

6. The paired-sample t-test is a popular analysis technique for comparing two experimental treatments in situations where units are available in natural pairs (or blocks of size 2). The experimental layout can be thought of as a p-row (for pairs) by two-column (for treatments) table, and the two-sided t-test is equivalent to the F-test for equal column effects, after accounting for rows, assuming no row-by-column interaction. For an unspecified value of $p > 1$:

(a) write a partitioned model for this situation

(b) fully characterize $\mathbf{X}_1$ and $\mathbf{X}_2$

(c) compute $\mathbf{H}_1$, $\mathbf{X}_{2|1}$, and $\mathcal{I}_{2|1}$.

7. A t-by-t symmetric, positive semi-definite matrix $\mathbf{M}$ has t eigenvalues $\{\lambda_1, \lambda_2, \lambda_3, \ldots, \lambda_t\}$, and t corresponding t-dimensional eigenvectors $\{\mathbf{e}_1, \mathbf{e}_2, \mathbf{e}_3, \ldots, \mathbf{e}_t\}$. Some of their properties are:

$$\lambda_i \geq 0, \quad \mathbf{e}_i'\mathbf{e}_i = 1, \quad \mathbf{e}_i'\mathbf{e}_j = 0, \quad i \text{ and } j = 1, 2, 3, \ldots, t, \ i \neq j.$$

The *eigenvalue-eigenvector decomposition* of $\mathbf{M}$ is:

$$\mathbf{M} = \sum_{i=1}^{t} \lambda_i \mathbf{e}_i \mathbf{e}_i'.$$

Also, if $\mathbf{M}$ is of full rank,

$$\mathbf{M}^{-1} = \sum_{i=1}^{t} \lambda_i^{-1} \mathbf{e}_i \mathbf{e}_i'$$

and if $\mathbf{M}$ is not of full rank, a generalized inverse is given by the same sum including only terms for which $\lambda_i \neq 0$. Suppose that two designs, denoted A and B, have information matrices $\mathcal{I}^A_{2|1}$ and $\mathcal{I}^B_{2|1}$ for the parameters of interest, and that these matrices have the same eigenvectors. Suppose also that $Var(\epsilon) = \sigma^2$ would be the same for either design. Finally, note that a given linear contrast of the treatment parameters is estimable under a given design if and only if the vector of coefficient weights can be expressed as a linear combination of the eigenvectors of $\mathcal{I}_{2|1}$ associated with nonzero eigenvalues.

(a) What must be true of the two sets of eigenvalues if design A offers precision that is at least as good as design B for all estimable functions of the parameters of interest? Prove this.

(b) What must be true of the two sets of eigenvalues if, under design A, the noncentrality parameter associated with the test:

$$\mathrm{Hyp_0} : \mathbf{y} = \mathbf{X}_1\boldsymbol{\theta}_1 + \epsilon$$

$$\mathrm{Hyp_A} : \mathbf{y} = \mathbf{X}\boldsymbol{\theta} + \epsilon$$

is at least as large as it is for design B, regardless of the value of $\boldsymbol{\theta}$? Prove this.

8. Recall R.A. Fisher's tea-tasting experiment (Chapter 1, exercise 1, part (d)). Suppose that, rather than stating her guess as to which ingredient was put into the cup first, the lady responded to each cup with a real-valued number (a response) reflecting her judgment of the tea's taste, and that we are willing to adopt the model:

$$y_{ijk} = \alpha_i + \beta_j + \epsilon_{ijk}$$

where

$$i = 1 \text{ for porcelain cups}$$
$$2 \text{ for china cups}$$
$$j = 1 \text{ for milk-before-tea preparations}$$
$$2 \text{ for tea-before-milk preparations}$$

and k indexes the first ($=1$) or second ($=2$) beverage made for the indicated values of i and j. Assuming that inferences about β_1 and β_2 are of greatest interest:

(a) construct a partitioned model for the experiment,

(b) compute $\mathbf{H}_1$, the hat matrix for the model including only cup-type parameters and random noise, and

(c) compute $\mathbf{X}_{2|1}$ and $\mathcal{I}_{2|1}$.

9. Continuing exercise 8, suppose now that one of the china cups was broken, and one of the tea-before-milk preparations was not made. For the resulting seven-trial experiment, find:

   (a) $\mathbf{H}_1$

   (b) $\mathbf{X}_{2|1}$ and $\mathcal{I}_{2|1}$

   (c) the expected squared length of a 90% confidence interval for $\beta_1 - \beta_2$ (in terms of σ^2).

10. Once again, continuing exercise 8, suppose only *three* beverages are prepared:

    • china cup, with milk-before-tea

    • porcelain cup, with milk-before-tea

    • porcelain cup, with tea-before-milk

    Show that the data value collected from the beverage prepared in the china cup is not used in estimating $\beta_1 - \beta_2$. Explain clearly, and without using mathematics, why this is reasonable.

## CHAPTER 3

Completely randomized designs

3.1 Introduction

The simplest of the experimental designs we shall discuss are the *Completely Randomized Designs* (CRD) for comparing t treatments, using N experimental units. Here (and through Chapter 8) we will be considering experimental designs in the context of *unstructured* treatments, by which we simply mean a discrete collection of experimental conditions, not necessarily related by factors, nesting relationships, specification by common controlled variable values, et cetera. Some of the treatments may be "special" from the standpoint of the experimental context, such as a condition regarded as a *control* against which other treatments are to be compared, but these particulars are generally not used in our characterization of the treatment structure. The name of this class of designs stems from the idea that the particular experimental units identified for application of any one of the treatments are selected from the available units in an unrestricted (or "completely") random manner. More precisely, we generally determine *a priori* the number of experimental units to be assigned to each treatment, $n_1, n_2, n_3, \ldots n_t$, such that their sum is N, the number of units we plan to use. Then we may randomly select n_1 of the units for allocation to the first treatment, n_2 of the remaining units for the second treatment, et cetera, without any *additional* restrictions on the process of random assignment. In particular, unlike the designs to be described in Chapters 4 and 5, a CRD does not contain *blocks* of units purposefully selected to be especially similar; rather all units are viewed as having been selected from a single collection of available units, and while we know they cannot be exactly identical, any pair of them is viewed as being related in the same way as any other pair.

3.1.1 Example: radiation and rats

As part of a study designed to investigate the effects of whole-body X-irradiation on the nervous system of rats, Matsuu et al. (2005) carried out a small experiment involving four experimental treatments. The units in this study were 20 Wistar-Kyoto rats. Treatments numbered 2 through 4 were defined as exposure to a standard dose of radiation, followed by 4, 8, or 24 hours, respectively, before sacrifice and analysis. Group 1 was a control group; rats in this group were housed and handled as those in groups 2-4, but they received no radiation. Upon sacrifice, the adrenal gland of each rat was removed, and

Table 3.1 Epinephrine Levels (Grams, g) in Rats Treated with Whole Body X-Irradiation, from Matsuu et al. (2005)

Treatment			
1	2	3	4
9.934	8.675	10.509	8.829
9.819	10.720	8.067	10.484
10.693	10.040	9.027	8.632
10.106	9.894	9.680	8.352
9.139	11.912	8.967	9.323

the amount of epinephrine in the gland determined, resulting in one data value per rat. Interest in this experiment lay in comparing the groups – especially the control group with each of groups 2 through 4 – for possible differences in the response that could be attributed to the treatments. Table 3.1 contains hypothetical data consistent with the results reported from this experiment. (The table entries were actually reconstructed from graphs of summary statistics published in the paper; as is often the case in research papers, the individual data values were not reported.) Proper experimental practice would call for randomized application of the four treatments to the 20 available rats in such a manner that every group of five rats has the same probability of being assigned to any of the treatments. One way this could have been done in this study is through arbitrary labeling of rats with numbers 1-20, and random selection without replacement of 20 tags from a container, where five of the tags are labeled "treatment 1," five labeled "treatment 2," et cetera. The ith tag drawn determines the treatment to be applied to rat i. In experiments of this type the technicians performing the necropsies often do so "blinded," that is, without knowing the treatment applied to each rat, so as to avoid intentional or unintentional technician bias. As an additional precaution, the temporal order in which the 20 adrenal glands are processed might also be randomized.

3.2 Models

In describing the structure of data from a CRD, models used in one-way analysis of variance are often appropriate. The *cell means model* sometimes used in this setting can be written as:

$$y_{ij} = \mu_i + \epsilon_{ij},$$
$$i = 1 \ldots t, \; j = 1 \ldots n_i,$$
$$\epsilon_{ij} \text{ i.i.d. with } E(\epsilon_{ij}) = 0 \text{ and } Var(\epsilon_{ij}) = \sigma^2 \tag{3.1}$$

for the response recorded from the jth unit allocated to the ith treatment. μ_i is the fixed-but-unknown expectation of responses associated with treatment i, and ϵ represents the independent, random component of each data value. In fact, ϵ often represents variation associated with multiple sources, including:

- unique and generally unknown physical characteristics of each unit,
- unreproducible "errors" associated with application of a treatment to a unit,
- any subsampling of experimental material for analysis, such as selection of an *aliquot*,
- sources of "error" associated with each application of the measurement process, or unrepeatable "noise" associated with the measurement instruments and techniques.

A more extensive model might express these sources of variation individually with multiple terms in place of ϵ, but there is little practical reason to do this in analyzing data from a CRD because these effects cannot be meaningfully separated. Because we want to use an analysis based on an assumption that the ϵ's are independent, it is important that the experiment be carried out so that this assumption is physically realistic, by:

- assuring that treatment-to-unit assignments are made randomly and independently for each unit (within the constraint imposed by the fixed sample sizes),
- applying each treatment individually and independently to each of its allocated units,
- carrying out any material handling or subsampling processes required for response evaluation independently for each unit, and
- applying the measurement process independently for each unit.

The cell means model has intuitive appeal and simple form, but also has an interpretive shortcoming when viewed from the standpoint of controlled experimentation. Recall that, in order to eliminate as much variation as possible, an experiment is performed under conditions that are as closely controlled as possible, so that any differences due to the intentionally varied treatment conditions may be detected. Often, this means that experimental runs receiving the same treatment are actually *artificially similar* compared to what might be expected in nonexperimental settings. For example, an industrial experiment may be carried out to compare quantities of chemicals produced under different versions of a process in a tightly controlled development laboratory, when the *real* interest is in understanding how these quantities would differ in a much larger and more variable production environment. In statistical language, the data collected in an experiment are realizations of random variables *conditioned on* all the specific circumstances held constant in the process of controlling the experiment. Hence, μ_i actually represents the expectation of responses associated with treatment i, *collected on the specific day of the experiment, using the specific batch of raw material employed in*

the experiment, by the single laboratory technician who carried out the exper-
iment ..., that is, with far more detail specified than would be involved in
realistic questions about the treatments.

Because only the experimental treatments, and not the particular circumstances of experimental execution, are of primary interest, the *effects model*:

$$y_{ij} = \alpha + \tau_i + \epsilon_{ij},$$
$$i = 1 \ldots t, \; j = 1 \ldots n_i,$$
$$\epsilon_{ij} \text{ i.i.d. with } E(\epsilon_{ij}) = 0 \text{ and } Var(\epsilon_{ij}) = \sigma^2, \qquad (3.2)$$

has some interpretive advantages. In the effects model, α is a nuisance parameter reflecting the contributions of the relatively uninteresting constant details
imposed on the experiment to create a controlled setting, such as the effects
of being *collected on the specific day of the experiment, using the specific batch
of raw material employed in the experiment, by the single laboratory technician who carried out the experiment* ..., and τ_i is the deviation, from α, of
the mean response associated with treatment i. If we *assume* that this deviation would have been the same regardless of the day/batch/technician and
the effects of all other operational details represented by α, then τ_i represents
information *only* about treatment i relative to the other treatments.

3.2.1 Graphical logic

Because CRDs with unstructured treatments are relatively simple, the form
of a graphical analysis to present experimental results can also be simple. The
fundamental questions are generally about whether the distributions of data
from the various treatment groups, especially their means or other measures of
central tendency, are different. A set of parallel *boxplots* (introduced by Tukey
(1977) and further described by Frigge, Hoaglin, and Iglewicz (1989)) of the
measured data, one boxplot generated from the data from each treatment
group, is a useful presentation for this purpose. It is an easy way to compare
the location and shape of the distributions associated with the various treatments, and so can be helpful in discovering treatment differences that are more
subtle than those easily detected with the standard analysis of variance. An
example of such a plot is displayed in Figure 3.1 for the data from the example
of subsection 3.1.1 (R3.1). Note that, following the logic used in modeling the
data, only the *relative* distances along the vertical axis are meaningful. Since
the common component of all the data is the experiment-specific α, the parallel
boxplot should be viewed as a device for comparing groups – as the formality of
estimable functions described in the next section suggests. Hence, the shapes
of the distributions and relative vertical positions are taken as meaningful,
but the direct comparison of any boxplot to the vertical axis is not.

If the boxplots are reordered by increasing or decreasing means, the plot
can be helpful in diagnostic checking for a relationship between the means and
variances of data from each treatment group. In many instances, the practical concern is for detecting increases in spread associated with increases in

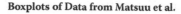

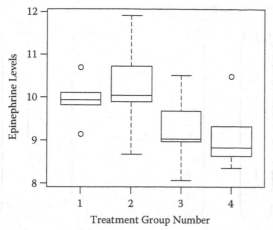

Figure 3.1 Boxplots of reconstructed data of Matsuu et al. (2005)(R3.1).

location, suggesting the need for a data transformation to satisfy the equality-of-variance assumption of the standard analysis of variance as discussed in Chapter 6.

3.3 Matrix formulation

A matrix expression of the cell means model representing all data in the experiment can be written as:

$$
\begin{pmatrix} y_{1,1} \\ \cdots \\ y_{1,n_1} \\ y_{2,1} \\ \cdots \\ y_{2,n_2} \\ \cdots \\ y_{t,1} \\ \cdots \\ y_{t,n_t} \end{pmatrix} = \begin{pmatrix} 1 & 0 & \cdots & 0 \\ \cdots & \cdots & \cdots & \cdots \\ 1 & 0 & \cdots & 0 \\ 0 & 1 & \cdots & 0 \\ \cdots & \cdots & \cdots & \cdots \\ 0 & 1 & \cdots & 0 \\ \cdots & \cdots & \cdots & \cdots \\ 0 & 0 & \cdots & 1 \\ \cdots & \cdots & \cdots & \cdots \\ 0 & 0 & \cdots & 1 \end{pmatrix} \begin{pmatrix} \mu_1 \\ \mu_2 \\ \cdots \\ \mu_t \end{pmatrix} + \begin{pmatrix} \epsilon_{1,1} \\ \cdots \\ \epsilon_{1,n_1} \\ \epsilon_{2,1} \\ \cdots \\ \epsilon_{2,n_2} \\ \cdots \\ \epsilon_{t,1} \\ \cdots \\ \epsilon_{t,n_t} \end{pmatrix} \tag{3.3}
$$

or in more compact form, $\mathbf{y} = \mathbf{X}\boldsymbol{\mu} + \boldsymbol{\epsilon}$, where $E(\boldsymbol{\epsilon}) = \mathbf{0}$ and $Var(\boldsymbol{\epsilon}) = \sigma^2 \mathbf{I}$, and

$$
\mathbf{X} = \begin{pmatrix} \mathbf{1}_{n_1} & \mathbf{0}_{n_1} & \cdots & \mathbf{0}_{n_1} \\ \mathbf{0}_{n_2} & \mathbf{1}_{n_2} & \cdots & \mathbf{0}_{n_2} \\ \cdots & \cdots & \cdots & \cdots \\ \mathbf{0}_{n_t} & \mathbf{0}_{n_t} & \cdots & \mathbf{1}_{n_t} \end{pmatrix}.
$$

The effects model may be written in partitioned matrix form to represent data from the entire experiment as:

$$
\begin{pmatrix} y_{1,1} \\ \cdots \\ y_{1,n_1} \\ y_{2,1} \\ \cdots \\ y_{2,n_2} \\ \cdots \\ y_{t,1} \\ \cdots \\ y_{t,n_t} \end{pmatrix} = \begin{pmatrix} 1 \\ \cdots \\ 1 \\ 1 \\ \cdots \\ 1 \\ \cdots \\ 1 \\ \cdots \\ 1 \end{pmatrix} \alpha + \begin{pmatrix} 1 & 0 & \cdots & 0 \\ \cdots & \cdots & \cdots & \cdots \\ 1 & 0 & \cdots & 0 \\ 0 & 1 & \cdots & 0 \\ \cdots & \cdots & \cdots & \cdots \\ 0 & 1 & \cdots & 0 \\ \cdots & \cdots & \cdots & \cdots \\ 0 & 0 & \cdots & 1 \\ \cdots & \cdots & \cdots & \cdots \\ 0 & 0 & \cdots & 1 \end{pmatrix} \begin{pmatrix} \tau_1 \\ \tau_2 \\ \cdots \\ \tau_t \end{pmatrix} + \begin{pmatrix} \epsilon_{1,1} \\ \cdots \\ \epsilon_{1,n_1} \\ \epsilon_{2,1} \\ \cdots \\ \epsilon_{2,n_2} \\ \cdots \\ \epsilon_{t,1} \\ \cdots \\ \epsilon_{t,n_t} \end{pmatrix} \quad (3.4)
$$

or more compactly, $\mathbf{y} = \mathbf{X}_1 \alpha + \mathbf{X}_2 \tau + \epsilon$, with

$$
\mathbf{X}_1 = \begin{pmatrix} \mathbf{1}_{n_1} \\ \mathbf{1}_{n_2} \\ \cdots \\ \mathbf{1}_{n_t} \end{pmatrix}, \quad \mathbf{X}_2 = \begin{pmatrix} \mathbf{1}_{n_1} & \mathbf{0}_{n_1} & \cdots & \mathbf{0}_{n_1} \\ \mathbf{0}_{n_2} & \mathbf{1}_{n_2} & \cdots & \mathbf{0}_{n_2} \\ \cdots & \cdots & \cdots & \cdots \\ \mathbf{0}_{n_t} & \mathbf{0}_{n_t} & \cdots & \mathbf{1}_{n_t} \end{pmatrix}.
$$

Note that $\mathbf{X}_2$ in the partitioned effects model is exactly the same matrix as $\mathbf{X}$ in the cell means model. Clearly there is no new "structure" added to the data analysis with the second model. The single column in $\mathbf{X}_1$ is, in fact, the sum of columns in $\mathbf{X}_2$, and so the two models are exactly equivalent mathematical statements about the data. However, by framing our analysis around the partitioned model, we force ourselves to consider what information is actually available about the treatments, apart from, or "correcting for," the collection of influences that are common to all data throughout the experiment.

Recall (Section 2.5) that any linear combination of treatment parameters estimable under the effects model must be $\mathbf{c}'\tau$ such that $\mathbf{c}'$ can be written as a linear combination of the rows of $\mathbf{X}_{2|1} = (\mathbf{I} - \mathbf{H}_1)\mathbf{X}_2$. For a CRD:

$$
\mathbf{X}_1'\mathbf{X}_1 = N, \qquad (\mathbf{X}_1'\mathbf{X}_1)^{-1} = \frac{1}{N}, \qquad \mathbf{H}_1 = \frac{1}{N}\mathbf{J}_{N \times N},
$$

$$
\mathbf{X}_{2|1} = \mathbf{X}_2 - \frac{1}{N}(n_1\mathbf{1}_N | n_2\mathbf{1}_N | \ldots | n_t\mathbf{1}_N).
$$

Notice that $\mathbf{X}_{2|1}$ is a "column-centered" version of $\mathbf{X}_2$, that is, each column of $\mathbf{X}_{2|1}$ is formed as the corresponding column of $\mathbf{X}_2$ *minus* the average of elements in that column. Notice also that each row of $\mathbf{X}_{2|1}$ has a zero sum, that is, $1 - \frac{n_1}{N} - \frac{n_2}{N} - \cdots - \frac{n_t}{N} = 0$, and it immediately follows that this must also be true of $\mathbf{c}'$ for any estimable $\mathbf{c}'\tau$. Hence, the only linear combinations of τ's that are estimable are *contrasts* – those for which $\mathbf{c}'\mathbf{1} = 0$.

It is worth spending a bit more effort carefully comparing this simple result to the functions of parameters that are estimable under the cell means model. Since there are no nuisance parameters in the cell means model, $c'\mu$ is estimable if c' is a linear combination of the rows of X. It is immediately obvious that this is no restriction at all; *any* real-valued vector c' can be formed as a linear combination of the rows of X. But why, if $X = X_2$, are there vectors c for which $c'\mu$ is estimable, but $c'\tau$ is not? The answer to this question is that μ and τ, while closely related sets of parameters, are not the same. The relationship between these two vectors is

$$\mu = \tau + \alpha 1.$$

Let c be any t-element vector; $c'\mu$ is estimable, and so is the equivalent expression $c'\tau + \alpha c'1$, but the latter is not $c'\tau$ *unless* $c'1 = 0$, the condition for estimable functions of τ derived above. Put more intuitively, in order to eliminate any experiment-wide effects *common to the means of all observations*, the only estimable functions of treatment-specific parameters are contrasts, because these are the only linear functions that eliminate the common nuisance parameter (α) in the expectation through cancellation.

Continuing the derivation of the design information matrix for the effects model, we drop the subscript "2|1" used for partitioned models, with the understanding that all subsequent design information matrices presented will be for parameters associated with treatments (here τ), controlling for nuisance parameters (here α), and find:

$$\mathcal{I} = X'_{2|1}X_{2|1} = \text{diag}(n) - \frac{1}{N}nn' \qquad (3.5)$$

where $n = (n_1, n_2, \ldots, n_t)'$, and $\text{diag}(n)$ is the square diagonal matrix defined by this vector. In the special case of equal sample sizes for each treatment ($n_i = n, i = 1, 2, 3, \ldots, t$), this reduces to:

$$\mathcal{I} = n\left(I - \frac{1}{t}J\right).$$

Substitution of the specific matrix forms given above yields the reduced normal equations for the CRD:

$$\left[\text{diag}(n) - \frac{1}{N}nn'\right]\hat{\tau} = X'_2 y - n\bar{y}_{..}, \qquad (3.6)$$

where $\bar{y}_{..}$ is the average of all elements of y. Further reduction shows that the ith scalar equation in this set is equivalent to:

$$\hat{\tau}_i - \hat{\bar{\tau}}_w = \bar{y}_{i.} - \bar{y}_{..} \qquad (3.7)$$

where $\hat{\bar{\tau}}_w$ is the weighted average of treatment parameter estimates $\frac{1}{N}\sum_i n_i\hat{\tau}_i$, and $\bar{y}_{i.}$ is the average of the n_i data values associated with treatment i. For estimable functions $c'\tau$, the least-squares estimate is formed as the linear combination of these scalar equations in which the ith equation receives weight

c_i; since the sum of these weights must be zero if $\mathbf{c}'\boldsymbol{\tau}$ is to be estimable, both $\hat{\bar{\tau}}_w$ and $\bar{y}_{..}$ are eliminated from the resulting equation:

$$\sum_{i=1}^{t} c_i \widehat{\tau_i} = \sum_{i=1}^{t} c_i \bar{y}_i. \tag{3.8}$$

so that the least-squares estimate of any contrast of treatment parameters is the same contrast in the corresponding treatment data averages.

3.4 Influence of the design on estimation

The variance of the estimate displayed in equation (3.8) is especially simple to derive because the estimate is a linear combination of independent sample means with known variances (apart from the value of σ^2), but it is instructive to see how it follows the general form discussed in Section 2.7:

$$\sigma^2 \mathbf{c}' \mathcal{I}^{-} \mathbf{c} = \sigma^2 \mathbf{c}' [\mathbf{X}_2'(\mathbf{I} - \mathbf{H}_1)\mathbf{X}_2]^{-} \mathbf{c}.$$

In this case, we can simplify the expression by demonstrating that $\mathrm{diag}(\mathbf{n})^{-1}$ is a generalized inverse of $\mathcal{I} = [\mathbf{X}_2'(\mathbf{I} - \mathbf{H}_1)\mathbf{X}_2]$. Using equation (3.5),

$$\mathcal{I}\mathrm{diag}(\mathbf{n})^{-1}\mathcal{I} = \left[\mathrm{diag}(\mathbf{n}) - \frac{1}{N}\mathbf{n}\mathbf{n}'\right]\mathrm{diag}(\mathbf{n})^{-1}\left[\mathrm{diag}(\mathbf{n}) - \frac{1}{N}\mathbf{n}\mathbf{n}'\right]$$

$$= \mathrm{diag}(\mathbf{n}) - 2\frac{1}{N}\mathbf{n}\mathbf{n}' + \frac{1}{n^2}\mathbf{n}\mathbf{n}'\mathbf{1}\mathbf{n}'$$

$$= \mathrm{diag}(\mathbf{n}) - \frac{1}{N}\mathbf{n}\mathbf{n}'$$

$$= \mathcal{I}. \tag{3.9}$$

Hence for estimable $\mathbf{c}'\boldsymbol{\tau}$,

$$Var(\widehat{\mathbf{c}'\boldsymbol{\tau}}) = \sigma^2 \mathbf{c}' \mathrm{diag}(\mathbf{n})^{-1}\mathbf{c} = \sigma^2 \sum_{i=1}^{t} c_i^2/n_i.$$

Similarly, for any collection of estimable functions $\mathbf{C}\boldsymbol{\tau}$, the variance matrix is:

$$Var(\widehat{\mathbf{C}\boldsymbol{\tau}}) = \sigma^2 \mathbf{C}\mathrm{diag}(\mathbf{n})^{-1}\mathbf{C}'. \tag{3.10}$$

Following the general form discussed in Section 2.7, the expected squared length of a $(1 - \alpha)100\%$ two-sided confidence interval for estimable $\mathbf{c}'\boldsymbol{\tau}$ is

$$4t_{1-\frac{\alpha}{2}}^2(N - \mathrm{rank}(\mathbf{X}))\sigma^2 \mathbf{c}'\mathcal{I}^{-}\mathbf{c} = 4t_{1-\frac{\alpha}{2}}^2(N - t)\sigma^2 \sum_{i=1}^{t} c_i^2/n_i$$

in this case. Ordinarily, one would want to design the experiment so that the square root of this quantity is small relative to anticipated values of $\mathbf{c}'\boldsymbol{\tau}$, i.e., so that

$$2t_{1-\frac{\alpha}{2}}(N-t)\sigma\sqrt{\sum_{i=1}^{t}c_i^2/n_i} << \sum_{i=1}^{t}c_i\tau_i.$$

There are obvious practical limitations in doing this, because the parameter values are not known in advance. However, for any (even unknown) value of $\sum_{i=1}^{t}c_i\tau_i/\sigma$ (what might be called a signal-to-noise ratio), designs that lead to relatively small values of

$$t_{1-\frac{\alpha}{2}}(N-t)\sqrt{\sum_{i=1}^{t}c_i^2/n_i}$$

either through large overall sample size (and so a relatively small t-quantile), or through allocation of relatively more units to groups for which the corresponding $|c_i|$ is large for contrasts of interest, are generally preferable.

3.4.1 Allocation

Because the CRD is a relatively simple and flexible experimental design, it is a good setting in which to introduce the idea of *optimal allocation*. While good experimental practice requires that treatments be applied to individual units randomly, we often have the freedom to select the number of units to be used in each treatment group (n_i) so long as the operational constraints such as limits on total number of units or cost are satisfied. It is often assumed that a CRD that allocates equal numbers of available units to each treatment must necessarily be optimal. A more careful consideration of the allocation problem requires that we take experimental goals into consideration. Specifically, what estimate or test properties are we trying to optimize through allocation? The CRD with equal group sizes has good overall properties, and is in fact optimal for many – but not all – experimental goals.

Allocation problems are formulated as constrained optimization problems in which the quantity to be optimized is, for example, the variance of an estimator, or a reasonable function of more than one such variance. The constraint most often reflects the total number of units allowed. So, for example, if an experiment involving five treatments is being conducted with the sole aim of estimating $\tau_1 - \tau_2$ (admittedly not very realistic), and 50 units may be used altogether, we may wish to select the five sample sizes so as to:

$$\text{minimize } \sigma^2\left(\frac{1}{n_1}+\frac{1}{n_2}\right) \qquad \text{subject to } \sum_{i=1}^{t}n_i = 50.$$

With a little thought, it should be clear that the solution to this problem is $n_1 = n_2 = 25$ and $n_3 = n_4 = n_5 = 0$ for any positive value of σ^2.

Most realistic experiments involve estimation of more than one quantity. Suppose the goal of the experiment is to estimate p linear contrasts of τ, $\mathbf{C}\tau$.

The variance matrix of the least-squares estimate is:

$$Var(\widehat{\mathbf{C}\boldsymbol{\tau}}) = \sigma^2 \mathbf{C}\mathcal{I}^-\mathbf{C}' = \sigma^2 \mathbf{C}\text{diag}(\mathbf{n})^{-1}\mathbf{C}' \tag{3.11}$$

from equation (3.10). If we elect to minimize the *average* variance of these estimates (not the only sensible summary measure, but one that is often useful), this is equivalent to minimization of:

$$\text{trace}(\mathbf{C}\text{diag}(\mathbf{n})^{-1}\mathbf{C}') = \text{trace}[(\mathbf{C}'\mathbf{C})\text{diag}(\mathbf{n})^{-1}] \tag{3.12}$$

regardless of the (in practice, unknown) value of σ^2. Note that the matrix product in the last expression represents a partitioning of the objective into two factors, the first ($\mathbf{C}'\mathbf{C}$) determined by the experimental goals, and the second ($\text{diag}(\mathbf{n})^{-1}$) determined by the design. The objective function we want to minimize can also be written in scalar terms as:

$$\sum_{k=1}^{p}\sum_{i=1}^{t}\{\mathbf{C}\}_{ki}^2 \frac{1}{n_i}. \tag{3.13}$$

Unconstrained optimization of this quantity is obvious; simply choose the largest possible value for each n_i. The solution of the more realistic constrained problem is not so obvious, but can be found using the Method of Lagrangian Multipliers.

Method of Lagrangian multipliers

The allocation of units in a CRD is summarized by the values of integer-valued quantities, $n_1, n_2, \ldots, n_t$. In practice, it is often easier to solve an optimization problem expressed in terms of continuous variables, because this allows the use of standard techniques from calculus. Hence, optimal design problems are often solved as if $n_1, n_2, \ldots, n_t$ were actually continuous variables, and the resulting solution is "rounded" to integer values if necessary. There is no guarantee that the rounded design will actually *be* optimal among exact (integer-valued $\mathbf{n}$) designs in all cases, but it is usually very close to the optimal arrangement.

Constrained optimization of a function of real-valued arguments can be accomplished via the *Method of Lagrangian Multipliers*. Briefly, suppose we wish to maximize or minimize a differentiable function $f(\mathbf{n})$ with respect to $\mathbf{n}$, a real-valued vector of t arguments, subject to the constraint $g(\mathbf{n}) = G$ for a specified differentiable function g and scalar value G. Now introduce a new scalar variable L, and define a function of $t + 1$ arguments:

$$h(\mathbf{n}, L) = f(\mathbf{n}) + L(g(\mathbf{n}) - G). \tag{3.14}$$

The technique calls for solving the set of $t + 1$ simultaneous equations:

$$\frac{\partial}{\partial n_1}h = 0 \quad \frac{\partial}{\partial n_2}h = 0 \quad \ldots \quad \frac{\partial}{\partial n_t}h = 0 \quad \frac{\partial}{\partial L}h = 0. \tag{3.15}$$

Hence, this is an extension of the widely used technique from basic calculus for unconstrained optimization, based on the solution of:

$$\frac{\partial}{\partial n_1} f = 0 \qquad \frac{\partial}{\partial n_2} f = 0 \qquad \ldots \qquad \frac{\partial}{\partial n_t} f = 0.$$

As with this simpler version of the method, a solution found using the Method of Lagrangian Multipliers can actually be any stationary point, and so careful use also requires checking to see that it is of the desired type (e.g., maximizer or minimizer). The "Lagrangian Multiplier" L is not a quantity of direct interest in the specification of the design, but its introduction enforces the desired constraint through the last equation:

$$\frac{\partial}{\partial L} h = g(\mathbf{n}) - G = 0.$$

Applying the Method of Lagrangian Multipliers to the allocation problem, if we wish to minimize the average variance of contrasts $\mathbf{C}\bar{\mathbf{y}}$, where $\bar{\mathbf{y}}$ is the t-element vector of treatment-specific response averages, controlling the total number of units to be a specified value N,

$$h(\mathbf{n}, L) = \sum_{k=1}^{p} \sum_{i=1}^{t} \{\mathbf{C}\}_{ki}^2 n_i^{-1} - L \left(\sum_{i=1}^{t} n_i - N \right) \tag{3.16}$$

$$\frac{\partial}{\partial n_i} h = -\sum_{k=1}^{p} \{\mathbf{C}\}_{ki}^2 n_i^{-2} - L = 0, \qquad i = 1 \ldots t \tag{3.17}$$

$$\frac{\partial}{\partial L} h = N - \sum_{i=1}^{t} n_i = 0. \tag{3.18}$$

The t equations described by (3.17) yield:

$$n_i \propto \sqrt{\sum_{k=1}^{p} \{\mathbf{C}\}_{ki}^2}, \qquad i = 1 \ldots t.$$

As is often the case with the Method of Lagrangian Multipliers, the last equation may not need to be formally solved at all. Here, we know that the sample sizes must total N, and so the solution for the (continuous) optimal design is:

$$n_i = \frac{\sqrt{\sum_{k=1}^{p} \{\mathbf{C}\}_{ki}^2}}{\sum_{j=1}^{t} \sqrt{\sum_{k=1}^{p} \{\mathbf{C}\}_{kj}^2}} N = \frac{\sqrt{\{\mathbf{C}'\mathbf{C}\}_{ii}}}{\sum_{j=1}^{t} \sqrt{\{\mathbf{C}'\mathbf{C}\}_{jj}}} N, \qquad i = 1 \ldots t. \tag{3.19}$$

As an example, consider the experiment of Matsuu et al. (2005) described in subsection 3.1.1, in which 20 rats were used to compare four treatments. Since the first treatment was an experimental control, interest might be focused on estimating the differences $\tau_2 - \tau_1$, $\tau_3 - \tau_1$, and $\tau_4 - \tau_1$, leading to:

$$\mathbf{C} = \begin{pmatrix} -1 & 1 & 0 & 0 \\ -1 & 0 & 1 & 0 \\ -1 & 0 & 0 & 1 \end{pmatrix}.$$

From equation (3.19),

$$n_1 = \frac{\sqrt{3}}{\sqrt{3}+3}20 = 7.32, \qquad n_2 = n_3 = n_4 = \frac{1}{\sqrt{3}+3}20 = 4.23,$$

suggesting the CRD allocation defined by the integer values $n_1 = 8$ and $n_2 = n_3 = n_4 = 4$. (Note that the values cannot be independently rounded since this would lead to a sum of 21 units.) This imbalance reflects the fact that treatment 1 is involved in all comparisons of interest, while the remaining treatments are each involved in only one comparison.

A useful general tutorial on the Method of Lagrangian Multipliers can be found in Edwards (1994).

3.4.2 Overall experiment size

In the allocation problem described above, for given experimental goals, optimal values of n_i, $i = 1, 2, 3, \ldots, t$, are derived under the constraint of a specified value for N. We can also consider the complimentary problem; suppose the *proportion* of units to be used with each treatment has been determined, and call these p_i, $i = 1, 2, 3, \ldots, t$, $\sum_i p_i = 1$. How large would the experiment have to be, that is, how large would N have to be, to assure that the experimental objectives are met?

Suppose the design information matrix $\mathcal{I}$ has been computed for a CRD based on an arbitrary value of N total units, with the required proportions assigned to each treatment. We can denote a *per-observation design informa- tion matrix* as $\mathcal{I}^1 = \mathcal{I}/N$. Note that for CRDs with fixed proportions of units assigned to each treatment, $\mathcal{I}^1$ would be the same matrix regardless of the value of N used to compute $\mathcal{I}$. (Proof of this is left to the reader; see Exercise 6 at the end of this chapter.) In fact, from equation (3.5) we see that

$$\mathcal{I}^1 = \text{diag}(\mathbf{p}) - \mathbf{p}\mathbf{p}'$$

where $\mathbf{p} = (p_1, p_2, p_3, \ldots, p_t)'$, and following equation (3.9), a generalized inverse is:

$$\mathcal{I}^{1-} = \text{diag}(\mathbf{p})^{-1}.$$

Then, for *any* value of N,

$$\mathcal{I} = N\mathcal{I}^1 = N[\text{diag}(\mathbf{p}) - \mathbf{p}\mathbf{p}'],$$

$$\mathcal{I}^- = \frac{1}{N}\mathcal{I}^{1-} = \frac{1}{N}\text{diag}(\mathbf{p})^{-1}.$$

If a particular linear contrast $\mathbf{c}'\boldsymbol{\tau}$ is of interest, the variance of its estimate is a function of N:

$$Var(\widehat{\mathbf{c}'\boldsymbol{\tau}}) = \sigma^2 \mathbf{c}' \mathcal{I}^- \mathbf{c} = \frac{\sigma^2}{N} \mathbf{c}' \text{diag}(\mathbf{p})^{-1} \mathbf{c} = \frac{\sigma^2}{N} \sum_{i=1}^{t} c_i^2 / p_i.$$

We want a design for which the square root of this variance will be small relative to $\mathbf{c}'\boldsymbol{\tau}$, hence N should be large enough to make

$$\frac{\mathbf{c}'\boldsymbol{\tau}}{\sqrt{Var(\widehat{\mathbf{c}'\boldsymbol{\tau}})}} = \sqrt{N} \frac{\mathbf{c}'\boldsymbol{\tau}/\sigma}{\sqrt{\mathbf{c}' \text{diag}(\mathbf{p})^{-1} \mathbf{c}}}$$

acceptably large. Use of this equation requires a value be proposed for the unitless signal-to-noise ratio $\psi = \mathbf{c}'\boldsymbol{\tau}/\sigma$; given this, we can easily determine the experiment size needed to, for example, result in a specified value of $\Psi = \mathbf{c}'\boldsymbol{\tau}/\sqrt{Var(\widehat{\mathbf{c}'\boldsymbol{\tau}})}$:

$$N = \Psi^2 \frac{\mathbf{c}' \text{diag}(\mathbf{p})^{-1} \mathbf{c}}{\psi^2}.$$

3.5 Influence of design on hypothesis testing

The comparison of experimental interest is between variation associated with $\boldsymbol{\tau}$, and variation associated with $\boldsymbol{\epsilon}$. This comparison is made formally using the F-statistic associated with the test of the null hypothesis Hyp_0 : $\tau_1 = \tau_2 = \tau_3 = \ldots = \tau_t$, computed as the ratio of:

$$MST = \sum_{i=1}^{t} n_i(\bar{y}_{i.} - \bar{y}_{..})^2/(t-1) \tag{3.20}$$

and

$$MSE = \sum_{i,j}(y_{ij} - \bar{y}_{i.})^2/(N-t) \tag{3.21}$$

with critical value $F_{1-\alpha}(t-1, N-t)$, where α is the selected level or size of the test. For a given value of $\boldsymbol{\tau}$, the noncentrality parameter associated with this test is:

$$\lambda = \boldsymbol{\tau}'\mathcal{I}\boldsymbol{\tau}/\sigma^2 = \boldsymbol{\tau}' \left[\text{diag}(\mathbf{n}) - \frac{1}{N}\mathbf{n}\mathbf{n}' \right] \boldsymbol{\tau}/\sigma^2 = \sum_i n_i(\tau_i - \bar{\tau}_w)^2/\sigma^2, \tag{3.22}$$

where $\bar{\tau}_w$ is the weighted average of the elements of $\boldsymbol{\tau}$, $\frac{1}{N}\mathbf{n}'\boldsymbol{\tau}$. In particular, the power of the test of equal treatment effects, for given values of $\boldsymbol{\tau}$ and σ^2, is

$$\text{Prob}\{W > F_{1-\alpha}(t-1, N-t)\},$$

where

$$W \sim F'\left(t-1, N-t, \sum_i n_i(\tau_i - \bar{\tau}_w)^2/\sigma^2\right).$$

For example, in the design of the experiment of Matsuu et al. described in subsection 3.1.1, investigators might have wanted to know the power of the test of equal treatment effects, under hypothetical conditions in which $\tau' = (0, 0, -1, -2)$, and $\sigma = 0.75$. For the design specified (five rats assigned to each of the four groups), the critical value of a 0.01-level test would be:

$$F_{0.99}(3, 16) = 5.292$$

and the resulting power would be based on a noncentral F-variate, W, with noncentrality parameter 24.444,

$$\text{Prob}\{W > 5.292\} = 0.845,$$

(R3.2).

3.6 Conclusion

Completely Randomized Designs are the simplest and least restrictive class of experimental designs. They are appropriate when the experimental material is homogeneous, e.g., not acquired or processed in batches, within which the material is especially consistent, and where treatments can be applied to experimental units without restriction. In fact, different numbers of units can easily be allocated to treatments in a CRD; this flexibility is more difficult to accommodate in many other standard classes of designs. Where the available experimental material *does* have structure, such as batches or groups of especially similar units, other classes of designs take advantage of this fact by allowing "correction for" variation that might be attributable to block differences before assessing the effects of treatments.

3.7 Exercises

1. Consider a completely randomized design with four treatment groups, with $n_i > 0$ units assigned to treatment $i = 1, 2, 3, 4$.

   (a) One way to model data from such an experiment is with the effects model:
   $$y_{ij} = \alpha + \tau_i + \epsilon_{ij} \quad i = 1, 2, 3, 4; \quad j = 1 \ldots n_i$$
   Under this model, show why each of the following is estimable or nonestimable.
   $$\tau_3 \qquad \tau_3 - \tau_2 \qquad \tau_3 + \tau_2$$

   (b) Now define a different model for the same experiment, as:
   $$y_{1,j} = \mu_1 + \epsilon_{1,j} \quad j = 1 \ldots n_1$$
   $$y_{ij} = \mu_1 + \theta_i + \epsilon_{ij} \quad i = 2, 3, 4; \quad j = 1 \ldots n_i$$
   Under this model, show why each of the following is estimable or nonestimable.
   $$\theta_3 \qquad \theta_3 - \theta_2 \qquad \theta_3 + \theta_2$$

(c) Clearly explain in words the difference between the meaning of τ_2 in the first model and θ_2 in the second.

(d) Is there any four-element vector c such that $c'\tau$ is estimable under model 1, but such that $c'\theta$ is not estimable under model 2? If yes, give an example of such a linear combination; if no, show why this is so.

2. Consider a completely randomized design with five treatment groups, in which a total of $N = 50$ units are to be used. Although it won't be explicitly used in the analysis model, treatments 1 through 5 actually represent increasing concentrations of one component in an otherwise standard chemical compound, and the primary purpose of the experiment is to understand whether certain measurable properties of the compound change with this concentration. The investigator decides to address these questions by estimating four quantities:

$$\tau_2 - \tau_1, \ \tau_3 - \tau_2, \ \tau_4 - \tau_3, \ \tau_5 - \tau_4,$$

where each τ_i is a parameter in the standard effects model. Find the optimal allocation for the 50 available units (i.e., values for $n_1 \ldots n_5$) that minimizes the average variance of estimates of the four contrasts of interest. Do this as a constrained, continuous optimization problem, then round the solution to integer values that are consistent with the required constraint.

3. Continue working with the experimental design described in exercise 2. Suppose the experiment-specific treatment means in this problem, as would be expressed in the cell means model, are actually:

μ_1	μ_2	μ_3	μ_4	μ_5
10	11	12	12	12

and $\sigma = 2$. What is the power of the standard F-test for the hypothesis $\tau_1 = \tau_2 = \tau_3 = \tau_4 = \tau_5$, at $\alpha = 0.05$:

(a) if all $n_i = 10$?

(b) under the optimal sample allocation you found in problem 2?

(c) Derive *an* optimal allocation for the F-test of equal treatment effects, i.e., the sample sizes (totaling 50) that would result in the greatest power, if in reality the experiment-specific means are $\mu_1 = 10$ and $\mu_2 = \mu_3 = \mu_4 = \mu_5 = 8$.

4. Consider a situation in which an investigator plans to execute a CRD to compare $t = 3$ treatments using N units, assigned to treatments in proportions $n_1 = N/2$, $n_2 = N/4$, and $n_3 = N/4$. The experiment he envisions is rather small (relatively small value of N), but could be accomplished with reasonably tight experimental control. However, he might also be able to conduct a larger experiment, continuing to use sample sizes in

the same proportions, if he were willing to use experimental units from a larger source. The difficulty is that the units in the alternative source are more heterogeneous, and should reasonably be expected to result in a value of σ^2 about four times the size of the experimental variance he would encounter with his present plan.

(a) How much larger would the alternative experiment have to be so that the precision of point estimators would be the same under the two plans?

(b) If the present plan is to use $N = 4$ units, what is the smallest possible number of units that would be needed in the larger plan so that the expected squared length of 95% confidence intervals would be reduced?

5. A completely randomized design is to be used in an experiment to compare $t = 3$ treatments. The contrasts of greatest interest to the investigator are:

$$\mu_1 - \mu_2, \quad \mu_1 - \mu_3, \quad \text{and} \quad \mu_2 - \mu_3$$

The investigator would like to minimize the average of the variances of the least squares estimators of these three quantities, i.e.,

$$\frac{1}{3}\left(Var(\widehat{\mu_1 - \mu_2}) + Var(\widehat{\mu_1 - \mu_3}) + Var(\widehat{\mu_2 - \mu_3})\right)$$

However, the costs are not the same for the three treatments. The cost of each observation is:

$$
\begin{aligned}
\text{treatment 1} &: \quad \$1 \\
\text{treatment 2} &: \quad \$2 \\
\text{treatment 3} &: \quad \$3
\end{aligned}
$$

Further, the total experimental budget is fixed at $100. Given this constraint, what is the optimal design (i.e., values of the three sample sizes) of this study? (Treat the n_i as continuous variables in this problem.)

6. As described in subsection 3.4.2, show that for any CRD in which the proportion of units assigned to treatment i is $p_i, i = 1, 2, 3, \ldots, t$, $\mathcal{I}/N = \text{diag}(\mathbf{p}) - \mathbf{p}\mathbf{p}'$.

7. A forensics researcher is interested in comparing the performance of a fingerprint expert to that of a new automated computer system for assessing whether two presented fingerprints "match." Both the expert and the automated system judge the quality of the match by the "number of points of agreement" they find, reported as a single number. An experiment is designed in which the expert and the automated system each evaluate N pairs of fingerprints, some of which are known to match (i.e., were made by the same finger) and the others known not to match. For each pair, both the expert's score and the automated system's score will be the recorded responses.

Discuss this experimental situation and think about how it does or does not fit the pattern of a CRD. Specifically, what are the units and treatments here? Given access to all the fingerprints you want, including information on how each was made, how would you randomize this study and why? How might you analyze the data if the score differences could be assumed to be approximately normally distributed?

8. A chemical engineer is interested in comparing three different versions of a reaction process, labeled A, B, and C, with respect to "percent conversion of feedstock." In a preliminary experiment, she applied each process to four batches of raw material, using appropriate randomization of the 12 available batches to the three treatments, and collected the percent conversion values presented in the following table.

Treatment		
A	B	C
27.3	41.9	36.5
34.6	36.8	39.2
31.8	38.2	35.1
35.4	38.4	34.7

Assuming the data are independent and can be reasonably modeled as:

$$y_{ij} = \mu_i + \epsilon_{ij}, \quad E(\epsilon_{ij}) = 0, \quad Var(\epsilon_{ij}) = \sigma^2:$$

(a) Estimate σ^2 and test the $\text{Hyp}_0 : \mu_1 = \mu_2 = \mu_3$.

(b) Using your estimate of σ^2 as if it were the true parameter value, how large would a follow-up experiment (with equal sample sizes) have to be so that the 0.05-level confidence interval for each $\mu_i - \mu_j$ would have expected squared width $(5\%)^2$?

9. One popular model used in the statistical characterization of measurement methodologies is

$$y = x + \beta + \epsilon, \quad E(\epsilon) = 0, \quad Var(\epsilon) = \sigma^2,$$

where x is the "measurand" (the true value of interest), β characterizes measurement *bias*, σ characterizes measurement *precision*, and y is the measurement. Large values of $|\beta|$ correspond to poor *accuracy*. Repeated measurements of the same physical quantity have variance σ^2, and the average of many measurements of the same quantity has error approaching β.

Suppose you compare five measurement methods by using each to measure the same unknown quantity ten times, denoting the bias parameter for method i by β_i, $i = 1, 2, 3, 4, 5$. Assume appropriate randomization of measurement order has been carried out.

(a) Can this experiment be used to support estimation of $\beta_i - \beta_j$?

(b) Does this experiment provide information about which measurement method is most accurate, i.e., has smallest $|\beta_i|$?

(c) If the model is extended to allow a different variance for each method, does this experiment provide information about which measurement method is most precise, i.e., has smallest σ_i^2?

10. Continuing exercise 9, suppose the sample means and standard deviations are:

	Method				
	1	2	3	4	5
$\bar{y}$:	10.12	10.43	10.27	9.98	10.08
s:	0.151	0.113	0.176	0.082	0.121

where units are in grams (weight).

(a) Assuming all methods are equally precise, test:

$$\text{Hyp}_0 : \beta_1 = \beta_2 = \beta_3 = \beta_4 = \beta_5$$

(b) Assuming $x = 10$, and that all methods are equally precise, test

$$\text{Hyp}_0 : \beta_1 = \beta_2 = \beta_3 = \beta_4 = \beta_5 = 0$$

(c) Assuming methods 1, 2, 3, and 5 are equally precise, test

$$\text{Hyp}_0 : \sigma_1^2 = \sigma_4^2$$

CHAPTER 4

Randomized complete blocks and related designs

4.1 Introduction

The completely randomized designs described in Chapter 3 are the simplest general class of experimental designs we shall discuss. A fundamental premise of the CRD is that the available collection of experimental units is homogeneous, and that the treatments can be randomly applied in such a way that any subsample of n_i units have the same probability of being assigned to the ith treatment. Hence, no predictable or systematic differences are expected in the collected data other than those that are attributable to the treatments.

While the CRD is simple, popular, and frequently used – especially in smaller experiments – its application is unrealistic or impractical in many settings. For example, an experiment might be performed to compare the resilience of surface finish resulting from the use of five different additives mixed in a standard base paint. An experimental unit includes the section of surface to be painted, along with the physical quantity of base paint used in any one test application. The standard paint contained in a "batch" (say, a commercially produced one-gallon container) may be very uniform in its properties, but different batches might be known to vary somewhat in consistency and other physical properties that could have some (hopefully small) effect on surface resilience. If a single batch of base paint contains only enough material to prepare five test formulations for application, this means that the five experimental units made from any one batch should be regarded as being more similar than a collection of units made from five different batches. Because every pair of units used in a CRD must be regarded as having the same "similarity" relationship, this leaves only two unpleasant choices for applying such a design here:

- Use only homogeneous units from the same batch, implying that N can be no greater than 5.

- Use only one unit from each batch, and so intentionally use units that are less homogeneous than those from a common batch.

In practice, of course, all five units from each of several batches of paint would be used in such an experiment, even though the units taken from one batch would be expected to be more homogeneous than those taken from different batches. Because a CRD is not appropriate in this application, the experiment is executed in *blocks*, or "sub-experiments," each using only the experimental material from one batch. In our example, the five units prepared from one batch of material are referred to as a block. If each of five treatments is applied once using the material in each block, making a *complete* unreplicated sub-experiment in our paint study, and this pattern is repeated using b such blocks, the entire experimental plan is called a *Complete Block Design* (CBD). Fisher is generally credited with developing the idea of blocks from a statistical perspective, first for agricultural experiments (1926) and later for more general settings (1971); see, e.g., Preece (1990) for a historical account of this and other of Fisher's early contributions to experimental design.

In a *Randomized* Complete Block Design, treatments are randomly applied to units within each block, but these random assignments cannot be made in the same manner as in a CRD because they are restricted to balance across the units within each block. So, for example, after dividing a batch of paint into five quantities, additive 1 might be applied to the first such unit selected at random, but after this assignment is made, none of the remaining quantities from that batch could be used with additive 1. Note that if treatments are sequentially applied to units randomly selected from those still available in the block, the final assignment is automatic, that is, completely determined after the other four units have been allocated.

4.1.1 Example: structural reinforcement bars

Kocaoz, Samaranayake, and Nanni (2005) performed a laboratory experiment to compare the effects of four coatings on the tensile strength of steel reinforcement bars of the type used in concrete structures. Three of the coatings were formed from a common matrix of Engineering Thermoplastic Polyurethane (ETPU), embedded with glass fibers, carbon fibers or aramid fibers, respectively; the fourth coating consisted of ETPU only (i.e., no added fibers) and served as an experimental control. The $N = 32$ specimens (coated bars) were:

> "... *prepared in eight groups of four, with each bar type represented in each of the eight groups. The groups act as the 'blocks' in a randomized complete block ... design, thus adjusting for systematic trends in environmental factors or testing conditions across time. The bars with(in) each group were prepared in random order ...*"

The prepared bars were tested (destructively) for strength in a set-up requiring each bar to be anchored in a pipe filled with grout. The bars from a given block were tested together:

> "*Since all four bars in a group were tested within a short period of time (1h) it is assumed that the test conditions within a group were similar. Also, for each group, a single batch of cementitious grout was prepared, thus eliminating any variation due to grout differences among the bars within each group.*"

Data reported on bar tensile strength are presented in Table 4.1, by block and coating type.

Table 4.1 Tensile Strength (Kilograms Per Square Inch, ksi) of Steel Reinforcement Bars, from Kocaoz et al. (2005)

Block	Coating			
	1	2	3	4
1	136	147	138	149
2	136	143	122	153
3	150	142	131	136
4	155	148	130	129
5	145	149	136	139
6	150	149	147	144
7	147	150	125	140
8	148	149	118	145

4.2 A model

The structure of a CBD suggests two potential systematic patterns in the data. One is associated with the applied treatments, and so is of primary interest to the investigator. The other is associated with blocks of units. These block differences are generally not of interest but, because they can lead to systematic patterns in the data, must be taken into account in modeling and analysis. An effects model that accommodates patterns of variation due to both t treatments and b blocks is:

$$y_{ij} = \alpha + \beta_i + \tau_j + \epsilon_{ij},$$

$$i = 1 \ldots b, \; j = 1 \ldots t,$$

$$\epsilon_{ij} \text{ i.i.d. with } E(\epsilon_{ij}) = 0 \quad \text{and} \quad Var(\epsilon_{ij}) = \sigma^2 \tag{4.1}$$

for the response for the jth treatment from the ith block. As with the effects model presented in Chapter 3 for the CRD, α represents the influences common to all runs in the experiment, and τ_j represents the systematic (common) contribution of the jth treatment to its responses. Similarly, β_i represents any systematic contribution unique to the units in the ith block. It is assumed that *all* of the common, systematic influences of the units in block i are represented by α and β_i, just as it was assumed that all of the common, systematic influences shared by all units in a CRD are represented by α. Given this accommodation for blocks, we assume that any further unit influence is random, and independent from unit to unit.

Two further important points should be made about model (4.1) before we proceed. The first is that in the present context, blocks are treated as fixed effects, symmetric to treatments in the statement of the model, even if this symmetry does not extend to the scientific interest of the investigation. In some settings, the block effect may be appropriately regarded as random instead; we shall address the use of random blocks in Chapter 8. The second point is that model (4.1) includes no block-by-treatment interaction terms.

We sometimes rephrase this to say that we assume the effect of blocks is *additive*. This is a straightforward but critical assumption for the usual analysis of CBD data, since inclusion of a block-by-treatment interaction would lead to a variance decomposition with no residual degrees of freedom (and so no standard F- or t-tests for treatment differences). Careful analysis of data from a CBD should include a diagnostic check of the validity of this assumption.

4.2.1 Graphical logic

As we saw in Chapter 3, parallel boxplots of response values from CRDs can be useful graphical aids in understanding the systematic differences between treatment groups and random variation within treatment groups. The only reservation in interpretation of these graphs was the understanding that the origin of the common axis isn't really meaningful since the (essentially arbitrary and uninteresting) experiment-, technique-, and material-influences common to all data values could well be different in, say, a second experiment run in a different laboratory on a different day.

Boxplots of data from a CBD should be constructed with somewhat more care because each data value contains a contribution from a specific block as well as a specific treatment. So, a simple boxplot of the data associated with a given treatment will display variation originating both with random noise *and* with blocks. In fact, parallel treatment-specific boxplots constructed from a CBD in which blocks contribute a large proportion of the total variation are likely to be quite uninformative, because each boxplot will be artificially broad.

Instead, consider a boxplot of "block-corrected" observations for each treatment group. For treatment j, summarize data by a boxplot of b values:

$$y_{ij}^* = y_{ij} - \bar{y}_{i.}, \quad i = 1, 2, 3, \ldots, b. \tag{4.2}$$

Since, according to the model, $y_{ij} = \alpha + \beta_i + \tau_j + \epsilon_{ij}$, and $\bar{y}_{i.} = \alpha + \beta_i + \bar{\tau}_. + \bar{\epsilon}_{i.}$, then

$$y_{ij}^* = (\tau_j - \bar{\tau}_.) + (\epsilon_{ij} - \bar{\epsilon}_{i.}). \tag{4.3}$$

Hence the systematic component of y_{ij}^* reflects only the contribution of treatment j, relative to the average effect of all treatments. As with the plots of uncorrected data for CRDs, we lose any sense of the origin on the measurement axis, but retain meaningful comparisons between groups. The random component of y_{ij}^* has variance $\sigma^2(1 - \frac{1}{t})$ under the assumed model, and so underestimates the variation that would be seen in repeated application of a single treatment to units from a common batch, but this reduction is minor if t is not small. This "block-corrected" boxplot can be easily generated in a statistical computing environment by constructing the treatment-specific boxplots of residuals from a model containing only block effects; an example based on the data from Table 4.1 is displayed in Figure 4.1 (R4.1).

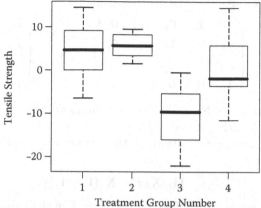

Block–Corrected Data from Kocaoz et al.

Figure 4.1 Boxplots of block-corrected data of Kocaoz et al. (2005)(R4.1).

Mild block-by-treatment interaction is not easily detected in a CBD; however, a similar "correction" strategy can be used to generate a graphical check of the assumed additive effects of treatments and blocks. Just as $y_{ij} - \bar{y}_{i.}$ "corrects" y_{ij} for a block effect, consider the block-and-treatment-corrected data, or equivalently, the residuals from the fit of model (4.1) to the data:

$$r_{ij} = y_{ij} - \bar{y}_{i.} - \bar{y}_{.j} + \bar{y}_{..} = y_{ij} - \hat{y}_{ij}.$$

A little algebra reveals that, under the assumed model:

$$E(r_{ij}) = 0 \qquad Var(r_{ij}) = \sigma^2 \frac{(t-1)(b-1)}{tb}.$$

These values are correlated; however, if t and b are not too small, they should roughly "look like" a random sample with constant mean and variance. Any outliers appearing in a boxplot of these values may be indicators of model inadequacy, such as possible treatment-block interaction. But they *could* also simply indicate experimental runs that were unusual for other reasons. Familiarity with the details of how the experiment was actually carried out is critical to interpreting this plot (and diagnostic plots in general).

4.3 Matrix formulation

Following the scalar representation of model (4.1), we may use matrix notation to write a partitioned model for all data from the experiment:

$$\mathbf{y} = \mathbf{X}_1 \boldsymbol{\beta} + \mathbf{X}_2 \boldsymbol{\tau} + \boldsymbol{\epsilon} \qquad \boldsymbol{\epsilon} \sim N(\mathbf{0}, \sigma^2 \mathbf{I}), \tag{4.4}$$

where $\boldsymbol{\beta}$ is the $(b+1)$-element vector of nuisance parameters α and β_i, $i = 1, 2, 3, \ldots, b$, $\boldsymbol{\tau}$ is the t-element vector of treatment parameters, and $\mathbf{y}$ and $\boldsymbol{\epsilon}$ are $N(= b \times t)$-element vectors of responses and random "errors," respectively.

If the elements of $\mathbf{y}$ are ordered by block, and by treatments within each block, the model matrices are:

$$\mathbf{X}_1 = \begin{pmatrix} \mathbf{1}_t & \mathbf{1}_t & \mathbf{0}_t & \cdots & \mathbf{0}_t \\ \mathbf{1}_t & \mathbf{0}_t & \mathbf{1}_t & \cdots & \mathbf{0}_t \\ \cdots & \cdots & \cdots & \cdots & \cdots \\ \mathbf{1}_t & \mathbf{0}_t & \mathbf{0}_t & \cdots & \mathbf{1}_t \end{pmatrix} \qquad \mathbf{X}_2 = \begin{pmatrix} \mathbf{I}_{t \times t} \\ \mathbf{I}_{t \times t} \\ \cdots \\ \mathbf{I}_{t \times t} \end{pmatrix},$$

where $\mathbf{1}_t$, $\mathbf{0}_t$, and $\mathbf{I}_{t \times t}$ refer to t-element column vectors of 1's and 0's, and the $t \times t$ identity matrix, respectively.

Starting with the general form of the reduced normal equations from Chapter 2, we have

$$\mathbf{X}_2'(\mathbf{I} - \mathbf{H}_1)\mathbf{X}_2\hat{\boldsymbol{\tau}} = \mathbf{X}_2'(\mathbf{I} - \mathbf{H}_1)\mathbf{y}.$$

Some of these matrices have different structure than their counterparts in the CRD. In particular:

$$\mathbf{X}_1'\mathbf{X}_1 = \begin{pmatrix} N & t\mathbf{1}_b' \\ t\mathbf{1}_b & t\mathbf{I}_{b \times b} \end{pmatrix}.$$

This matrix is of rank b; a generalized inverse can be constructed by omitting the first row and column, inverting the remainder, and "padding" the result with zeros corresponding to the omitted row and column:

$$(\mathbf{X}_1'\mathbf{X}_1)^- = \begin{pmatrix} 0 & \mathbf{0}_b' \\ \mathbf{0}_b & \frac{1}{t}\mathbf{I}_{b \times b} \end{pmatrix},$$

and it follows that:

$$\mathbf{H}_1 = \frac{1}{t} \begin{pmatrix} \mathbf{J}_{t \times t} & \mathbf{0}_{t \times t} & \cdots & \mathbf{0}_{t \times t} \\ \mathbf{0}_{t \times t} & \mathbf{J}_{t \times t} & \cdots & \mathbf{0}_{t \times t} \\ \cdots & \cdots & \cdots & \cdots \\ \mathbf{0}_{t \times t} & \mathbf{0}_{t \times t} & \cdots & \mathbf{J}_{t \times t} \end{pmatrix},$$

$$\mathbf{X}_{2|1} = (\mathbf{I} - \mathbf{H}_1)\mathbf{X}_2 = \mathbf{X}_2 - \frac{1}{t}\mathbf{J}_{N \times t},$$

where $\mathbf{0}$ and $\mathbf{J}$ are matrices of 0's and 1's of the indicated dimension.

In fact, note here that for the purpose of inferences about $\boldsymbol{\tau}$, we might have omitted α from model (4.1), simply using β_i as the "intercept" for runs in the ith block. The result would have been a full-rank $\mathbf{X}_1$ of b columns, leading to exactly the same $\mathbf{H}_1$ and $\mathbf{X}_{2|1}$ as are given above. What is important here is that the columns of $\mathbf{X}_1$ span the same vector space whether the (linearly redundant) column associated with α is included or not. It is customary to include α in such models; for example, most computer programs are set up to "correct the data for the mean" first, leaving a sum of squares for blocks after removal of the common intercept.

Note that for this model, $\mathbf{X}_2$ and $\mathbf{X}_{2|1}$ are *the same as would be obtained with a CRD with b units in each treatment group.* As a result, it should be obvious that the reduced normal equations for the CBD take the same form as those for the CRD:

$$\hat{\tau}_j - \bar{\hat{\tau}} = \bar{y}_{.j} - \bar{y}_{..} \quad j = 1, \dots, t. \tag{4.5}$$

(Unlike the general reduced normal equations presented for the CRD in Chapter 3, this equation does not indicate the need for a *weighted* average of $\hat{\tau}_j$ on the left side, because each treatment is applied to the same number of units in a CBD.) This result is not mathematically complex, but it is quite profound. The implication is that the least-squares point estimators of estimable functions of treatment effects can be constructed for CBDs by simply ignoring the blocks and treating the data as if they were collected using a CRD. We shall discuss why this is so in somewhat more depth in Section 4.6; for now it suffices to say that it is critically related to the symmetry properties of the CBD, specifically, that each treatment is applied to exactly one unit in each block. Furthermore, since $\mathcal{I} = \mathbf{X}'_{2|1}\mathbf{X}_{2|1}$ is the same matrix as would be seen with a CRD with $n_i = b$, $i = 1, 2, 3, \dots, t$, we have

$$\mathcal{I} = b\mathbf{I} - \frac{b}{t}\mathbf{J} = b\left(\mathbf{I} - \frac{1}{t}\mathbf{J}\right), \tag{4.6}$$

and one generalized inverse of this matrix is

$$\mathcal{I}^- = \frac{1}{b}\mathbf{I}. \tag{4.7}$$

The shared structure of the reduced normal equations for CBD and CRD also implies that estimable linear combinations $\mathbf{c}'\boldsymbol{\tau}$ must be such that $\mathbf{c}'$ can be written as:

$$\mathbf{c}' = \mathbf{l}'(\mathbf{I} - \mathbf{H}_1)\mathbf{X}_2 = \mathbf{l}'\left(\mathbf{X}_2 - \frac{1}{t}\mathbf{J}_{N \times t}\right)$$

once again implying that $\mathbf{c}'\mathbf{1} = 0$ since all rows of $(\mathbf{I} - \mathbf{H}_1)\mathbf{X}_2$ have zero sums.

4.4 Influence of design on estimation

The least-squares estimator of any set of estimable functions $\mathbf{C}\boldsymbol{\tau}$ is:

$$\widehat{\mathbf{C}\boldsymbol{\tau}} = \mathbf{C}\begin{pmatrix} \bar{y}_{.1} \\ \bar{y}_{.2} \\ \dots \\ \bar{y}_{.t} \end{pmatrix}. \tag{4.8}$$

The variance matrix for this estimate is easily constructed due to the simple form of the estimator, or equivalently through the simple generalized inverse available for the design information matrix:

$$Var(\widehat{\mathbf{C}'\boldsymbol{\tau}}) = \mathbf{C}(\sigma^2\mathcal{I}^-)\mathbf{C}' = \frac{\sigma^2}{b}\mathbf{C}\mathbf{I}\mathbf{C}' = \frac{\sigma^2}{b}\mathbf{C}\mathbf{C}'. \tag{4.9}$$

In particular, for a single estimable function,

$$Var(\widehat{\mathbf{c}'\boldsymbol{\tau}}) = \frac{\sigma^2}{b} \sum_{j=1}^{t} c_j^2. \tag{4.10}$$

While these variance functions have the same form as those for the CRD in which $n = b$ units are assigned to each treatment, it is important to remember that σ^2 represents uncontrolled variation *among all units* in a CRD, while it represents only uncontrolled variation *among units from a common block* in a CBD. Recall that all operational effects of blocks, including block-to-block differences in batches of units, are represented by the elements of $\boldsymbol{\beta}$. Hence the sampling variance of treatment contrasts may be substantially smaller for a CBD than for a CRD of the same size if blocking is "effective," that is, if it results in greater homogeneity among units-within-blocks than can be expected within a larger collection of unblocked units.

In some cases, a CBD is chosen as the design for an experiment because of operational requirements; it may simply not be possible to select enough units from a homogeneous source to allow execution of a CRD, while smaller groups of units from different homogeneous sources may be available. In other cases, this may be a choice the investigator is free to make; an experiment *can* be executed as either a CRD or a CBD, and the question is which would be better for the specific purposes of the experiment. Suppose that in a given situation, the variance associated with ϵ for the CRD is σ_{CRD}^2, while that associated with the CBD is σ_{CBD}^2. When either is operationally possible, we would generally expect $\sigma_{CBD}^2 < \sigma_{CRD}^2$ since the CBD involves a greater degree of experimental control. For any particular estimable contrast $\mathbf{c}'\boldsymbol{\tau}$, the expected squared length of the associated confidence interval is, for a CRD and CBD that both call for application of each treatment to N/t units:

$$4t_{1-\alpha/2}^2(N-t)\sigma_{CRD}^2 \, \mathbf{c}'\mathbf{c} \, t/N$$

$$4t_{1-\alpha/2}^2(N-b-t+1)\sigma_{CBD}^2 \, \mathbf{c}'\mathbf{c} \, t/N,$$

respectively. Regardless of the value of $\mathbf{c}$, the ratio of these two quantities is:

$$[t_{1-\alpha/2}^2(N-t)\sigma_{CRD}^2] \, / \, [t_{1-\alpha/2}^2(N-b-t+1)\sigma_{CBD}^2] \tag{4.11}$$

and so in this sense the CBD can be expected to yield more precise intervals if:

$$\sigma_{CBD}/\sigma_{CRD} \, < \, t_{1-\alpha/2}(N-t)/t_{1-\alpha/2}(N-b-t+1). \tag{4.12}$$

The ratio of t-quantiles on the right side of inequality (4.12) is always at least somewhat less than 1 because the number of degrees of freedom is larger in the numerator quantile. This suggests that at least *some* reduction in unit-to-unit variability must be expected as a result of blocking, compared to the variability expected from selecting all units from the same source without restriction, if blocking is to be considered a reasonable option. For larger values of N, this is generally not a practical constraint since the ratio of t-quantiles approaches one for larger designs (where blocking is generally expected to be most effective in reducing variation).

4.4.1 Experiment size

If each block contains one unit assigned to each treatment, the overall size of the experiment is determined by the number of blocks, and for any estimable function $\mathbf{c}'\boldsymbol{\tau}$, the variance of the estimate is reduced in inverse proportion to b:

$$Var(\widehat{\mathbf{c}'\boldsymbol{\tau}}) = \frac{\sigma^2}{b} \sum_{j=1}^{t} c_j^2.$$

For any treatment contrast of interest, we generally want

$$\Psi = \frac{\mathbf{c}'\boldsymbol{\tau}}{\sqrt{Var(\widehat{\mathbf{c}'\boldsymbol{\tau}})}}$$

to be acceptably large. For a given signal-to-noise ratio $\psi = \mathbf{c}'\boldsymbol{\tau}/\sigma$ and desired value of Ψ, we can solve for the required number of blocks:

$$b = \Psi^2 \frac{\mathbf{c}'\mathbf{c}}{\psi^2}.$$

4.5 Influence of design on hypothesis testing

While the form of the least-squares estimates of estimable functions of treatment effects is the same for CRDs and CBDs, this does *not* mean that the entire analysis of data from a CBD can be carried out ignoring blocks. In particular, the sum of squares for residuals will be too large in the CRD variance decomposition of a CBD, because the sum of squares associated with block differences will not be taken into account. Specifically, while SSE for a CRD is the within-group sum of squares, for a CBD:

$$SSE = \sum_{i=1}^{b}\sum_{j=1}^{t}(y_{ij} - \bar{y}_{..})^2 - \sum_{i=1}^{b}t(\bar{y}_{i.} - \bar{y}_{..})^2 - \sum_{j=1}^{t}b(\bar{y}_{.j} - \bar{y}_{..})^2$$

$$= \sum_{i=1}^{b}\sum_{j=1}^{t}(y_{ij} - \bar{y}_{i.} - \bar{y}_{.j} + \bar{y}_{..})^2.$$

However, for $\text{Hyp}_0 : \tau_1 = \tau_2 = \tau_3 = \ldots = \tau_t$, the noncentrality parameter associated with the F-test is:

$$\boldsymbol{\tau}'\boldsymbol{\mathcal{I}}\boldsymbol{\tau}/\sigma^2 = \boldsymbol{\tau}'\left[b\mathbf{I} - \frac{b^2}{N}\mathbf{1}\mathbf{1}'\right]\boldsymbol{\tau}/\sigma^2 = \sum_j b(\tau_j - \bar{\tau})^2/\sigma^2. \qquad (4.13)$$

Note that in comparison to the corresponding result from Chapter 3, the matrices (and hence the quadratic form) are identical to those for a CRD with equal numbers of units assigned to each treatment, and $\bar{\tau}$ need not be a weighted average here since all treatments are applied to the same number of units. Comparisons between a CRD and a CBD of the same size can be made in the context of the power of the test of equal treatment effects. For either design and a fixed hypothetical value of τ, the numerator of the noncentrality parameter associated with the F-test is $\tau'\mathcal{I}\tau = b\sum_j(\tau_j - \bar{\tau})^2$. Hence the power of the F-test for equality of treatments is, for these two designs:

$$\text{Prob}\{W_{CRD} > F_{1-\alpha}(t-1, N-t)\}$$
$$\text{where } W_{CRD} \sim F'(t-1, N-t, \tau'\mathcal{I}\tau/\sigma_{CRD}^2) \qquad (4.14)$$
$$\text{Prob}\{W_{CBD} > F_{1-\alpha}(t-1, N-b-t+1)\}$$
$$\text{where } W_{CBD} \sim F'(t-1, N-b-t+1, \tau'\mathcal{I}\tau/\sigma_{CBD}^2), \qquad (4.15)$$

respectively. Again, the trade-off is between the degrees of freedom (favoring the CRD) and the size of the noncentrality parameter (favoring the CBD if $\sigma_{CBD} < \sigma_{CRD}$). As with the precision of estimates, unit-to-unit variation must be at least somewhat smaller with the CBD to justify its use.

4.6 Orthogonality and "Condition E"

Return now to more carefully consider the equivalence of normal equations and design information matrices for the CBD and CRD with $n_j = b$, $j = 1, 2, 3, \ldots, t$. A general statement of similarity between two designs that leads to this result follows; for convenience we will refer to the two requirements of this statement as "Condition E" (for "Equivalent"):

Consider two designs, each constructed to accommodate t treatments in N runs. Data can be modeled via a partitioned linear model in each case:

$$\mathbf{y} = \mathbf{X}_1\boldsymbol{\beta} + \mathbf{X}_2\boldsymbol{\tau} + \boldsymbol{\epsilon},$$

where $\mathbf{X}_2$ is an $N \times t$ matrix for either design, but $\mathbf{X}_1$ may have a different number of columns for the two designs, depending on the blocking strategy used. Say that the two designs satisfy Condition E if, for some ordering of rows (experimental runs):

- $\mathbf{X}_2$ *is the same matrix for each design, and*
- $\mathbf{H}_1\mathbf{X}_2$ *is the same matrix for each design.*

An immediate consequence is that $\mathbf{X}_{2|1} = (\mathbf{I}-\mathbf{H}_1)\mathbf{X}_2$ is the same for any two designs that jointly satisfy Condition E. Recall from Chapter 2 that $(\mathbf{I}-\mathbf{H}_1)$ is a projection matrix associated with the orthogonal compliment of the column space of $\mathbf{X}_1$ in $\mathcal{R}^N$. When Condition E is satisfied by two designs, they are in a sense *equivalent* after the respective corrections for the terms represented by $\mathbf{X}_1$ have been made in each case. As a result, the solutions to the reduced normal equations for treatment parameter estimates are of exactly the same form. In particular, since a CRD and a CBD that assign the same number of units to each treatment jointly satisfy Condition E, the estimable contrasts in the elements of τ can be estimated in a CBD ignoring blocks, just as they can be estimated in a CRD ignoring α. This is often summarized by saying that "*treatments are orthogonal to blocks*" in a CBD, or that "*a CBD is an orthogonally blocked design.*" These phrases are also appropriate for any other blocked design which, together with a CRD, satisfies Condition E.

Among designs that partition units into a single set of blocks, CBDs are probably the most popular, but they are not the only such arrangements for which treatments and blocks are orthogonal. Suppose an experiment is organized in b blocks, but that these blocks may be of different size, and the number of units to which any treatment is applied may be different in each block. What conditions must hold so that $\mathbf{X}_2$ and $\mathbf{H}_1\mathbf{X}_2$ are as they would be in a CRD of the same size? Let $n_1, n_2, \ldots, n_t$ be the number of units associated with the various treatments in each design (thereby satisfying the first part of Condition E). Recall that $\mathbf{H}_1 = \frac{1}{N}\mathbf{J}$ for a CRD; this means that each row of $\mathbf{H}_1\mathbf{X}_2$ must be $(\frac{n_1}{N}, \frac{n_2}{N}, \frac{n_3}{N}, \ldots, \frac{n_t}{N})$ for this design.

Next, suppose that for the blocked design, blocks are of size $m_1, m_2, \ldots, m_b$ units, so that

$$\sum_i m_i = \sum_j n_j = N.$$

Assume for simplicity that we work with a parameterization in which a term for each block is included (i.e., β_i), but that we omit the redundant intercept (α). $\mathbf{X}_1$ for the blocked design is such that each row contains a single 1 and the remaining elements 0, and the ith column has sum m_i, so that

$$(\mathbf{X}_1'\mathbf{X}_1)^{-1} = \text{diag}(\mathbf{m})^{-1}, \tag{4.16}$$

where $\mathbf{m} = (m_1, m_2, m_3, \ldots, m_b)'$. $\mathbf{X}_1'\mathbf{X}_2$ is called an *incidence matrix*; its (i, j) element is the number of times treatment j appears in block i. Hence

$$(\mathbf{X}_1'\mathbf{X}_1)^{-1}\mathbf{X}_1'\mathbf{X}_2 \tag{4.17}$$

contains, in the ith row, the *proportion* of units assigned to each treatment in the ith block:

$$(p_{i,1}, p_{i,2}, \ldots, p_{i,t}), \tag{4.18}$$

where p_{ij} is the proportion of units in the ith block assigned to treatment j. Next, let $B(i)$ be the column number of the entry in the ith row of $\mathbf{X}_1$ that is 1, i.e., the number of the block containing the ith experimental run. Then the (i, j) entry of $\mathbf{X}_1[(\mathbf{X}_1'\mathbf{X}_1)^{-1}\mathbf{X}_1'\mathbf{X}_2] = \mathbf{H}_1\mathbf{X}_2$ must be $p_{B(i),j}$. So, $\mathbf{H}_1\mathbf{X}_2$ will be the same matrix as would be calculated for a CRD if and only if

$$p_{B(i),j} = \frac{n_j}{N} \quad \text{for all } i = 1, 2, \ldots, N, \quad j = 1, 2, \ldots, t. \tag{4.19}$$

That is, any specific treatment must be applied to the same proportion of units in each block (although these proportions do not need to be the same for each treatment).

Note that this condition cannot be met for all possible integer values of t, $m_1, m_2, \ldots, m_b$; however, it can be easily satisfied when all blocks contain the same number of units. In most practical experimental layouts, each block contains the same number of units, and this general result is useful in cases where the common block size can be greater than t. Such designs were called *augmented complete block designs* by Federer (1955 and 1961), since they allow at least some treatments to be applied to more than one unit in the same block. For example, in situations where it is desirable to have a relatively larger group of control units (e.g., as in the discussion of subsection 3.4.1), an augmented complete block design can often be constructed to accommodate this.

While orthogonally blocked designs are attractive because they result in *simple* reduced normal equations, are such designs really *statistically superior* to other blocked arrangements? In one important sense, the answer to this question is "yes." Consider again two designs, each of which assigns n_j units to treatment $j = 1, 2, 3, \ldots, t$. For the moment, assume that the error variance is the same in both experiments, that is that $Var(\mathbf{y}) = \sigma^2 \mathbf{I}$ for either design. Design 1 is a CRD, so the variance of $\widehat{\mathbf{c}'\boldsymbol{\tau}}$ under this design is $\sigma^2 \sum_{j=1}^{t} \frac{c_j^2}{n_j}$. Design 2 is a blocked design for which we write a general partitioned model as:

$$\mathbf{y} = \mathbf{X}\theta = \mathbf{X}_1\beta + \mathbf{X}_2\boldsymbol{\tau} + \epsilon$$

where $\theta = (\beta', \boldsymbol{\tau}')'$. Under this design, the variance of $\widehat{\mathbf{c}'\boldsymbol{\tau}}$ is $\sigma^2 \mathbf{c}'(\mathbf{X}'\mathbf{X})_{22}^{-}\mathbf{c}$, where $(\mathbf{X}'\mathbf{X})_{22}^{-}$ is the lower right $t \times t$ submatrix of a generalized inverse of $\mathbf{X}'\mathbf{X}$. One convenient re-expression of $(\mathbf{X}'\mathbf{X})_{22}^{-}$ for this situation is

$$(\mathbf{X}_2'\mathbf{X}_2)^{-1}$$
$$+ (\mathbf{X}_2'\mathbf{X}_2)^{-1}\mathbf{X}_2'\mathbf{X}_1(\mathbf{X}_1'\mathbf{X}_1 - \mathbf{X}_1'\mathbf{X}_2(\mathbf{X}_2'\mathbf{X}_2)^{-1}\mathbf{X}_2'\mathbf{X}_1)^{-}\mathbf{X}_1'\mathbf{X}_2(\mathbf{X}_2'\mathbf{X}_2)^{-1}$$

or more conveniently, $(\mathbf{X}_2'\mathbf{X}_2)^{-1} + \mathbf{Q}$, where the generalized inverse is replaced with a unique inverse if one exists, and $\mathbf{Q}$ is positive semi-definite in any case. Using this result, and realizing that $\mathbf{X}_2'\mathbf{X}_2 = \text{diag}(\mathbf{n})$, we have that the variance of $\widehat{\mathbf{c}'\boldsymbol{\tau}}$ under design 2 can be written as

$$Var(\widehat{\mathbf{c}'\boldsymbol{\tau}}) = \sigma^2 \sum_{j=1}^{t} \frac{c_j^2}{n_j} + \sigma^2 \mathbf{c}'\mathbf{Q}\mathbf{c}$$

where the second term must be non-negative. Hence, apart from its effect on the value of σ^2, blocking cannot improve the variance properties of a design, and *can* make them much worse (depending on the values of $\mathbf{c}$ and $\mathbf{Q}$). On the other hand, the variance properties are not degraded for blocked designs which, together with a CRD of the same size, satisfy Condition E; for these designs, $\mathbf{c}'\mathbf{Q}\mathbf{c} = 0$ for all estimable $\mathbf{c}'\boldsymbol{\tau}$.

One caveat should be added to the above argument. We have shown that with the same value of σ^2, the best that can be expected of a blocked design is that it yields the same precision of estimation as a CRD of the same size, and that designs that are equivalent to CRDs in the sense of Condition E attain this bound. The equal-variances requirement in this claim is usually unrealistic in practice since we expect blocking to reduce random noise in the data. Hence designs that are *not* orthogonally blocked may sometimes also actually reduce estimation variance through better control of noise. The real point to be made here is that when blocking is required or desired, where two plans result in the same value of σ^2 and one is Condition E-equivalent to a CRD while the other is not, the first should be preferred for its superior precision properties.

4.7 Conclusion

In blocked experiments, treatment comparisons are made as precise/powerful as possible by arranging treatment applications within groups of especially similar units. In Complete Block Designs, each treatment is applied to exactly one unit in each such group or block; every pair of experimental observations in a CBD differs in the applied treatment, the block of the design, or both. Standard analysis requires that treatments and blocks be assumed to have additive effects (i.e., do not interact), to provide for an indirect estimate of σ^2. When this assumption is justified and blocking is "effective," CBDs yield more precise/powerful inference than unblocked experimental designs of the same size. Because blocks and treatments are orthogonal in a CBD, the reduced normal equations are identical in form to those associated with a CRD in which each treatment is applied to $n = b$ units. Non-orthogonally blocked experiments can be no more efficient than CBDs of the same size, and for some treatment contrasts must be less efficient.

4.8 Exercises

1. Consider a modification of the usual randomized complete block design. Suppose there are b blocks, and that each block contains units allocated to each of t treatments, as in the CBD. However, while each treatment is assigned a unit in each block (as with a CBD), $r > 0$ *additional* units are allocated to treatment 1 in each block. (This is an *augmented* CBD, as discussed in Section 4.6.) That is, each block contains $t + r$ units, $r + 1$ of these are allocated to treatment 1, and one of these is allocated to each of treatments 2 through t. There are thus $N = b(t + r)$ units used (and observations recorded) in the entire experiment. Answer the following questions, using the usual notation for an effects model parameterization and assuming that blocks and treatment effects do not interact (e.g., as with the usual analysis of CBDs).

   (a) Write expressions for the partitioned model matrices, $\mathbf{X}_1$ and $\mathbf{X}_2$. (Use general symbolic expressions like $\mathbf{1}_t$ and $\mathbf{J}_{N \times b}$ to do this; segments of the matrices that are filled with zeros can be left blank.)

   (b) Write expressions for the matrices $\mathbf{H}_1$, $\mathbf{X}_{2|1}$, and $\mathcal{I}$.

   (c) Give a scalar expression (i.e., not in terms of matrices) for the variance of $\widehat{\tau_1 - \tau_2}$.

   (d) Similarly, give a scalar expression for the variance of $\widehat{\tau_2 - \tau_3}$.

   (e) Does this design and a CRD with $n_1 = b(r + 1)$ and $n_j = b$, $j = 2, 3, 4, \ldots, t$ satisfy Condition E?

(f) For this design, σ^2 can be unbiasedly estimated even if treatments and blocks interact. Write this estimate, a quadratic form in the data for which the rank of the central matrix is rb, and show that it is statistically independent of the least-squares estimate of any estimable $\mathbf{c}'\boldsymbol{\tau}$.

2. Now consider a different kind of balanced block design consisting of $b = t$ blocks. As in the design of exercise 1, each block contains $t + r$ units with $r > 0$, and in each block, one of the units is paired with each treatment as in a CBD. However, here the r additional units in block i are allocated to treatment i, so that each treatment is allocated to the same number of units $(t + r)$ overall.

   (a) Write expressions for the partitioned model matrices, $\mathbf{X}_1$ and $\mathbf{X}_2$. Use general symbolic expressions like $\mathbf{1}_t$ and $\mathbf{J}_{N \times b}$ to do this; segments of the matrices that are filled with zeros can be left blank.

   (b) Write expressions for the matrices $\mathbf{H}_1$, $\mathbf{X}_{2|1}$, and $\mathcal{I}$.

   (c) Give a scalar expression (i.e., not in terms of matrices) for the variance of $\widehat{\tau_1 - \tau_2}$.

   (d) Does this design and a CRD with $n_i = t + r$, $i = 1, 2, 3, \ldots, t$, satisfy Condition E?

3. Consider a very small designed experiment in which the experimental units are grouped in two blocks; one block contains four units and the other contains two units:

block 1   block 2

For t treatments, we are willing to assume that blocks and treatments do not interact; the data will be modeled as:

$$y_{ijk} = \alpha + \beta_i + \tau_j + \epsilon_{ijk} \quad i = 1, 2 \quad j = 1 \ldots t$$

where k indexes replication for any (i, j) pair.

   (a) Compute $\mathbf{H}_1$, the hat matrix corresponding to the blocks-only submodel.

   (b) Suppose $t = 2$. Is it possible to assign three units to each treatment in this experiment so that the reduced normal equations are the same as those for a completely randomized design with $n_1 = n_2 = 3$? Defend your answer without computing the entire reduced normal equations for either design.

   (c) Suppose $t = 3$ and that two units from block 1 will be assigned to treatment 1, one unit from each block will be assigned to treatment 2, and one unit from each block will be assigned to treatment 3. Are the reduced normal equations for this design the same as those for a completely randomized design with $n_1 = n_2 = n_3 = 2$? Defend your answer without computing the entire reduced normal equations for either design.

(d) Is $\tau_1 - \tau_2$ estimable using the design described in part (c)? Clearly defend your answer using your solution from part (c).

4. Perhaps the simplest use of blocking is the basic paired-comparison study, i.e., complete blocks to compare two treatments. Consider the expected squared length of the usual two-sided 95% confidence interval for $\mu_1 - \mu_2$ based on such a paired experiment, with 10 pairs of observations; call this EL_p^2. In contrast to this, consider also the expected squared length of an unpaired t-based 95% interval for the same quantity, assuming the same number of observations (20); call this EL_u^2. The variation associated with the paired-units study (σ_p^2) *should* be smaller than that associated with the unpaired study (σ_u^2), since it requires only that pairs of homogeneous units be identified. What is the largest possible value of σ_p/σ_u that results in $EL_p^2 < EL_u^2$?

5. Suppose an experimenter initially decides to test four treatments in a randomized complete block design, using five blocks of size 4. Within each block, she randomly assigns one unit to each treatment. However, before the experiment begins, she discovers that she has the opportunity to add two additional blocks each containing *eight* units to the experiment. How can the units in blocks 6 and 7 be assigned to treatments so that the resulting design and a CRD with $n_j = 9$, $j = 1, 2, 3, 4$ satisfy Condition E?

6. Continuing exercise 5, suppose the investigator strongly suspects that the relative effects of her treatments can be approximately expressed as

$$\tau_1 = -1, \quad \tau_2 = \tau_3 = 0, \quad \tau_4 = +1.$$

Given this information, how should treatments be assigned in blocks 7 and 8 so as to maximize the power of the F-test for

$$\text{Hyp}_0 : \tau_1 = \tau_2 = \tau_3 = \tau_4$$

Does this design and a CRD assigning each treatment to the same number of units satisfy Condition E?

7. Write a detailed set of instructions for how treatments could be properly randomized to experimental material in the experiment of Kocaoz et al., described in subsection 4.1.1. Assume that you have:

- 32 uncoated bars,
- sufficient material to coat the bars as described, but must do so one bar at a time,
- 8 pipes as described, each with 4 pre-drilled holes, one bar to be inserted in each hole, and
- sufficient grout (in a single batch) to accomplish the execution of the experiment as described.

8. An investigator wishes to design an experiment to compare three treatments using a CBD (of three units per block). From long experience with similar experiments, he knows that σ^2 will be very close to 2.4. He believes that there really is no difference between treatments 1 and 2, but that treatment 3 produces responses about 0.6 larger, on average, than these. Assuming this is true:

    (a) How many blocks should be included in the design to provide power of 0.8 for testing $\text{Hyp}_0 : \tau_1 = \tau_2 = \tau_3$ with type I error probability of 0.05?

    (b) If $b = 10$ blocks are used, what will be the expected squared length of a 95% confidence interval for $\frac{1}{2}(\tau_1 + \tau_2) - \tau_3$?

9. Continuing exercise 8, suppose that the investigator decides to perform a CBD in $b = 10$ blocks, but fails to randomize units to treatments appropriately. In each block, he always assigns treatment 1 to the first unit selected, measuring the response from it first, then evaluates treatment 2 using the second selected unit, and finally tests treatment 3 by assigning it to the third unit. Suppose also that, unknown to the investigator, his response measurement system has a temporal "drift" problem. It functions as it should for the first measurement in each block, but the second is always 0.3 units less than it should be, and the third is always 0.6 units less than it should be. That is, the statistical model for the data he generates *should* be:

$$y_{ij} = \alpha + \beta_i + \tau_j + \epsilon_{ij} \qquad \text{for the first measurement in each block}$$
$$= \alpha + \beta_i + \tau_j - 0.3 + \epsilon_{ij} \quad \text{for the second measurement in each block}$$
$$= \alpha + \beta_i + \tau_j - 0.6 + \epsilon_{ij} \quad \text{for the third measurement in each block}$$

If the investigator bases his inference on the usual model (i.e., does not account for measurement bias), $\sigma^2 = 2.4$, and a nominal type I error probability of 0.05 is used,

    (a) What is the actual probability of a type I error?

    (b) What is the real power of the test if $\tau_1 = \tau_2 = \tau_3 - 0.6$?

10. In subsection 4.2.1, it is suggested that the model residuals:

$$r_{ij} = y_{ij} - \bar{y}_{i.} - \bar{y}_{.j} + \bar{y}_{..}$$

may be used in a diagnostic analysis to check for possible block-by-treatment interactions, or other unanticipated effects associated with individual data values. For example, we might define:

$$t_{ij} = r_{ij}/\sqrt{MSE \times \frac{(t-1)(b-1)}{tb}}$$

and declare as "unusual" any value that has absolute value larger than the t-quantile with $N - b - t + 1$ degrees of freedom, associated with a selected probability.

(a) Explain why t_{ij} is not actually a t-statistic.

(b) Write a small computer program to compute the probability that for any one t_{ij},

$$|t_{ij}| > t_{.975}(N - b - t + 1)$$

Do this for any t and b by letting $i = j = 1$, and repeatedly generating normally distributed data that have the structure of the assumed model.

Hint: In this exercise, you may find it helpful to recall that $\mathbf{r} = (\mathbf{I} - \mathbf{H})\mathbf{y}$, and that $MSE = \mathbf{y}'(\mathbf{I} - \mathbf{H})\mathbf{y}/(N - b - t + 1)$.

CHAPTER 5

Latin squares and related designs

5.1 Introduction

Randomized Complete Block Designs, and the other orthogonally blocked designs described in Chapter 4, are motivated by the idea that units can be naturally grouped in blocks, within which uncontrolled unit-to-unit variation is reduced. These blocks form a partition of the units used in the experiment; each unit is an element of exactly one block. Even if all the physical sources of unit-to-unit variation are not fully understood, it is assumed that the single classification variable associated with blocks accounts for a substantial proportion of it.

In some cases, the relationships among experimental units are more complicated, and a more complicated system may be required to describe potential systematic differences among them. *Row-Column Designs* (e.g., John and Williams, 1995) are used in settings where units can reasonably be sorted by two characteristics rather than one, and the most commonly used of these are *Latin Square Designs* (LSD). For example, suppose an agricultural experiment were to be performed in a square field, divided into "rows" and "columns" (as they would appear on a map) of smaller square experimental plots as displayed in Figure 5.1. The treatments will be different strains of corn, say, and a single strain of corn will be sown (i.e., the treatment applied) in each small square of land (i.e., unit). Now suppose that the field slopes gradually downhill from north to south, and that it is understood that this could have an effect on responses; the units in the northern-most row could absorb less rain water (depending on drainage characteristics) than the units in the next row to the south, and these in turn could be systematically different than those in the third row, et cetera. If this were our only concern regarding systematic differences among units, we might use a CBD, grouping the units in each east-west row as a separate block, randomly selecting one plot in each row for application of each of the strains of corn. But now also suppose that the prevailing winds in this area are west-to-east, and that this might cause the western-most "column" of units to be somewhat different in their response from those in the next column to the east, and that these in turn could be systematically different from those in the third column, et cetera. The result is a two-way classification of units based on two potential sources of nuisance variation, so the unit-to-unit relationships cannot be described independently within rows ignoring columns, or independently within columns ignoring rows.

↑ N   → prevailing wind

↓ downhill

2	1	4	3
1	4	3	2
3	2	1	4
4	3	2	1

Figure 5.1 Example of a Latin square; four treatments are denoted by numbers.

In order to ensure treatment-block balance comparable to that found in CBDs, it would seem minimally necessary that treatments should be associated with units in such a way that:

- the design is a CBD with respect to rows as blocks, ignoring columns, and
- the design is a CBD with respect to columns as blocks, ignoring rows.

This is, in fact, how a LSD is constructed. Note that there are a number of immediate implications of these requirements, including:

- The number of row-blocks of units must be t (the number of treatments), since each treatment must appear exactly once in each column-block.

- The number of column-blocks of units must be t, since each treatment must appear exactly once in each row-block.

- Therefore, a LSD must contain a total of t^2 units, t of which must be assigned to each treatment, and result in $N = t^2$ data values for each response variable.

The basic pattern of a LSD can be described by a $t \times t$ array of t symbols, where each treatment is associated with one of the symbols, in which each symbol appears once in each row and once in each column. For small values of t there are relatively few *unique* Latin squares, those not equivalent to others through reordering of rows, reordering of columns, and/or relabeling of symbols. But for some larger values of t, many such unique Latin squares exist. Latin square patterns are often tabulated in "standard form," with symbols appearing in increasing order across the top row and down the first column; this is a convenient way to clearly list each unique Latin square exactly once. Figure 5.2 contains examples of Latin squares in standard form for $t = 3$ through 6 (Beyer, 1968).

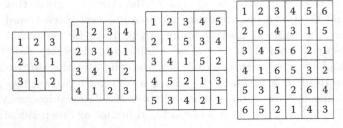

Figure 5.2 Examples of Latin squares in standard form for $t = 3, 4, 5$ and 6.

Since each experimental unit is contained in two blocks, randomization is somewhat less straightforward for LSDs than with CBDs. For a given value of t and corresponding Latin square in standard form (generally taken from a reference table or constructed by hand), the challenge is to randomly select one of the physical experimental layouts that conforms to this pattern. The critical aspects of the "pattern" are that each symbol appear once in each row and once in each column, but the physical order of placement of rows and columns, and the physical meaning attached to each symbol, can be randomized. Hence, for a selected pattern, randomization can be accomplished by:

- randomly shuffling the rows of the Latin square, so that each of the $t!$ row orderings is equally likely,

- randomly shuffling the columns of the Latin square, so that each of the $t!$ column orderings is equally likely, and

- randomly shuffling the association of symbols to treatments, so that each of the $t!$ assignments is equally likely.

The fullest possible (and therefore best) randomization for a LSD actually involves one further step. The process outlined above supposes that a standard-form Latin square has been selected. More completely, this starting point should be randomly selected from the collection of unique Latin squares of the desired size.

5.1.1 Example: web page links

Murphy, Hofacker and Mizerski (2006) used a series of Latin square designs to investigate the effect of link placement in a web page on the probability of a visitor "clicking" on that link. One study focused on a website maintained by a Florida restaurant, including seven links to other pages containing information on the restaurant's offerings, travel directions, and local attractions. Seven versions of this page were constructed in which the seven links appeared in different positions (top-to-bottom in the page) in such a way that each link was located at each position in exactly one version of the web page. Over the course of an eight-week period, the website was visited 18,134 times; visitors were randomly divided into seven groups, with visitors in the first group shown the first version of the web page, visitors in the second group shown the second version, et cetera. This arrangement led to a seven-by-seven table with rows associated with the seven links, and columns associated with the seven groups of visitors, in which each cell was associated with one of the seven positions on the web page (the treatment of interest in this experiment). In each of the 49 cells, the proportion of visitors from the associated group who clicked on the associated link was tabulated. The investigators used this Latin square arrangement to separate the possible effects on "click probability" of visitor group and specific link, from the effect of link position in the web page. The authors presented these overall click proportions for each location in a graph, from which the data of Table 5.1 have been extracted, and reported results

Table 5.1 Proportion of Visitors Clicking
on the Link at the Indicated Web Page Position, from Murphy et al. (2006)

Link Position	Proportion of Clicks
1 (top)	0.0462
2	0.0438
3	0.0392
4	0.0380
5	0.0356
6	0.0350
7 (bottom)	0.0368

of a formal analysis using a logistic regression model accounting for all three
possible sources of variation. The investigators interpreted their study as supporting previous research indicating a "primacy" effect (increased probability
that links located at the top of the page will be clicked), but also indicating
the possibility of a "recency" effect (increase in probability for links at the
bottom of the page).

5.2 Replicated Latin squares

The CRDs and CBDs discussed in Chapters 3 and 4, respectively, can easily
be "sized" to meet experimental requirements. CRDs allow the freedom to
select the number of units to be allocated to each treatment, and CBDs can
be adjusted in size by adding or removing complete blocks, or by expanding
blocks to include replication of some treatments. However, the structure of
the basic LSD is such that this isn't possible. For t treatments, a Latin square
can only be constructed for t^2 units, each classified by one row-block and one
column-block; increasing either the number of rows or columns would destroy
the Latin square structure. This would be a serious restriction if it meant that
the total sample size could not be increased as needed, to allow the investigator
to control the power of tests and the expected size of confidence intervals.

However, experiments designed as Latin squares can be adjusted in size
by adding additional replicates of *the entire basic design*; that is by combining r basic Latin squares in a design calling for a total of rt^2 units. These
replicates can be thought of as "superblocks," each containing all the experimental material for a single Latin square, each organized by row-blocks and
column-blocks. So, for example, in the hypothetical agricultural experiment
described in Section 5.1 carried out to compare four different varieties of corn,
we might increase the size of the study by using $r = 3$ complete Latin squares
on three different fields, as depicted in Figure 5.3. The effects associated with
row-blocks and column-blocks for each of these could be different, in part
because the slope of the ground might not be the same and/or the effect of
the prevailing wind might be different for each basic Latin square. But what-
ever systematic differences might be associated with east-west differences or

↑N

4	3	2	1
1	4	3	2
3	2	1	4
2	1	4	3

↑N

3	4	1	2
1	2	3	4
2	3	4	1
4	1	2	3

↑N

1	4	2	3
2	3	1	4
4	2	3	1
3	1	4	2

Figure 5.3 Example of a replicated Latin square; three replicates of a basic Latin square in four treatments.

north-south differences, these can be accounted for in each square in the replicated plan.

Where a replicated LSD is used, randomization should be performed independently for each replicate in the experiment.

5.3 A model

A basic LSD recognizes the possibility of three systematic sources of variation in the data related to rows, columns, and treatments. If these effects can be assumed to influence the data additively, a three-way main effects analysis of variance model can be used to describe the structure of the data:

$$y_{ijk} = \alpha + \beta_i + \gamma_j + \tau_k + \epsilon_{ijk},$$
$$i = 1 \ldots t, \quad j = 1 \ldots t, \quad k = 1 \ldots t,$$
$$\epsilon_{ijk} \text{ i.i.d. with } E(\epsilon_{ijk}) = 0 \quad \text{and} \quad Var(\epsilon_{ijk}) = \sigma^2 \tag{5.1}$$

where y_{ijk} is the data value observed for the unit appearing in the ith row-block and jth column-block. Because only one unit is included in the "intersection" of any row and column, the values of i and j are sufficient to identify any specific experimental run, but k is included in the indexing system to identify the effect of the treatment assigned to that unit. Hence, not all possible combinations of i, j, and k are represented in any specific Latin square arrangement.

Two-way interaction terms representing effects due to row-column combinations, row-treatment combinations, or column-treatment combinations, cannot be meaningfully accounted for in a Latin square design, and in fact must ordinarily be assumed to be zero in order to allow an analysis of the three first-order effects. To see this, consider an analysis that recognizes only effects due to rows and columns and their interaction. We might adopt the following model as a basis for this analysis:

$$y_{ijk} = \alpha + \beta_i + \gamma_j + (\beta\gamma)_{ij} + \epsilon_{ijk},$$
$$i = 1 \ldots t, \quad j = 1 \ldots t, \quad k = 1 \ldots t,$$
$$\epsilon_{ijk} \text{ i.i.d. with } E(\epsilon_{ijk}) = 0 \quad \text{and} \quad Var(\epsilon_{ijk}) = \sigma^2. \tag{5.2}$$

Note that the index k can be ignored here since it no longer indexes any fixed effect or replication (because only one value of k is included for each (i, j) pair). This leads to a two-way ANOVA of an unreplicated t-by-t table, and a standard decomposition of variation would assign $t - 1$ degrees of freedom to rows, $t - 1$ degrees of freedom to columns, and $(t - 1)^2$ degrees of freedom to the row-column interaction. A model containing an intercept (α) along with terms representing all these degrees of freedom is *saturated* – it represents *all* variation in any data set of this form. Adding model terms corresponding to treatments cannot improve, or change in any way, the fit of this model, because any variation that *might* have been attributable to treatments has been accounted to the row-column interaction. Put another way, the $t - 1$ degrees of freedom that might have explained variation due to treatments are completely confounded with $t - 1$ of the $(t - 1)^2$ degrees of freedom associated with the row-column interaction, and so the effects of both cannot be simultaneously assessed. The practical result is that, in a LSD, inferences can only be made about treatment effects under a model in which the contributions of treatments, rows, and columns are assumed to be additive.

If the basic Latin square pattern is replicated, an augmented model is needed to represent the additional sources of variation. Here we must be careful to think physically about how this replication is being carried out. First, consider the hypothetical experiment with four strains of corn described in Section 5.2, where the three replicate squares are physically unrelated. A model for this experiment can be written as:

$$y_{ijkl} = \alpha + \rho_l + \beta_{i(l)} + \gamma_{j(l)} + \tau_k + \epsilon_{ijkl},$$

$$i = 1 \ldots t, \quad j = 1 \ldots t, \quad k = 1 \ldots t, \quad l = 1 \ldots r,$$

$$\epsilon_{ijkl} \text{ i.i.d. with } E(\epsilon_{ijkl}) = 0 \text{ and } Var(\epsilon_{ijkl}) = \sigma^2. \tag{5.3}$$

In this model, a new additive effect due to replicates is denoted by ρ_l, where $l = 1 \ldots r$; note that there is no restriction on the number of replicates that can be included (as with row-blocks and column-blocks within each basic Latin square). Because each replicate Latin square is comprised of a physically different collection of row-blocks and column-blocks, these entities are *nested* within replicates. This is indicated by the indexing in the notation $\beta_{i(l)}$ and $\gamma_{j(l)}$. So, for example, the common effect of the third column-block may be entirely different in replicates 1 and 2, that is, $\gamma_{3(1)}$ and $\gamma_{3(2)}$ are different nuisance parameters in this model. Treatment effects are not nested within replicates, that is, we continue to use τ_k rather than $\tau_{k(l)}$, because they are assumed to be the same in each replicate.

But now recall that east-to-west column-blocking was actually implemented to protect against possible differences associated with prevailing winds. If all three replicates are physically placed in locations where these effects are known or assumed to be the same (i.e., with the same level of exposure to or protection from predominantly west winds), the nested pattern of column-blocks within replicates may not be the most realistic description of the data.

Instead, a model of form:

$$y_{ijkl} = \alpha + \rho_l + \beta_{i(l)} + \gamma_j + \tau_k + \epsilon_{ijkl},$$
$$i = 1 \ldots t, \quad j = 1 \ldots t, \quad k = 1 \ldots t, \quad l = 1 \ldots r,$$
$$\epsilon_{ijkl} \text{ i.i.d. with } E(\epsilon_{ijkl}) = 0 \quad \text{and} \quad Var(\epsilon_{ijkl}) = \sigma^2 \tag{5.4}$$

describes a situation in which, for example, the systematic effect of the third
column-block, γ_3, is the same in each replicate, while the effects of row-blocks
(north-to-south drainage) continue to be expressed differently for each replicate.

Finally, suppose that all three replicates are, in fact, executed on the same
plot of ground, but in three different growing seasons. If our concern continues
to be protection against the possible effects of prevailing wind and drainage
pattern, *and* we are willing to assume that these effects (if present) do not
change from year to year, then we might choose to model them with the same
set of parameters in each year (replicate):

$$y_{ijkl} = \alpha + \rho_l + \beta_i + \gamma_j + \tau_k + \epsilon_{ijkl},$$
$$i = 1 \ldots t, \ j = 1 \ldots t, \ k = 1 \ldots t, \ l = 1 \ldots r,$$
$$\epsilon_{ijkl} \text{ i.i.d. with } E(\epsilon_{ijkl}) = 0 \quad \text{and} \quad Var(\epsilon_{ijkl}) = \sigma^2. \tag{5.5}$$

Note that the progression through these three models actually represents
a *strengthening* of assumptions being made about the variation associated
with blocks. In model (5.3), we assume only that row-blocks and column-
blocks may have effects. In model (5.4), we *further* assume that the effects
of column-blocks are the same in each replicate. Finally, in model (5.5) we
add a similar assumption concerning the equivalence of row-blocks in each
replicate. In this sense, use of model (5.3) is always most conservative because
it requires the fewest assumptions. But it contains more nuisance parameters
than model (5.4), which contains more nuisance parameters than model (5.5).
This means that an analysis based on model (5.5) will result in more residual
degrees of freedom than one based on model (5.4), and so will provide more
power for tests and narrower confidence intervals (other things being equal).
Similarly, the statistical inferences based on model (5.4) – when it is an accurate representation of the system being studied – will be superior to those
of model (5.3). The important principle here is that the modeling must accurately represent the actual (physical) experimental situation in order to assure
a valid data analysis, but that given this, relatively fewer nuisance parameters
generally lead to more precision and power in the analysis of the data.

5.3.1 Graphical logic

Because the data collected from a Latin square contains variation associated
with row-blocks, column-blocks, and treatments, graphical displays of data
by treatment should be "adjusted" to remove both sets of nuisance effects.
Extending the logic we used in Chapter 4 for CBDs, consider adjusting each

datum from an unreplicated LSD by subtracting averages representing both kinds of nuisance effects and the overall mean:

$$y_{ijk}^* = y_{ijk} - (\bar{y}_{i..} - \bar{y}_{...}) - (\bar{y}_{.j.} - \bar{y}_{...}) - \bar{y}_{...} = \tau_k - \bar{\tau}. - \epsilon_{ijk} - \bar{\epsilon}_{i..} - \bar{\epsilon}_{.j.} + \bar{\epsilon}_{...}.$$

The expectation of each such value is $\tau_k - \bar{\tau}.$, and the variance is:

$$Var(y_{ijk}^*) = \sigma^2 \left(1 - \frac{1}{t}\right)^2.$$

As with CRDs, the variability is somewhat smaller than that associated with ϵ_{ijk}, especially when t is not large. But given this qualification, parallel boxplots of these adjusted data values provide, under the assumed model, relative comparisons of the responses associated with each treatment, after correcting for both kinds of blocks.

A similar data correction can be made for each of the three versions of replicated LSDs described previously. For models (5.3), (5.4), and (5.5), respectively, define:

$$y_{ijkl}^* = y_{ijkl} - (\bar{y}_{i..l} - \bar{y}_{...l}) - (\bar{y}_{.j.l} - \bar{y}_{...l}) - \bar{y}_{...l},$$

$$y_{ijkl}^* = y_{ijkl} - (\bar{y}_{i..l} - \bar{y}_{...l}) - (\bar{y}_{.j..} - \bar{y}_{....}) - \bar{y}_{...l},$$

$$y_{ijkl}^* = y_{ijkl} - (\bar{y}_{i...} - \bar{y}_{....}) - (\bar{y}_{.j..} - \bar{y}_{....}) - \bar{y}_{...l}.$$

Each quantity is actually a residual from the least-squares fit of the data to a model containing only the nuisance parameters, i.e., those associated with $\mathbf{X}_1$ for the appropriate model, and in each case the expectation of y_{ijkl}^* is $\tau_k - \bar{\tau}.$.

5.4 Matrix formulation

Beginning with model (5.1) for an unreplicated Latin square, and collecting all nuisance parameters in the first model partition and those associated with treatments in the second, we can write:

$$\mathbf{y} = \mathbf{X}_1\boldsymbol{\beta} + \mathbf{X}_2\boldsymbol{\tau} + \boldsymbol{\epsilon} \quad \boldsymbol{\epsilon} \sim (\mathbf{0}, \sigma^2\mathbf{I}), \tag{5.6}$$

where $\boldsymbol{\beta}$ is the $(2t+1)$-element vector of α and the block parameters, $\boldsymbol{\tau}$ is the t-element vector of treatment parameters, and $\mathbf{y}$ and $\boldsymbol{\epsilon}$ are $N(= t^2)$-element vectors of responses and random "errors," respectively. If the elements of $\mathbf{y}$ are sorted by row-blocks, and within each row-block by column-blocks (that is, as would be the case in "reading" the design pattern left-to-right and top-to-bottom), the model matrices can be written as:

$$\mathbf{X}_1 = \begin{pmatrix} \mathbf{1}_t & \mathbf{1}_t & \mathbf{0}_t & \cdots & \mathbf{0}_t & \mathbf{I}_{t\times t} \\ \mathbf{1}_t & \mathbf{0}_t & \mathbf{1}_t & \cdots & \mathbf{0}_t & \mathbf{I}_{t\times t} \\ \cdots & \cdots & \cdots & \cdots & \cdots & \cdots \\ \mathbf{1}_t & \mathbf{0}_t & \mathbf{0}_t & \cdots & \mathbf{1}_t & \mathbf{I}_{t\times t} \end{pmatrix} \quad \mathbf{X}_2 = \begin{pmatrix} \mathbf{P}_1 \\ \mathbf{P}_2 \\ \cdots \\ \mathbf{P}_t \end{pmatrix} \quad \text{where} \sum_{i=1}^{t} \mathbf{P}_i = \mathbf{J}.$$

$$\tag{5.7}$$

In this expression, $\mathbf{P}_i$, $i = 1 \ldots t$ are $t \times t$ *permutation matrices* containing zero elements at every position except for a single 1 in each row and column. This class of matrices gets its name from the fact that multiplication by any such matrix permutes the elements of a vector, e.g.:

$$\begin{pmatrix} 1 & 0 & 0 & 0 \\ 0 & 0 & 0 & 1 \\ 0 & 1 & 0 & 0 \\ 0 & 0 & 1 & 0 \end{pmatrix} \begin{pmatrix} v_1 \\ v_2 \\ v_3 \\ v_4 \end{pmatrix} = \begin{pmatrix} v_1 \\ v_4 \\ v_2 \\ v_3 \end{pmatrix}.$$

Note that the identity matrix is a permutation matrix (even though it corresponds to a fairly uninteresting permutation). A consequence of these being permutation matrices is that each treatment is applied exactly once in each row-block. The requirement that the permutation matrices sum to $\mathbf{J}$ implies that there is exactly one application of each treatment in any column-block.

For this model, a little matrix algebra shows that

$$\mathbf{X}_1'\mathbf{X}_1 = \begin{pmatrix} t^2 & t\mathbf{1}_t' & t\mathbf{1}_t' \\ t\mathbf{I}_t & \mathbf{J}_{t \times t} \\ & & t\mathbf{I}_{t \times t} \end{pmatrix}.$$

$\mathbf{X}_1$ is of rank $2t - 1$ because the sums of columns 2 through $t + 1$, and $t + 2$ through $2t + 1$, are each equal to the first column. As a direct result, $\mathbf{X}_1'\mathbf{X}_1$ is also of rank $2t - 1$, so two judiciously chosen rows and columns can be ignored in constructing a generalized inverse. However, in this case, we can take a somewhat less computationally burdensome route to the construction of $\mathbf{H}_1\mathbf{X}_2$, the key quantity in determining the form of the reduced normal equations. First, we repartition

$$\mathbf{H}_1\mathbf{X}_2 = [\mathbf{X}_1(\mathbf{X}_1'\mathbf{X}_1)^-][\mathbf{X}_1'\mathbf{X}_2],$$

and consider the second matrix factor in this expression. The inner products of columns from $\mathbf{X}_1$ and $\mathbf{X}_2$ "count" the number of positions in which both columns contain 1's. The first column of $\mathbf{X}_1$ contains all 1's, so the first row of $\mathbf{X}_1'\mathbf{X}_2$ records the number of times each treatment appears in the entire design; the result is t in each case. Inner products involving each of the other columns of $\mathbf{X}_1$ record the number of times a particular row-block or column-block contains a unit assigned to the treatment associated with the selected column of $\mathbf{X}_2$; the result is 1 in each case. As a result:

$$\mathbf{X}_1'\mathbf{X}_2 = \begin{pmatrix} t\mathbf{1}_t' \\ \mathbf{J}_{(2t) \times t} \end{pmatrix}.$$

Now, we can write another simple matrix product that has the same value, namely:

$$\mathbf{X}_1'\left(\frac{1}{t}\mathbf{J}_{t^2 \times t}\right) = \begin{pmatrix} t\mathbf{1}_t' \\ \mathbf{J}_{(2t) \times t} \end{pmatrix}.$$

Using this equivalence, we have:

$$\mathbf{H}_1\mathbf{X}_2 = [\mathbf{X}_1(\mathbf{X}_1'\mathbf{X}_1)^-]\left[\mathbf{X}_1'\left(\frac{1}{t}\mathbf{J}_{t^2 \times t}\right)\right] = \mathbf{H}_1\left(\frac{1}{t}\mathbf{J}_{t^2 \times t}\right).$$

But we know that multiples of $\mathbf{1}_{t^2}$ lie in the column space of $\mathbf{X}_1$ because $\mathbf{1}_{t^2}$ is the first column of this matrix. Therefore $\mathbf{H}_1(\frac{1}{t}\mathbf{1}) = \frac{1}{t}\mathbf{1}$, and so $\mathbf{H}_1(\frac{1}{t}\mathbf{J}_{t^2 \times t}) = \frac{1}{t}\mathbf{J}_{t^2 \times t}$. But note that this is the same as $\mathbf{H}_1\mathbf{X}_2$ for a CRD with t units allocated to each treatment, so an unreplicated LSD and a CRD with the same number of units assigned to each treatment jointly satisfy Condition E. It follows immediately that the reduced normal equations for a LSD are of form:

$$\hat{\tau}_k - \hat{\bar{\tau}} = \bar{y}_{..k} - \bar{y}_{...}, \quad k = 1, \ldots, t,$$

and that the design information matrix and one of its generalized inverses can be written as:

$$\mathcal{I} = t\mathbf{I} - \mathbf{J} = t\left(\mathbf{I} - \frac{1}{t}\mathbf{J}\right), \quad \mathcal{I}^- = \frac{1}{t}\mathbf{I}.$$

For replicated LSDs, these matrix arguments can be extended by adding the necessary columns of indicator variables for each replicate, and additional necessary columns if row-blocks or column-blocks are nested within replicates, to $\mathbf{X}_1$. For example, for an $r = 2$ replicated LSD of $t = 3$ treatments in which row-blocks are different for each replicate but column-blocks are not, the form of $\mathbf{X}_1\beta$, following the parameterization of model (5.4), is:

$$\begin{pmatrix} 1 & 1 & 0 & 1 & 0 & 0 & 0 & 0 & 0 & 1 & 0 & 0 \\ 1 & 1 & 0 & 1 & 0 & 0 & 0 & 0 & 0 & 0 & 1 & 0 \\ 1 & 1 & 0 & 1 & 0 & 0 & 0 & 0 & 0 & 0 & 0 & 1 \\ 1 & 1 & 0 & 0 & 1 & 0 & 0 & 0 & 0 & 1 & 0 & 0 \\ 1 & 1 & 0 & 0 & 1 & 0 & 0 & 0 & 0 & 0 & 1 & 0 \\ 1 & 1 & 0 & 0 & 1 & 0 & 0 & 0 & 0 & 0 & 0 & 1 \\ 1 & 1 & 0 & 0 & 0 & 1 & 0 & 0 & 0 & 1 & 0 & 0 \\ 1 & 1 & 0 & 0 & 0 & 1 & 0 & 0 & 0 & 0 & 1 & 0 \\ 1 & 1 & 0 & 0 & 0 & 1 & 0 & 0 & 0 & 0 & 0 & 1 \\ \hline 1 & 0 & 1 & 0 & 0 & 0 & 1 & 0 & 0 & 1 & 0 & 0 \\ 1 & 0 & 1 & 0 & 0 & 0 & 1 & 0 & 0 & 0 & 1 & 0 \\ 1 & 0 & 1 & 0 & 0 & 0 & 1 & 0 & 0 & 0 & 0 & 1 \\ 1 & 0 & 1 & 0 & 0 & 0 & 0 & 1 & 0 & 1 & 0 & 0 \\ 1 & 0 & 1 & 0 & 0 & 0 & 0 & 1 & 0 & 0 & 1 & 0 \\ 1 & 0 & 1 & 0 & 0 & 0 & 0 & 1 & 0 & 0 & 0 & 1 \\ 1 & 0 & 1 & 0 & 0 & 0 & 0 & 0 & 1 & 1 & 0 & 0 \\ 1 & 0 & 1 & 0 & 0 & 0 & 0 & 0 & 1 & 0 & 1 & 0 \\ 1 & 0 & 1 & 0 & 0 & 0 & 0 & 0 & 1 & 0 & 0 & 1 \end{pmatrix} \begin{pmatrix} \alpha \\ \rho_1 \\ \rho_2 \\ \beta_{1(1)} \\ \beta_{2(1)} \\ \beta_{3(1)} \\ \beta_{1(2)} \\ \beta_{2(2)} \\ \beta_{3(2)} \\ \gamma_1 \\ \gamma_2 \\ \gamma_3 \end{pmatrix}$$

where the number of columns is 12, and the number of linear dependencies among columns is 4, because linear combinations of columns associated with

- $\rho_1 + \rho_2 - \alpha$,
- $\beta_{1(1)} + \beta_{2(1)} + \beta_{3(1)} - \rho_1$,
- $\beta_{1(2)} + \beta_{2(2)} + \beta_{3(2)} - \rho_2$, and
- $\gamma_1 + \gamma_2 + \gamma_3 - \alpha$

each sum to $\mathbf{0}$. Hence $\text{rank}(\mathbf{X}_1) = \text{rank}(\mathbf{H}_1) = 12 - 4 = 8$. Each of the three forms of replicated LSDs discussed in Section 5.3 is Condition E-equivalent to a CRD with rt units assigned to each treatment. The reduced normal equations are then

$$\hat{\tau}_k - \bar{\hat{\tau}} = \bar{y}_{..k.} - \bar{y}_{....}, \quad k = 1, \ldots, t,$$

and the design information matrix and one of its generalized inverses can be written as:

$$\mathcal{I} = rt\left(\mathbf{I} - \frac{1}{t}\mathbf{J}\right), \quad \mathcal{I}^- = \frac{1}{rt}\mathbf{I}.$$

5.5 Influence of design on quality of inference

Because Latin square designs are balanced and orthogonally blocked, the reduced normal equations for treatment effects take the same form as those for complete block designs, and the results of Sections 4.4 and 4.5 hold for Latin squares after an appropriate adjustment for the number of times each treatment is included in the design, and the residual degrees of freedom. In particular, for a t-treatment Latin square design replicated r times (including $r = 1$):

- The residual degrees of freedom are found by subtracting from $N = rt^2$, the number of fixed parameters in the appropriate model:

  - 1 for the experiment-wide nuisance parameter, α,
  - $r - 1$ for replicates,
  - $t - 1$ for unnested column-blocks, or $r(t - 1)$ for nested column-blocks,
  - $t - 1$ for unnested row-blocks, or $r(t - 1)$ for nested row-blocks,
  - $t - 1$ for treatments.

  In the following, we use "df" to stand for this quantity.

- The residual sum of squares is found by subtracting from the corrected total sum of squares, $\sum_l (\bar{y}_{ijkl} - \bar{y}_{....})^2$, the orthogonal sums of squares associated with each set of parameters in the model:

  - $\sum_l t^2 (\bar{y}_{...l} - \bar{y}_{....})^2$ for replicates if $r > 1$,
  - $\sum_i rt(\bar{y}_{i...} - \bar{y}_{....})^2$ for row-blocks if $r = 1$ or if they are physically the same in each replicate,

- $\sum_{il} t(\bar{y}_{i..l} - \bar{y}_{...l})^2$ for row-blocks if $r \neq 1$ and they are physically different (nested) in each replicate,

- $\sum_{j} rt(\bar{y}_{.j..} - \bar{y}_{....})^2$ for column-blocks if $r = 1$ or if they are physically the same in each replicate,

- $\sum_{jl} t(\bar{y}_{.j.l} - \bar{y}_{...l})^2$ for column-blocks if $r \neq 1$ and they are physically different (nested) in each replicate,

- $\sum_{k} rt(\bar{y}_{..k.} - \bar{y}_{....})^2$ for treatments.

- The variance of an estimable function (i.e., contrast) of treatment parameters is

$$Var(\widehat{\mathbf{c}'\boldsymbol{\tau}}) = \frac{\sigma^2}{rt} \sum_{k=1}^{t} c_k^2. \tag{5.8}$$

- The Latin square design can be expected to yield better estimation precision than a CRD with rt units assigned to each treatment if

$$\sigma_{LSD}/\sigma_{CRD} < t_{1-\alpha/2}(rt^2 - t)/t_{1-\alpha/2}(df), \tag{5.9}$$

and better estimation precision than a CBD in rt blocks if

$$\sigma_{LSD}/\sigma_{CBD} < t_{1-\alpha/2}(rt^2 - rt - t + 1)/t_{1-\alpha/2}(df), \tag{5.10}$$

where σ_{CRD}, σ_{CBD}, and σ_{LSD} are the standard deviations associated with ϵ in each of the three designs.

- For a given estimable function $\mathbf{c}'\boldsymbol{\tau}$ and signal-to-noise ratio $\psi = \mathbf{c}'\boldsymbol{\tau}/\sigma$, a desired $\Psi = \mathbf{c}'\boldsymbol{\tau}/\sqrt{Var(\mathbf{c}'\hat{\boldsymbol{\tau}})}$ can be obtained with

$$r = \Psi^2 \frac{\mathbf{c}'\mathbf{c}}{\psi^2 t} \tag{5.11}$$

replicates of the Latin square.

- For testing $\text{Hyp}_0 : \tau_1 = \tau_2 = \tau_3 = \ldots = \tau_t$, the noncentrality parameter associated with the F-test is:

$$\boldsymbol{\tau}'\boldsymbol{\mathcal{I}}\boldsymbol{\tau}/\sigma^2 = \sum_{k=1}^{t} rt(\tau_k - \bar{\tau})^2/\sigma^2, \tag{5.12}$$

and the power of this test at level α for given values of $\boldsymbol{\tau}$ and σ^2 is

$$\text{Prob}\{W > F_{1-\alpha}(t-1, df)\} \text{ where } W \sim F'\left(t - 1, df, \sum_{k=1}^{t} rt(\tau_k - \bar{\tau})^2/\sigma^2\right). \tag{5.13}$$

5.6 More general constructions: Graeco-Latin squares

A slightly more general version of the algebraic argument described in Section 5.4 can be used to demonstrate that some even more complicated designs are orthogonally blocked. Consider any blocked design in which N

units are evenly distributed among t treatments. Suppose the nuisance parameters are represented by the model matrix $\mathbf{X}_1$ of N rows and p columns, and that the column totals (the number of 1's in each column) of this matrix are $s_1, s_2, s_3, \ldots, s_p$. The requirements of the more general result are that:

- **1** be in the column space of $\mathbf{X}_1$, that is, **1** can be expressed as a linear combination of the columns of $\mathbf{X}_1$, and
- $\mathbf{X}_2'\mathbf{X}_1 = \frac{1}{t}(s_1\mathbf{1}|s_2\mathbf{1}|s_3\mathbf{1}|\ldots|s_p\mathbf{1})$.

Note that the first of these conditions is immediately satisfied if the model contains an intercept term. It follows then that:

- $\mathbf{X}_1'\mathbf{X}_2 = \mathbf{X}_1'(\frac{1}{t}\mathbf{J})$ because the total of elements in the ith row of $\mathbf{X}_1'$ is s_i, and so
- $\mathbf{H}_1\mathbf{X}_2 = \mathbf{H}_1(\frac{1}{t}\mathbf{J}) = \frac{1}{t}\mathbf{J}$, because **1** is in the space spanned by the columns of $\mathbf{X}_1$,

that is, the conditions for the design to be Condition E-equivalent to a CRD of the same size with units divided equally among treatments are satisfied. The key to this result is the requirement that the inner produce of the ith column of $\mathbf{X}_1$, and *any* column of $\mathbf{X}_2$, be s_i/t. That is, any blocking arrangement for which the units in each block are evenly divided among the treatments is an orthogonally blocked design. One interesting point is that this says absolutely nothing about how the blocks must relate to each other; the condition is entirely characterized by how the treatments are assigned within each block individually.

This result saves us substantial effort in analyzing the *Graeco-Latin Square Design* (GLSD), a direct generalization of the LSD. Suppose we have three sources of "nuisance" variation with which we must deal, rather than the two accounted for by the rows and columns of a Latin square. If we are willing to accept the no-interaction assumption required by the Latin square, we can construct a design in $N = t^2$ units for this situation by "superimposing" two *orthogonal* Latin squares of this size. Two Latin squares are said to be orthogonal if, when superimposed, each treatment symbol in one Latin square is paired with every treatment symbol in the other Latin square in exactly one cell. Figure 5.4 displays the 4-by-4 Latin square we used to introduce the basic design in Section 5.1, with the four treatments indicated as numbers 1–4.

2A	3B	4D	1C
3D	2C	1A	4B
4C	1D	2B	3A
1B	4A	3C	2D

Figure 5.4 Example of a Graeco-Latin square; four treatments are denoted by numbers.

A second Latin square is superimposed, using letters A-D for clarity, arranged so that each number 1-4 appears in exactly one cell with each letter A-D, hence the two Latin squares are orthogonal. Three overlapping types of blocking are now represented by tabular rows and columns (as in a Latin square) *and* letters, while numbers continue to denote treatments. As with Latin square designs, straightforward analysis requires an assumption that interactions are not necessary in the response model.

We can extend the notation used in defining a matrix model for a Latin square by adding t new columns to $\mathbf{X}_1$ to represent t new blocks represented by the symbols of the second Latin square:

$$\mathbf{X}_1 = \begin{pmatrix} \mathbf{1}_t & \mathbf{1}_t & \mathbf{0}_t & \cdots & \mathbf{0}_t & \mathbf{I}_t & \mathbf{R}_1 \\ \mathbf{1}_t & \mathbf{0}_t & \mathbf{1}_t & \cdots & \mathbf{0}_t & \mathbf{I}_t & \mathbf{R}_2 \\ \cdots & \cdots & \cdots & \cdots & \cdots & \cdots \\ \mathbf{1}_t & \mathbf{0}_t & \mathbf{0}_t & \cdots & \mathbf{1}_t & \mathbf{I}_t & \mathbf{R}_t \end{pmatrix} \qquad \mathbf{X}_2 = \begin{pmatrix} \mathbf{P}_1 \\ \mathbf{P}_2 \\ \cdots \\ \mathbf{P}_t \end{pmatrix}. \qquad (5.14)$$

Here, $\mathbf{R}_i, i = 1, 2, 3, \ldots, t$ are a second set of order-t permutation matrices which must be chosen so that:

- $\sum_{i=1}^{t} \mathbf{R}_i = \mathbf{J}$, requiring that the "letters" form a second Latin square with respect to the tabular rows and columns, and

- $\sum_{i=1}^{t} \mathbf{P}'_i \mathbf{R}_i = \mathbf{J}$, requiring that the two Latin squares be orthogonal.

With some effort, it would be possible to work out the algebraic form of the reduced normal equations for these specific model matrices. However, we note that *all* blocks in the GLSD (denoted by tabular rows, tabular columns, and letters associated with the second Latin Square pattern) are of size t, and contain each treatment exactly once. Hence by the result discussed at the beginning of this section, this is also an orthogonally blocked experimental design, and the reduced normal equations and design information matrix for treatments take the same form as those for CRDs, CBDs, and LSDs in the same number of units for each treatment.

It is important to remember that while the reduced normal equations are the key to point estimates, the form of the full model determines the precision with which the variance of ϵ can be estimated. Writing the model matrices explicitly, as in equation (5.14), helps us see the number of degrees of freedom associated with the model (the column rank of matrices $\mathbf{X}_1$ and $\mathbf{X}_2$ combined). In the case of the GLSD, our representation of $\mathbf{X}_1$ includes a leading column of 1's, followed by three groups of t columns each; the sum of each of these sets of columns is $\mathbf{1}$, indicating the presence of three linear dependencies in the columns of this matrix. In addition, the sum of the columns of $\mathbf{X}_2$ is also $\mathbf{1}$, indicating a fourth linear dependency among the columns of the full model matrix $(\mathbf{X}_1 | \mathbf{X}_2)$. The rank of this combined model matrix is then $4t + 1$ (columns) minus 4 dependencies, or $4t - 3$, so the number of degrees of freedom available for estimating σ^2 is t^2 (data values) minus $4t - 3$ (rank of the full model matrix), or $t^2 - 4t + 3$.

A GLSD can be randomized by randomly selecting a pair of orthogonal Latin squares in normalized form, independently randomizing each of them as described in Section 5.1 (where symbols in the second Latin square are being associated with blocks in the third system, rather than to treatments), and overlaying the resulting randomized arrays as in Figure 5.4. For larger experiments, Graeco-Latin squares can be replicated an arbitrary number of times, where any combination of row-blocks, column-blocks, and second-square-blocks can be regarded as either common to each replicate or nested within replicates, according to the physical detail of the experiment.

5.7 Conclusion

The Latin square design is an extension of the complete block design in which two systems of blocks account for two possible sources of variation, and each unit is contained in exactly one block from each system. Like CBDs, LSDs are orthogonally blocked and so have reduced normal equations and design information matrices that are identical to those of CRDs of the same size and with the same number of units assigned to each treatment. The basic Latin square contains t^2 units, but the size of a LSD can be increased by replicating the basic pattern any number of times. In addition to the standard assumption of no block-by-treatment interaction in CBDs, LSDs also require that the effects of the two blocking systems be additive, that is, no row-block-by-column-block interactions, for clear interpretation of treatment effects; Hunter (1989) discusses the potential for misleading analysis if these assumptions are violated. The Latin square structure can be further extended to three blocking systems by combining two orthogonal Latin squares to form a Graeco-Latin square.

5.8 Exercises

1. Consider an experiment in which an unreplicated Latin square is constructed in $t + 1$ row-blocks and column-blocks, anticipating $t + 1$ treatments. However, since there are only t treatments to be compared, treatment 1 is applied each place a "1" or a "$t + 1$" appears in the Latin Square layout; that is, treatment 1 actually appears *twice* in each row-block and each column-block.

   (a) Derive the $t \times t$ design information matrix for this design. (Hint: Remember that $\mathbf{H}_1$ is the same for this design as for a Latin Square with $t + 1$ treatments).

   (b) Does this design, along with a CRD with treatment 1 assigned to $2(t + 1)$ units and each of treatments 2-t assigned to $t + 1$ units, satisfy Condition E?

2. Discuss the meaning of *experimental unit* in the web page link-clicking experiment of Murphy et al. (2006), described in subsection 5.1.1.

3. Consider the following (rather unusual) experimental design in three treatments, with nine units organized in four overlapping blocks:

Treatment	Block 1	Block 2	Block 3	Block 4
1	•			•
2	•		•	
3	•		•	
1		•	•	
2		•		•
3		•		•
1		•		•
2		•		•
3		•		•

Write (without using any algebra at all) the reduced normal equations for treatments for this design, and explain (using only words) why your answer is correct for this design.

4. (a) Construct a Graeco-Latin square of order 3 (i.e., for three treatments), and explain why an unreplicated design of this form is of limited practical value in most experimental settings.

(b) Explain why a Graeco-Latin square of order 2 cannot be constructed.

5. An experiment was set up to compare the wear characteristics of four kinds of automobile tires. Sample tires of each kind were tested under "live conditions" by mounting on fleet cars; wear was determined by a measurement on each tire after a specified number of miles in use. However, it is known that the position of the tire (e.g., left-rear, et cetera) can also have an effect on wear, as can the specific automobile. To control for these uninteresting sources of variation, a 4-by-4 Latin square design was used in which automobiles were thought of as "rows," tire positions were thought of as "columns," and the four kinds of tires were the treatments of interest. Before execution of the study, however, the investigator decided that a larger experiment should be performed, so he enlarged the experimental plan to include two complete Latin squares as described above, using four different automobiles in each (i.e., total of eight automobiles).

(a) With the usual model for analysis of data from a Latin square design, how many degrees of freedom should appear in the indicated lines of the ANOVA table?

- replicates
- automobiles
- positions
- tire types
- residual

(b) Suppose that in fact, $\tau_1 = +1$, $\tau_2 = -1$, $\tau_3 = \tau_4 = 0$, and $\sigma = 2$. What is the value of the noncentrality parameter associated with the

F-test of the hypothesis of equality among treatments? (Hint: Recall that Latin squares have the same $\mathbf{H_1X_2}$ as completely randomized designs with the same number of units assigned to each treatment – this is also true of replicated Latin squares.)

6. Consider a Latin square design for comparing three treatments, specifically:

1	2	3
2	3	1
3	1	2

where rows in this figure correspond to one kind of block, columns to another kind of block, and numbers refer to the associated treatments.

(a) How many degrees of freedom are available for estimating σ^2 in this design, if row-blocks, column-blocks, and treatments are assumed to have additive treatments (i.e., no interactions)?

(b) What is $\mathbf{H_1X_2}$ for this design? Are the least-squares estimates of estimable contrasts of τ's the same for this design as they would be for a completely randomized design with three units assigned to each treatment? Why or why not?

(c) Apart from the factor of σ^2, what is $Var(\widehat{\tau_1 - \tau_2})$ for this design?

7. Suppose the investigator who was planning to use the design in exercise 6 decided at the last minute that she was not interested in treatment 3 after all. However, she had already arranged to use units which would be appropriate for a Latin square of order 3. So, she considered simply not using those units which *would* have been assigned treatment 3. That is, she considered the experimental design which might be described as:

1	2	-
2	-	1
-	1	2

where as before, rows in the figure correspond to one kind of block, columns another kind of block, and numbers refer to the associated treatments, but where the units corresponding to the cells containing "-" were simply not used.

(a) How many degrees of freedom are available for estimating σ^2 in this design, if row-blocks, column-blocks, and treatments are assumed to have additive treatments (i.e., no interactions)?

(b) What is $\mathbf{H_1X_2}$ for this design? Are the least-squares estimates of estimable contrasts of τ's the same for this design as they would be for a completely randomized design with three units assigned to each of treatments 1 and 2? Why or why not?

(c) What is the design information matrix for this design? Apart from the factor of σ^2, what is $Var(\widehat{\tau_1 - \tau_2})$ for this design?

8. In the example of Murphy et al. (2006) described in subsection 5.1.1, the response variable was a binomial proportion in each cell of the Latin square (or Bernoulli for each user within each cell). If sample sizes are large enough and proportions are not extreme, approximate inferences can be based on standard linear models. Suppose a smaller experiment was conducted with four treatments, using a LSD, resulting in the following response values, each a proportion from a sample size of 100:

trt 1: 0.51	trt 2: 0.56	trt 3: 0.52	trt 4: 0.53
trt 2: 0.52	trt 3: 0.59	trt 4: 0.55	trt 1: 0.43
trt 3: 0.53	trt 4: 0.58	trt 1: 0.42	trt 2: 0.45
trt 4: 0.48	trt 1: 0.47	trt 3: 0.41	trt 3: 0.38

(a) Although a "normalizing" transformation might be applied, since all of these proportions are similar and none especially near 0 or 1, ANOVA-based inference should be adequate. Compute the ANOVA decomposition and perform the F-test for equality of treatments.

(b) Because all proportions are similar, an estimate of the near-common variance for each is $\frac{1}{100}\bar{p}(1 - \bar{p})$, where $\bar{p}$ is the average of the 16 tabulated proportions. Based on this estimate, and regarding it as a "constant" since the accumulated sample size is 1600, develop a χ^2-test based on the residual sum of squares to determine whether the assumed model is adequate.

9. Revise and extend each of the points made in Section 5.5, for Graeco-Latin squares.

10. Consider the following (highly artificial) data set for a Latin square in four treatments:

trt 1: 1.23	trt 2: 1.59	trt 3: 1.37	trt 4: 1.28
trt 2: 1.83	trt 3: 1.71	trt 4: 7.22	trt 1: 1.64
trt 3: 1.17	trt 4: 1.65	trt 1: 1.18	trt 2: 1.44
trt 4: 1.36	trt 1: 1.27	trt 2: 1.52	trt 3: 1.03

(a) Present a careful argument supporting why you think that this data set is or is not consistent with the model usually used in analysis of data from a LSD.

(b) If you argued that the data are consistent with the standard model, perform a test for equality of the four treatment effects. If you argued otherwise, give at least two possible explanations, i.e., specific model failures that could result in the observed data pattern.

Some data analysis for CRDs and orthogonally blocked designs

6.1 Introduction

The emphasis of this book is on experimental designs and the statistical rationale for their use, in particular, the impact of the design on the precision of estimators and power of hypothesis tests in the context of linear models. However, since the properties of analysis are the foundation for our motivation to study experimental design, it is fitting to spend some effort discussing the ideas upon which some of these analytical techniques are based. This is only a very brief summary of a very few widely-applicable techniques. For more information on analysis methods, the reader should consult some of the many excellent books that have been written on this topic.

This chapter is placed at this location in the book because the structure of CRDs and orthogonally blocked designs leads to especially simple forms for many popular analytical techniques. That being said, some of the methods discussed here are applicable in a far wider variety of settings. The material is presented in the order in which it might actually be used (at least sometimes) in analyzing data; model diagnostics to check the validity of the intended assumptions, transformation of the data to more closely meet those assumptions, basic statistical inference to answer questions about treatment effects, and more specific techniques for controlling overall risk when many estimates are required.

6.2 Diagnostics

6.2.1 Residuals

Any application of linear models depends on assumptions about the form of the model representing the mean and the statistical behavior of the observable realizations. The most commonly encountered assumptions of the latter type are that the data have equal variances and are statistically independent. Taken together, these imply:

$$E(\mathbf{y}) = \mathbf{X}\boldsymbol{\theta}, \quad Var(\mathbf{y}) = \sigma^2 \mathbf{I} \tag{6.1}$$

where the assumptions do not include specification of σ^2 or $\boldsymbol{\theta}$. In some cases, an additional assumption of normality is added:

$$\mathbf{y} \sim N(\mathbf{X}\boldsymbol{\theta}, \sigma^2 \mathbf{I}).$$

Validity of these assumptions is often checked by examining the *residuals* from the fitted mean model:

$$\mathbf{r} = \mathbf{y} - \hat{\mathbf{y}} = (\mathbf{I} - \mathbf{H})\mathbf{y}.$$

If the assumptions about the mean structure are correct, the implication is that

$$E(\mathbf{r}) = \mathbf{0}, \quad Var(\mathbf{r}) = \sigma^2(\mathbf{I} - \mathbf{H}).$$

If $\mathbf{y}$ is also normally distributed,

$$\mathbf{r} \sim N(\mathbf{0}, \sigma^2(\mathbf{I} - \mathbf{H})).$$

Further, since $\mathbf{r}$ is a linear combination of $\mathbf{y}$, the Central Limit Theorem often provides justification for treating $\mathbf{r}$ as being approximately normally distributed even when $\mathbf{y}$ is not. If the observed residual vector is a credible realization from this distribution for *some* value of σ^2, this may be interpreted as evidence in support of the assumptions; a residual vector that would be an unusual realization from such a distribution for *any* value of σ^2 suggests that one or more of the modeling assumptions is questionable.

The simplest of these residual-based assumption checks are plots or indices computed from the residuals, designed to indicate whether they have the appearance of an i.i.d. sample from a normal distribution of mean zero and unknown variance. In some cases, the assumed common standard deviation is estimated and the *standardized residuals* – original residuals divided by an estimate of σ – are examined instead. This does nothing to destroy any pattern or curious feature that might be apparent in the residuals because it is a common rescaling of all values, but it does eliminate the need to consider "any" or "some" value of σ^2 because it puts all residual values on a scale that should be approximately appropriate for $N(0, 1)$ data.

A more subtle issue, that is sometimes important, is that the residuals are not really an i.i.d. sample even when the assumptions are correct, because $Var(\mathbf{r}) = \sigma^2(\mathbf{I} - \mathbf{H})$. The residuals *would* be i.i.d. draws from $N(0, \sigma^2)$ if $\mathbf{H}$ were $\mathbf{0}$, and in well-designed experiments in which N is much larger than rank$(\mathbf{X})$, this is often approximately (but never entirely) true. *Studentized residuals*, unlike standardized residuals, are formed by normalizing relative to an estimate of the actual standard deviation of each:

$$\frac{r_i}{\sqrt{MSE(1 - h_{ii})}}, \quad i = 1, 2, 3, \ldots, N,$$

where r_i is the ith element of $\mathbf{r}$ and h_{ii} is the ith diagonal element of $\mathbf{H}$.

Note that while studentized residuals are approximately normally distributed with zero mean and of equal variance (if the model assumptions hold), they are not statistically independent because $\mathbf{H}$ is not a diagonal matrix except under very unusual circumstances. (See exercise 1.) Since $\mathbf{H}$ is known, it *is* possible to linearly transform $\mathbf{r}$ to a vector of independent random variables with equal variances, through a linear transformation:

$$\mathbf{r}^* = \mathbf{A}_{N-p \times N}\mathbf{r} \quad \text{such that} \quad \mathbf{A}(\mathbf{I} - \mathbf{H})\mathbf{A}' = \mathbf{I}_{N-p \times N-p}$$

where $p = \text{rank}(\mathbf{X})$. However, this sacrifices much of the intuitive appeal of the residuals as diagnostics because there is no one-to-one relationship between the elements of $\mathbf{r}^*$ and the experimental observations; one or a few unusual elements of $\mathbf{r}^*$ could not generally be easily traced back to one or a few experimental observations at which something might have "gone wrong."

Hence, simple listings of residuals, standardized residuals, or studentized residuals can be a useful screening device to indicate which, if any, individual observations might be regarded as suspicious. Aggregated plots of these quantities, grouped by treatment or block, can also be useful diagnostic checks of the validity of model assumptions. For example, a treatment group in which the sample variance of residuals is substantially larger may indicate a treatment that influences both the mean and variance of the response. The most effective diagnostic checks are often relatively simple graphical displays such as these, because the intelligent consideration of how assumptions *can* fail is usually highly context-specific. A *pattern* of suspicious residuals can often be the basis of a fruitful discussion between an investigator and a statistician. This is often a process of discovery rather than of formal model comparison and testing. Still, there are contexts in which it is useful to have a "standard" diagnostic procedure at hand for specific kinds of assumption failures. The analysis methods described in the remainder of this section are useful for detecting inequality of variance in CRDs, and interaction between blocks and treatments in CBDs.

6.2.2 Modified Levene test

Consider data collected from an experiment executed as a CRD (or any other modeling scenario in which one-way ANOVA is the default analysis) in which there is concern over the assumption of equal variances in each treatment group. Following the notation of Chapter 3, let y_{ij} denote the jth observation of the ith (treatment) group, where all groups need not be of the same size. The modified Levene test for equality of group variances, introduced by Levene (1960) and shown to be superior to a number of competing procedures by Conover, Johnson, and Johnson (1981), is very simple and is performed as follows:

1. For each group (i), compute the *median* of data values, $\tilde{y}_i$, $i = 1, 2, 3, \ldots, t$.

2. For each data value, compute the absolute difference between y_{ij} and the associated group median:

$$z_{ij} = |y_{ij} - \tilde{y}_i|, \qquad i = 1, 2, 3, \ldots, t, \qquad j = 1, 2, 3, \ldots, n_i.$$

3. Perform an F-test (one-way ANOVA) for equality of *means*, using the transformed data z_{ij}.

The intuition for why the procedure works is revealed in step 2; any group with an unusually large *spread* of data values (y_{ij}) will tend to have relatively large transformed data (z_{ij}), and the *average* value of the group will tend to be relatively large. While the test is not exact, an extensive simulation study by Conover et al. showed it to approximately maintain the nominal type I error level, and have reasonable power for moderate nonhomogeneous variance.

6.2.3 General test for lack of fit

A key assumption in all blocked experiments we have discussed is that blocks and treatments do not interact. In most of these designs, the information available to check this assumption is limited. However, designs that have been enlarged to include "true replication" yield data in which the variation of model residuals can be compared to the variation within groups of runs with common treatment and block, via a formal test for lack of fit (subsection 2.7.1). Such designs include augmented complete block designs (Section 4.6), extended complete block designs (to be discussed in subsection 7.5.1), and any other blocked design in which multiple units are assigned to the same treatment and block for at least some treatment/block combinations.

For example, consider the small augmented CRD for $t = 3$ treatments in $b = 3$ blocks, in which treatment 1 is assigned to two units within each block (of size four). Extending the notation of Section 4.3:

$$\mathbf{X} = (\mathbf{X}_1 | \mathbf{X}_2) = \begin{pmatrix} 1 & 1 & 0 & 0 & 1 & 0 & 0 \\ 1 & 1 & 0 & 0 & 1 & 0 & 0 \\ 1 & 1 & 0 & 0 & 0 & 1 & 0 \\ 1 & 1 & 0 & 0 & 0 & 0 & 1 \\ 1 & 0 & 1 & 0 & 1 & 0 & 0 \\ 1 & 0 & 1 & 0 & 1 & 0 & 0 \\ 1 & 0 & 1 & 0 & 0 & 1 & 0 \\ 1 & 0 & 1 & 0 & 0 & 0 & 1 \\ 1 & 0 & 0 & 1 & 1 & 0 & 0 \\ 1 & 0 & 0 & 1 & 1 & 0 & 0 \\ 1 & 0 & 0 & 1 & 0 & 1 & 0 \\ 1 & 0 & 0 & 1 & 0 & 0 & 1 \end{pmatrix}.$$

In this case rank$(\mathbf{X}) = 5$ since the sum of columns 2 through 4 and the sum of columns 5 through 7 each are equal to column 1. SSE therefore is associated with $12 - 5 = 7$ degrees of freedom, and can be decomposed into $SSPE$ with 3 degrees of freedom (N - the number of unique rows of $\mathbf{X}$), and $SSLOF$ with 4 degrees of freedom. $SSPE$ is the "within-group" sum of squares for the three groups associated with rows 1 and 2, rows 5 and 6, and rows 9 and 10, of $\mathbf{X}$, and a formal test for adequacy of the no-interaction model can be carried out by comparing $MSLOF/MSPE$ to $F_{1-\alpha}(4, 3)$.

6.2.4 Tukey one-degree-of-freedom test

Unless a design contains "true replication" – multiple units in at least some blocks that are assigned to the same treatment – the general test for lack of fit described in subsection 6.2.3 is not available. For example, a CBD is typically analyzed as an unreplicated two-way table, in which b rows represent blocks and t columns represent treatments. Because the table contains no replication, variation that cannot be ascribed to row differences or column differences *could* be due to row-column interaction, random noise, or both. A standard analysis is to assume that any residual variation associated with these $(t-1)(b-1)$ degrees of freedom is random noise.

One test for interaction, introduced by Tukey (1949), partitions this variation into one single degree of freedom component associated with one particular kind of possible interaction, and the remaining $(t-1)(b-1)-1$ degree-of-freedom component which is assumed to represent random noise. Using the notation of Chapter 4 in which y_{ij} denotes the observation associated with treatment j in block i, Tukey's procedure calls for computing an interaction mean square:

$$MSI^* = \frac{[\sum_{ij} y_{ij}(\bar{y}_{i.} - \bar{y}_{..})(\bar{y}_{.j} - \bar{y}_{..})]^2}{[\sum_{i}(\bar{y}_{i.} - \bar{y}_{..})^2][\sum_{j}(\bar{y}_{.j} - \bar{y}_{..})^2]}$$

and an adjusted error mean square:

$$MSE^* = \frac{SSE - MSI^*}{(b-1)(t-1)-1}$$

and tests for interaction by comparing the ratio MSI^*/MSE^* to $F_{1-\alpha}(1, (b-1)(t-1)-1)$ for selected α, interpreting a large statistic value as evidence that blocks and treatments do interact.

Tukey's formulation is designed to be sensitive to cases in which the interaction, or nonadditivity between rows and columns, is primarily due to an omitted term that is proportional to the product of the marginal row and column effects in each cell. Examination of the numerator of MSI^* shows that it is a squared, unnormalized "covariance" between the data, y_{ij}, and the product of these marginal effects – large values of this "covariance," either positive or negative, lead to large test statistics. Of course, this test will not be so sensitive to other patterns of nonadditivity, and it is perhaps for this reason that this procedure has not been referenced so heavily in recent years. As noted above, subject matter knowledge of the form interaction might reasonably take – if it *is* present – is the best clue concerning how data should be examined for model diagnosis.

6.3 Power transformations

When the variance of the response appears to be inconsistent across groups of units (defined by block or assigned treatments or both), a decision must be made as to whether the heterogeneity is severe enough to merit a modification in the form of the analysis. In many cases where data values are strictly non-negative, and the variance is inhomogeneous, careful examination of enough data reveals that the variance and mean are related. Many simple one-parameter probability distributions have this property, e.g., the exponential (mean equal to variance) and the Poisson (mean equal to standard deviation). But such relationships also sometimes occur with data that are at least approximately normally distributed; for example, in chemical or material science experiments where the standard deviation of replicated concentration measurements is often roughly proportional to mean concentration. In our context, the empirical selection of a response transformation that equalizes variances over experimental groups, and preserves the desired additive structure for the response mean as a function of treatments and blocks, is sometimes a practical challenge.

The power transformation is often useful in this context and has found heavy use in experimental analysis. A diagnostic graph of group means against group variances (or standard deviations) that suggests a monotonic relationship is evidence that power-transformed data may more nearly satisfy the equal-variances assumption of unweighted least-squares analysis. Suppose that our response variable y is actually such that the mean and variance are related through a power law

$$Var(y) = E(y)^q.$$

If we actually knew the value of q, we might select a power transform y^p with an appropriately selected value of p to make the variances more homogeneous. Using the delta method (e.g., the expectation of a Taylor Series expansion of y^p as a function of y), the approximate variance of our transformed variate is:

$$Var(y^p) \approx p^2 \times E(y)^{q+2p-2}.$$

Hence, selecting $p = (2 - q)/2$ would provide a scale on which the variance is approximately constant with respect to the mean.

The one-parameter power transformation as described by Box and Cox (1964) is:

$$y_p^* = \frac{y^p - 1}{p}.$$

A convenient characteristic of this parameterization is that values of p close to 0 correspond to transformations that are "close to" logarithmic because:

$$lim_{p \to 0} \frac{y^p - 1}{p} \to ln(y).$$

This form also facilitates the use of available data to empirically fit an appropriate value of p. Suppose that there actually is some value of p, say π, for which y_π^*:

• has mean structure as described in the model we are fitting, and

• has errors that are i.i.d., and *normally distributed*.

Under this model, the maximum likelihood estimate (MLE) of π can be numerically found as follows:

1. Compute the geometric mean of the untransformed data, i.e., $\tilde{y} = [\prod_{i=1}^{N} y_i]^{\frac{1}{N}}$.
2. For a collection of values of p, fit $y_p^{**} = y_p^*/\tilde{y}^p$ to the intended model.
3. The value of p that minimizes SSE is the MLE of π.

Note, for example, that this approach will not work without normalizing the transformed data by the geometric mean since SSE values would then be expressed in different physical units for each value of p and so would not be directly comparable. In practice, values of p between 0 and 2 are generally of most interest, and many investigators limit attention to $p \in \{0, \frac{1}{2}, 1, 2\}$ unless the data set is large enough to support accurate resolution over a finer grid, or there is a good physical reason to consider other specific values.

When the selected value of p is treated as a known constant, tests for equality of treatments can generally proceed with usual procedures applied to the transformed data. However, interpretation of estimates derived under power-transformed data require more attention. To see this, consider a CRD in which the response data are measured in physical units of seconds (time). The transformed data are modeled as:

$$y_{ij}^* = \alpha^* + \tau_i^* + \epsilon_{ij}^*$$
$$i = 1 \ldots t, \qquad j = 1 \ldots n_i.$$

Here, the modeled response is in units of seconds^p; due to the linear form of the model, these units also apply to α^*, τ_i^*, and ϵ_{ij}^* (and hence also to the standard deviation of ϵ^*). These physically meaningless units also apply to $\mathbf{c}'\tau$ and $\widehat{\mathbf{c}'\tau}$ where the elements of $\mathbf{c}$ are (unitless) weights in an interesting treatment contrast.

The simplest approach to partially addressing this problem of interpretation is through use of the reverse transformation. The data model implies that

$$E\left(\frac{y_{ij}^p - 1}{p}\right) = \alpha^* + \tau_i^*.$$

The approximate relationship found by replacing $E(\frac{y_{ij}^p - 1}{p})$ by $\frac{E(y_{ij})^p - 1}{p}$ leads to

$$E(y_{ij}) \approx [p\alpha^* + p\tau_i^* + 1]^{\frac{1}{p}}.$$

So linear contrasts in treatment means can be estimated (although not unbiasedly except in the trivial case of $p = 1$) as:

$$\sum_{i=1}^{t} c_i E(y_i) = \overbrace{\sum_{i=1}^{t} c_i [p\hat{\alpha}^* + p\hat{\tau}_i^* + 1]^{\frac{1}{p}}}.$$

Note that α^*, which would not play a role in any estimable treatment *contrast* on the transformed scale, *does* have an influence in this estimate due to the nonlinear form of the transformation.

6.4 Basic inference

In Chapters 3-5 the two basic analysis tools discussed in motivating the structure of experimental designs are the F-test for equivalence of treatments, and t-based confidence intervals for specified linear contrasts of treatment effects. Formulae for these procedures are especially simple for the designs considered to this point because analysis for the CRD is based on one-way ANOVA, and since CBDs and LSDs are orthogonally blocked designs, they share much of this simplicity of analysis. We assume that the reader understands how these procedures would be performed in the general (unbalanced) case; the following paragraph is a brief statement of the relevant formulae as they apply to data from experiments designed as CRDs, CBDs, or LSDs.

For balanced CRDs and orthogonally blocked designs, let T_i be the average of all observations associated with treatment i, $\mathbf{T}$ denote the t-element vector of these treatment averages, r denote the number of times each treatment appears in the design, MSE be the error (residual) mean square from an ANOVA determined by the design and model used, and df be its associated degrees of freedom. In each case, the mean square associated with treatments can be written as:

$$MST = \sum_{i=1}^{t} r(T_i - \bar{T}.)^2/(t-1).$$

To test for equality of treatment means, the ratio of MST to MSE is compared to $F_{1-\alpha}(t-1, df)$ for a selected value of the type I error probability α, and the hypothesis of no treatment difference is rejected if the statistic is larger than the F-quantile. The power of this test is $P\{W > F_{1-\alpha}(t-1, df)\}$ where $W \sim F'(t-1, df, \sum_i r(\tau_i - \bar{\tau})^2/\sigma^2)$.

For any estimable function $\mathbf{c}'\boldsymbol{\tau}$, the t-based two-sided $(1-\alpha)100\%$ confidence interval can be written as:

$$\mathbf{c}'\mathbf{T} \pm t_{1-\alpha/2}(df)\sqrt{\mathbf{c}'\mathbf{c} \times MSE/r}.$$

6.5 Multiple comparisons

In experiments performed to compare a large number of treatments, there may be a need to estimate or test hypotheses about a large number of estimable parameter contrasts. In these situations, the problem of controlling the *experiment-wise error probability* may be of concern. Suppose that 95% confidence intervals are to be constructed for several specified linear contrasts. For example, in a screening experiment designed to compare 10 treatments, it might be of interest to construct confidence intervals on all 45 pairwise differences of two treatment parameters. If conventional t-based intervals are used and all the required assumptions are valid, each interval has a probability of 0.95 (before the experiment is executed) of containing its target parameter contrast value, or a 0.05 probability of failing to contain it. However, the probability that *at least one* of several such intervals fails to contain its target may be substantially greater because the 0.05-risk is taken several times; in the 10-treatment example the probability that at least one interval for a treatment difference will fail is approximately 0.64. Multiple comparisons confidence intervals are constructed as modifications of the usual forms, so that the pre-execution probability of *any* of them being incorrect (i.e., not containing its target parameter) is less than or equal to a user-supplied experiment-wise error probability we denote by α_E. Similar adjustments are available for multiple hypothesis tests. Miller (1981) is a popular in-depth treatment of the general multiple comparisons problem.

In this section, we briefly describe four procedures for constructing collections of confidence intervals that maintain a selected experiment-wise type I error probability. Each is a modification of the t-based confidence interval for a linear contrast of treatment parameters. In each case described, the procedure calls for constructing multiple intervals of standard form, but replacing the $t_{1-\alpha/2}$ quantile by the $\alpha_E/2$ quantile from a different, related distribution.

6.5.1 Tukey intervals

In many experiments, the treatment contrasts of greatest interest are the pairwise differences between treatment parameters, i.e., $\mathbf{c}_{ij}'\boldsymbol{\tau}$ where $\mathbf{c}_{ij}$ contains all zeros except for $\{\mathbf{c}_{ij}\}_i = +1$, and $\{\mathbf{c}_{ij}\}_j = -1$, for all $1 \leq i < j \leq t$. If all such comparisons are of interest, the number of inferences is $\binom{t}{2}$. ($\mathbf{c}_{ij}$ and $\mathbf{c}_{ji}$ are not both considered, since one is simply the negative of the other.) If individual t-based 95% confidence intervals are used for each comparison, the expected number of intervals that are incorrect (i.e., do not contain their respective target parameter contrast values) when $t = 7$ is $\binom{7}{2} \times 0.05 = 1.05$, and when $t = 20$ is $\binom{20}{2} \times 0.05 = 9.50$.

Tukey (1953) described how quantiles from the *studentized range* distribution could be used as the basis for simultaneous inference in this case. Let $u_1, u_2, u_3, \ldots, u_t$ and $v_1, v_2, v_3, \ldots, v_{df+1}$ be independent random variables following a common normal distribution, and let $R = u_{max} - u_{min}$ be the sample range of the first sample, and $S = \sqrt{\frac{1}{df}\sum_i(v_i - \bar{v})^2}$ be the sample standard deviation of the second sample. Then the ratio $\frac{R}{\sqrt{2}S}$ follows the studentized range distribution, which is fully characterized by the values of t and df. Let $q_{1-\alpha_E/2}(t, df)$ denote the $1-\alpha_E/2$ quantile of this distribution; then the Tukey simultaneous pairwise confidence intervals are of form:

$$T_i - T_j \pm q_{1-\alpha_E/2}(t, df)\sqrt{2 \times MSE/r}$$

constructed for all $1 \leq i \leq j \leq t$.

6.5.2 Dunnett intervals

When one of the treatments is a control, reference condition, or well-understood "standard," the comparisons of greatest interest are often the $t - 1$ pairwise contrasts involving this special treatment (treatment 1, say) and each of the other treatments in turn, that is, $\mathbf{c}'_{1j}\boldsymbol{\tau}$ where $\mathbf{c}_{1j}$ contains all zeros except for $\{\mathbf{c}_{1j}\}_1 = +1$, and $\{\mathbf{c}_{1j}\}_j = -1$, for all $2 \leq j \leq t$. Dunnett (1964) developed a procedure for this situation which, like the Tukey procedure, requires substitution of a different distributional quantile for t in the standard confidence interval. As in our description of the Tukey method, let $u_1, u_2, u_3, \ldots, u_t$ and $v_1, v_2, v_3, \ldots, v_{df+1}$ be independent random variables following a common normal distribution, and let $S = \sqrt{\frac{1}{df} \sum_i (v_i - \bar{v})^2}$, but in this case define $D = max_{j=2,3,4,\ldots,t} |u_1 - u_j|$. Then the ratio $\frac{D}{\sqrt{2}S}$ follows a distribution characterized by t and df; let $d_{1-\alpha_E/2}(t, df)$ denote the $1 - \alpha_E/2$ quantile of this distribution. Then the simultaneous intervals are of form:

$$T_1 - T_j \pm d_{1-\alpha_E/2}(t, df)\sqrt{2 \times MSE/r}$$

constructed for all $2 \leq j \leq t$. For specified values of t and df, there are fewer Dunnett intervals than Tukey intervals, and so the Dunnett intervals can be somewhat smaller for a given experiment-wise error probability.

6.5.3 Simulation-based intervals for specific problems

The similarity of arguments given for the Tukey and Dunnett modifications to the general t-based intervals suggests a general procedure for generating joint confidence intervals for any specified collection of treatment comparisons. The quantiles required in each case are not analytically tractable quantities, and the tables originally published for implementing these methods were based on extensive and careful numerical analytic evaluation of the integrals. This effort is warranted for collections of contrasts that are apt to be of interest in many studies – here all paired differences, and all differences including a control, respectively.

However, the context of an experiment often suggests specific comparisons of particular interest. For example, in a study comparing $t = 5$ treatments, there might be four contrasts of primary interest, denoted by the rows of the matrix:

$$\mathbf{C} = \begin{pmatrix} +1 & -1 & 0 & 0 & 0 \\ +1 & 0 & -1 & 0 & 0 \\ 0 & 0 & +1 & -1 & 0 \\ 0 & 0 & +1 & 0 & -1 \end{pmatrix}.$$

This might be the case if, for example, treatments 1 and 3 are fundamentally different preparations, treatment 2 is a modification of treatment 1, and treatments 4 and 5 are modifications of treatment 3. This fits neither of the patterns for the Tukey or Dunnett procedures, but could in principle be solved the same way. Generally, suppose there are c contrasts in t treatment parameters, defined as $c_i'\tau, i = 1, 2, 3, \ldots, c$. Let $u_i, i = 1, 2, 3, \ldots, t$, and S be defined as described in subsections 6.5.1 and 6.5.2. Define the t-element vector u to have ith element u_i, and let $C = max_{i=1,2,3,\ldots,c} \frac{|c_i'u|}{\sqrt{c_i'c_i}S}$. Then the $1 - \alpha_E/2$ quantile of the distribution of C would be the appropriate factor for modifying standard confidence intervals in this case.

While it is probably impractical to compute this quantile by numerical integration techniques for each case that might arise, these quantities are quite easy to compute via stochastic simulation using most modern statistical computing packages. Edwards and Berry (1987) outlined a simple algorithm for doing this, and studied the precision of results obtained using it. Essentially, for given values of α_E, t, df, and contrasts C, their approach is to repeatedly (say M times, where M is a large number):

1. Generate u and v values as independent random draws from a normal distribution, say $N(0,1)$.

2. Use these to compute each of $c_i'u$, S, and C by the formulae described above.

3. Sort the resulting M values of C in ascending order, and use the $(M + 1)$ $(1 - \alpha_E/2)$-st of these as a Monte Carlo estimate of $f_{1-\alpha_E/2}(t, df)$, rounding or interpolating if needed.

The resulting value is used as the quantile required in the simultaneous intervals:

$$c_i'T \pm f_{1-\alpha_E/2}(t, df)\sqrt{c_i'c_i \times MSE/r}$$

constructed for all $i = 1, 2, 3, \ldots, c$. The procedure is "approximate" only in that $f_{1-\alpha_E/2}(t, df)$ is determined by simulation; for sufficiently large M it is essentially exact.

6.5.4 Scheffé intervals

The most general simultaneous confidence intervals we shall mention are the Scheffé (1953) intervals for simultaneous estimation of *any* collection of contrasts in the elements of τ. Because they are so generally applicable, they are wider than the more specialized intervals of Tukey and Dunnett, and so should generally not be used in situations for which the latter were developed. However, for several contrasts involving more than two treatments, especially when the form of some of these contrasts may result from a preliminary inspection of the data, Scheffé intervals provide an easy way to control experiment-wise error probability.

For any one contrast $c'\tau$, the Scheffé interval is yet another modification of the t-interval form:

$$c'T \pm \sqrt{(t - 1)F_{1-\alpha_E}(t - 1, df)}\sqrt{c'c \times MSE/r}.$$

Table 6.1 Example Data, Four-Treatment Latin Square Design

treatment 3 $y = 16.2$	treatment 2 $y = 18.3$	treatment 1 $y = 15.7$	treatment 4 $y = 25.0$
treatment 1 $y = 13.1$	treatment 4 $y = 21.9$	treatment 3 $y = 16.8$	treatment 2 $y = 21.0$
treatment 4 $y = 24.1$	treatment 3 $y = 18.1$	treatment 2 $y = 18.9$	treatment 1 $y = 18.8$
treatment 2 $y = 20.0$	treatment 1 $y = 18.2$	treatment 4 $y = 23.5$	treatment 3 $y = 22.7$

Regardless of the number of such intervals formed (and note that the interval form is not a function of this number), the pre-experiment probability that any of the intervals does not contain its target parameter contrast value is less than or equal to α_E. Unlike the intervals produced by the Tukey and Dunnett procedures, Scheffé intervals are conservative (and in some cases, very conservative); that is, they generally have pre-experiment probability of actual joint coverage of greater than $1 - \alpha_E$.

6.5.5 Numerical example

Table 6.1 presents data from a four-treatment experiment designed as an un-replicated Latin square. Because the blocks of a Latin square are orthogonal to treatments, least-squares estimates of estimable treatment contrasts are simply the corresponding contrasts of the treatment means, Table 6.2. The additional information needed to construct confidence intervals on these con-trasts is the error mean square from the full model, in this case containing additive effects for the row-blocks, column-blocks, and treatments (Table 6.3, R6.1). The quantiles needed to construct Tukey pairwise confidence inter-vals can be found in published tables of the studentized range distribution (e.g., Beyer, 1968), but many statistical computing packages contain routines that can calculate many kinds of simultaneous confidence intervals (Table 6.4, R6.2). In this case, the null hypothesis of no treatment difference would be rejected for even a very small selected type I error probability. Simultane-ous interval estimates of pairwise differences of means, using Tukey's proce-dure, indicate differences between each pair of treatments except for 1 and 3, and 2 and 3, while preserving overall confidence for all paired comparisons of 95%.

Table 6.2 Treatment Averages for Data of Table 6.1

Treatment	1	2	3	4
Average	16.45	19.55	18.43	23.63

Table 6.3 Analysis of Variance for Data of Table 6.1. (R6.1)

Source of Variation	Degrees of Freedom	Sum of Squares	Mean Square	F
Rows	3	20.0025	6.66750	
Columns	3	31.0475	10.34917	
Treatments	3	109.9025	36.63417	42.72
Residuals	6	5.1450	0.85750	

6.6 Conclusion

Most diagnostic checks of the assumptions made in analyzing experimental data are based on residuals from the fitted model. Many of these are graphical presentations designed to reveal residual values that would be unusual under the model assumptions. Analytical procedures described in this chapter are the modified Levene test for detecting heterogeneity of variance in a CRD, a formal test for lack of fit for blocked experiments in which some units in a common block receive the same treatment, and the Tukey one-degree-of-freedom test for detecting block-by-treatment interaction in a CBD. When the variance of observations is monotonically related to their mean, the Box-Cox procedure can sometimes be used to find a power transformation that preserves the desired form of the linear model while making variances more homogeneous.

A large number of simultaneous inference procedures have been developed and are applicable to data collected from designed experiments. These differ in the collection of tests or estimates of interest, and the approach taken to controlling the pre-experiment probability of making one or more errors. In this chapter, we have described two modifications of standard t-based confidence intervals for specific collections of pairwise treatment contrasts, a simulation-based generalization of these that can be used for any *a priori* collection of contrasts, and a conservative method that can be used for any collection of contrasts – even those determined after looking at the data – that control the experiment-wise error probability α_E. We note that the use of these methods does not come without some cost. While they do control the experiment-wise error probability, and so provide better control over the probability that any error is made, they do this by increasing the width of the interval estimates.

Table 6.4 Tukey 95% Simultaneous Confidence Intervals for Data of Table 6.1. (R6.2)

Treatment Comparison	Estimate	Lower Bound	Upper Bound
1-2	−3.10	−5.37	−0.833
1-3	−1.98	−4.24	0.292
1-4	−7.17	−9.44	−4.910
2-3	1.12	−1.14	3.390
2-4	−4.07	−6.34	−1.810
3-4	−5.20	−7.47	−2.930

In some cases, the resulting intervals may be so wide as to be of limited practical value to the investigator. An appropriate balance between relatively narrow intervals (and thus higher risk) versus lower risk (and thus wider intervals) is highly application-specific, but should be understood by all involved in interpreting the data.

6.7 Exercises

1. In the discussion of subsection 6.2.1, residual analysis is described as often being loosely based on the idea that the off-diagonal elements of the hat matrix are "small." However, they cannot all be zero in realistic experiments, and *can* easily be large even when N is substantial.

    (a) Determine the conditions under which a CRD results in a hat matrix **H** that is diagonal (i.e., has off-diagonal elements that are all exactly zero).

    (b) Consider designs for comparing two treatments using two blocks of 10 units each. Demonstrate that the off-diagonal elements of **H** are not necessarily "small" by computing this matrix for a (not very attractive) design that meets these specifications.

2. Using the data reconstructed from the experiment of Matsuu et al. (2005) (Table 3.1),

    (a) Perform the modified Levene test to check for equality of variance in the four treatment groups.

    (b) Assuming variances *are* equal in the four treatment groups, use the appropriate method to construct simultaneous confidence intervals for comparing the control condition to each of the other treatment groups; use $\alpha_E = 0.05$.

3. Using the data of Kocaoz et al. (2005) (Table 4.1),

    (a) Perform the Tukey one-degree-of-freedom test to check for interaction between the blocks and treatments.

    (b) Using the appropriate method, construct simultaneous confidence intervals for comparing all pairs of treatments; use $\alpha_E = 0.05$.

4. Write a computer program or script (in any language or package you like) to compute the quantile value necessary to perform the simulation-based simultaneous confidence intervals of Edwards and Berry. Your program will need (as input):

    - the number of treatments, t,
    - the number of contrasts, c, and
    - the $c \times t$ matrix of contrast coefficients, **C**.

5. Suppose Murphy et al. (2006)(subsection 5.1.1) are especially interested in comparing the proportion of "clicks" observed at the first web page location to the average proportion of clicks observed at locations 2 through 6, *and* the average proportion of clicks observed at the last web page location to the average proportion of clicks observed at locations 2 through 6. Use the program you wrote in exercise 4 to compute the quantile that should be used to construct simultaneous confidence intervals for these two contrasts, using $\alpha_E = 0.10$. (For this exercise, ignore the fact that the data were actually zeros and ones in this experiment, and that no information is given about a value of MSE.)

6. All simultaneous interval estimation procedures described in Section 6.5 are such that $\mathbf{c}'\boldsymbol{\tau}$ is estimated by $\widehat{\mathbf{c}'\boldsymbol{\tau}} \pm K\sqrt{\mathbf{c}'\mathbf{c}\,MSE/r}$, for an appropriate value of K. What value of K should be used if, based on the data presented in the example of subsection 6.5.5, intervals are to be constructed for 1000 values of the vector $\mathbf{c}$ and overall confidence of 80% is desired?

7. In most applications of the Box-Cox transform, the value of p is fitted as described in Section 6.3, but is subsequently treated as if it is a known constant. The degree to which this practice affects the quality of inferences depends on many things, perhaps especially the size of the samples. In order to get some understanding of this, perform a simulation study in which you repeatedly:

   • Generate $n_i = 10$ data values from normal distributions with $\mu_i = i \times 10$ and $\sigma_i = \sqrt{i} \times 10$, for $i = 1 \ldots 5$.

   • Use the Box-Cox method to estimate a power transform parameter p.

   • Use the transformed data to construct a 95% confidence interval on $\mu_2 - \mu_1$.

   Over a large number of simulations (say 1000), keep track of the proportion of times the constructed confidence interval contains the true mean difference of 10. Is this proportion significantly different from 0.95?

8. Compute the studentized residuals for the data presented in Table 6.1. Are any of these values large enough to raise suspicions about the standard assumptions?

## CHAPTER 7

Balanced incomplete block designs

7.1 Introduction

The Complete Block Designs and Latin Square Designs introduced in Chapters 4 and 5 share special structure that leads to simple and desirable characteristics in data analysis. In particular, they are orthogonally blocked designs, composed of blocks of the smallest size (t units) for which this is possible. However, effective designs of even smaller blocks can certainly be constructed, and are necessary in many applications. For example, in some settings, groups of two similar units form natural blocks; e.g., identical twins, left and right halves of a common plant leaf, opposite surfaces of a metal plate, and the two front tires mounted on the same vehicle. While such blocks may offer especially "tight" experimental control of noise, they cannot be used in a CBD or LSD in which more than two treatments are to be compared.

In such cases, unit-to-treatment assignments necessarily result in *incomplete blocks*, that is, blocks in which only a subset of treatments are assigned. We maintain the notation of t for the number of treatments and b for the number of blocks, but now denote the block size, or number of units in each block, by $k < t$. Note that this means the total number of units and observations N is bk rather than bt. In most cases, there are many ways in which a design of t treatments in blocks of size k *can* be constructed, but it should be clear that some arrangements are better than others.

In this chapter, we focus most of our attention on *Balanced Incomplete Block Designs* (BIBDs), a special class of designs which, as their name suggests, maintain statistically desirable "balance" properties despite the requirement that k be less than t. An incomplete block design is a BIBD when three requirements are met:

1. Each treatment is applied to at most one unit in each block.

2. Each treatment is applied to a unit in the same number of blocks. We refer to this as the *first-order balance requirement*. It follows that this common number of units per treatment must be $r = bk/t$, the number of units divided by the number of treatments.

3. Each pair of treatments is applied to two units in the same number of blocks. We refer to this as the *second-order balance requirement*. The common number of blocks in which each pair of treatments appear is $\lambda = r(k-1)/(t-1)$. This result follows from considering any one treatment, say the treatment labeled "1". Treatment 1 is assigned in r blocks, so there are

$r(k-1)$ units available *in these blocks* for allocation of the $t-1$ other treatments, and so $r(k-1)/(t-1)$ is the average number of within-block pairings of treatment 1 with any other treatment. But if the second-order balance requirement holds, the average number of within-block treatment pairings is also the common number of within-block pairings for any two treatments.

So, for example

1	1	1	1	1	1	2	2	2	3
2	2	2	3	3	4	3	3	4	4
3	4	5	4	5	5	4	5	5	5

is a BIBD for $t=5$ treatments in $b=10$ blocks of size $k=3$, with $r=6$ and $\lambda=3$, while

1	3	5	2	4	1	3	5	2	4
2	4	1	3	5	2	4	1	3	5
3	5	2	4	1	3	5	2	4	1

is not a BIBD, even though it meets requirements 1 and 2 above.

7.1.1 Example: drugs and blood pressure

Kraiczi, Hedner, Peker and Grote (2000) compared the effects of five drugs – atenolol, amlodipine, enalapril, hydrochlorothiazide, and losartan, each at a standard dose – on the blood pressure of patients suffering from both hypertension and obstructive sleep apnea. The 40 patients included in the study were males of age 25 to 70 years, each of whom was tested by objective criteria for both medical conditions. Each patient was randomly assigned to one of 20 treatment protocols. A treatment protocol consisted of a baseline period during which blood pressure measurements were made without medical therapy, followed by two 6-week treatment periods separated by a 3-week "washout" period. During each treatment period, the patient received daily oral doses of one of the five drugs. The 20 treatment protocols were identical except for the *ordered* sequence of two drugs used; 20 such sequences are possible, and two of the 40 patients were assigned to each of the 20 protocols. The primary response variable in this study was the average of three diastolic blood pressure measurements, taken 24 hours after the last dose of medication administered in each period.

The design of this study is actually somewhat more complex than a basic BIBD, because the investigators were also concerned about the possibility of an effect due to the order of drugs administered. Models containing additional terms to explain *carryover effects* are often used with *crossover designs* in this context. For our purposes, suppose order of administration is of no consequence, e.g., that we are willing to assume that any systematic difference between the effects of atenolol and amlodipine is not dependent on the order in which they are administered. With this assumption, and disregarding the data collected during the baseline period, this experiment can be viewed as a BIBD in $t = 5$ treatments (drugs), in $b = 40$ blocks (patients), each of which contains $k = 2$ units (treatment periods experienced by the specified patient). The 10 block patterns used correspond to the 10 (now unordered) combinations of two drugs from among five, and each of the basic patterns is used with four patients. Hence in the notation described above $r = kb/t = 2 \times 40/5 = 16$ and $\lambda = r(k-1)/(t-1) = 16 \times 1/4 = 4$. Under our simplified scenario, appropriate assignment of units to treatments could be accomplished by randomly dividing the 40 patients into 10 groups of equal size (each group corresponding to one pair of drugs), and then for each patient individually, flipping a coin to determine which of the two assigned treatments would be used in the first treatment period.

7.1.2 Existence and construction of BIBDs

The construction of CBDs is such that a design can be constructed in any number of blocks for any number of treatments. However, there are many combinations of values of t, k, and b for which a BIBD does not exist. A necessary, but not sufficient, condition for the existence of a BIBD follows immediately from the construction requirements: For given t, k, and b, a BIBD cannot exist unless:

- $r = bk/t$ is an integer, and
- $\lambda = r(k-1)/(t-1) = bk(k-1)/[t(t-1)]$ is an integer.

For example, a BIBD cannot exist for six treatments in five blocks, each of size 4, since the required number of units associated with each treatment would be $r = 3\frac{1}{3}$. For seven treatments, a design in 21 blocks, each of size 5 *may* exist since this would allow for $r = bk/t = 21 \times 5/7 = 15$ and $\lambda = r(k-1)/(t-1) = 15 \times 4/6 = 10$. In fact, such a BIBD does exist in this case, and it is easy to show that no BIBD for seven treatments in fewer blocks of size 5 can exist since $b = 21$ is the smallest integer satisfying the two conditions in this case.

Construction of the BIBD for the case of $t = 7$ and $k = 5$ just mentioned is actually simple. It can be accomplished by including one block with each of the $\binom{7}{5} = 21$ subsets of five treatments. In general, a BIBD for t treatments in $\binom{t}{k}$ blocks of size k can always be constructed, even though it is not always the smallest BIBD that is possible for the given values of t and k. It is easy to show that such all-possible-subsets BIBDs are characterized by $r = \binom{t-1}{k-1}$ and $\lambda = \binom{t-2}{k-2}$. Another simple and useful technique is to note that any BIBD in b blocks can be expanded to a BIBD in mb blocks where m is an integer greater than one, by including m "copies" of each required block; the values of r and λ for the (larger) BIBD are each also increased by a factor of m. The BIBD described in the blood pressure experiment of subsection 7.1.1 is an example of both techniques; all possible combinations of two from five treatments are used as block patterns, and each block pattern is replicated in four different patients.

Several more elaborate algebraic techniques have been developed for constructing BIBDs, either for specified values of t, k, and/or b, or, as in the case of the replicating technique mentioned above, by modifying other BIBDs. Some of these are discussed in John (1998). Colburn and Dinitz (1996) are editors of an extensive table of BIBDs, and Prestwich (2003) has described a construction algorithm.

7.2 A model

Aside from the values of indices that can occur together in describing a response, the form of an effects model for a BIBD is the same as that for a CBD:

$$y_{ij} = \alpha + \beta_i + \tau_j + \epsilon_{ij},$$
$$i = 1 \ldots b, \qquad j \in S(i),$$
$$\epsilon_{ij} \text{ i.i.d. with } E(\epsilon_{ij}) = 0 \quad \text{and} \quad Var(\epsilon_{ij}) = \sigma^2 \qquad (7.1)$$

for the response from a unit in block i that received treatment j, where $S(i)$ is the set of k treatments assigned to experimental units in block i. As with CBDs, standard analysis of data from a BIBD generally proceeds under the assumption that blocks and treatments do not interact.

7.2.1 Graphical logic

At first glance, it might seem that a graphical presentation of the response data collected from a BIBD could be designed using a similar strategy as that described in Chapter 4 for CBDs. There, each data value was altered by subtraction of its corresponding block average to remove variability that could reasonably be attributable to blocks before treatment-specific boxplots were constructed. However, that approach is not so reasonable in this case since each block (and so each block average) represents only a subset of treatments. With CBDs, "correction" for block totals is an adjustment for a *common* average effect over all treatments, within each block, leaving a corrected value that represents the deviation of a specific treatment from the average of all treatments. Applied to BIBDs, this correction represents a different subset of treatments in each block, and so does not produce the desired result.

One solution to this problem, which is of practical value primarily when t is relatively small, is to construct a boxplot of differences for each *pair* of treatments, using only data from the λ blocks in which both treatments have been applied to units:

$$d_{ijj'} = y_{ij} - y_{ij'}, \quad j \in S(i) \quad \text{and} \quad j' \in S(i), \ j \neq j'. \tag{7.2}$$

Based on our model, each $d_{ijj'}$ has mean $\tau_j - \tau_{j'}$ and variance $2\sigma^2$, and so reflects only relative characteristics of treatments j and j' and within-block variability assumed to be consistent throughout the experiment.

It should be noted that this collection of $\binom{t}{2}$ boxplots may not seem entirely consistent to an untrained observer. Since different subsets of data associated with a given treatment are used in comparisons to other treatments, it is likely that (for example) the average of λ values for treatments 1 and 3 is *not* equal to the difference of averages for treatments 1 and 2, and treatments 2 and 3. However, discrepancies of this sort that are large, relative to the standard deviation of $\sqrt{2\sigma^2/\lambda}$ for the average of values in each boxplot, may be indicators of a violation of modeling assumptions.

The plots just described, while sometimes useful, do not actually represent *all* information about treatment differences in a BIBD. For example, consider a BIBD in blocks of size 2, in which treatments 1 and 3 are applied in block 1, and treatments 2 and 3 are applied in block 2. While treatments 1 and 2 do not appear together in either block, it is easy to see that $y_{1,1} - y_{1,3} + y_{2,3} - y_{2,2}$ has mean $\tau_1 - \tau_2$, but variance larger than that of a difference between two data values from the same block, $2\sigma^2$. More generally, if blocks 1 and 2 each contain units assigned to several common treatments other than 1 and 2, the averages of all such values can be used in place of $y_{1,3}$ and $y_{2,3}$ to reduce the variance of the contrast while maintaining the expectation of $\tau_1 - \tau_3$. These multi-block contrasts can be added to graphical displays, but only with the understanding that they are less precise than differences computed within a single block.

7.2.2 Example: dishwashing detergents

John (1961) described an experiment performed to compare $t = 9$ dishwashing detergent formulations. Interest centers on how long the soapsuds remain in a standard solution of each detergent, and the recorded response is the number of plates (each soiled in a controlled, uniform way) that are washed before the "foam" disappears. The experiment is performed in three basins; the tests performed during the same time period are treated as a block so as to control for fluctuations in the temperature of the room and other unaccountable sources of variation that would affect all three simultaneous tests. In the data listed in Table 7.1, treatment 9 is actually regarded as a standard formulation or control, while treatments 1–8 are alternative detergent formulations of interest. Examination of the pattern of entries in the table shows that this is a BIBD, with $b = 12$ blocks each containing $k = 3$ units, for $t = 9$ treatments each of which appears in $r = 4$ blocks, and each pair of which appear together in $\lambda = 1$ block.

Table 7.1 Data (Number of Plates) from the Detergent Example of John (1961)

Treatment	Block 1	2	3	4	5	6	7	8	9	10	11	12	Treatment Total
1	19	-	-	20	-	-	20	-	-	20	-	-	79
2	17	-	-	-	17	-	-	21	-	-	17	-	72
3	11	-	-	-	-	15	-	-	13	-	-	14	53
4	-	6	-	7	-	-	-	-	7	-	6	-	26
5	-	26	-	-	26	-	26	-	-	-	-	24	102
6	-	23	-	-	-	23	-	23	-	24	-	-	93
7	-	-	21	20	-	-	-	21	-	-	-	21	83
8	-	-	19	-	19	-	-	-	20	19	-	-	77
9	-	-	28	-	-	31	31	-	-	-	29	-	119
Block Total	47	55	68	47	62	69	77	65	40	63	52	59	704

7.3 Matrix formulation

A matrix representation of all data from an experiment arranged as a BIBD also follows the general form of that for a CBD:

$$y = X_1\beta + X_2\tau + \epsilon \quad \epsilon \sim N(0, \sigma^2 I), \tag{7.3}$$

where β is the $(b+1)$-element vector of nuisance parameters α and β_i, $i = 1, 2, 3, \ldots, b$, τ is the t-element vector of treatment parameters and X_1 takes the form that would be used in a CBD with k (rather than t) treatments:

$$X_1 = \begin{pmatrix} 1_k & 1_k & 0_k & \cdots & 0_k \\ 1_k & 0_k & 1_k & \cdots & 0_k \\ \cdots & \cdots & \cdots & \cdots & \cdots \\ 1_k & 0_k & 0_k & \cdots & 1_k \end{pmatrix}.$$

It follows immediately that H_1 also takes the same form as it would in a CBD with blocks of size k:

$$H_1 = \frac{1}{k} \begin{pmatrix} J_{k\times k} & 0_{k\times k} & \cdots & 0_{k\times k} \\ 0_{k\times k} & J_{k\times k} & \cdots & 0_{k\times k} \\ \cdots & \cdots & \cdots & \cdots \\ 0_{k\times k} & 0_{k\times k} & \cdots & J_{k\times k} \end{pmatrix}. \tag{7.4}$$

The form of X_2 is more difficult to write simply, but can be characterized by noting that

- each row consists of zeros with the exception of a single 1,
- the total of elements in any column of X_2 is r,
- the inner product of any two columns of X_2 is λ, and
- the group of rows numbered $((i-1)k+1, (i-1)k+2, (i-1)k+3, \ldots ik)$, $i = 1, 2, 3, \ldots, b$, i.e., those rows coding treatment assignments in the ith block, contain 1's placed according to each element of $S(i)$.

It follows that the lth row of $\mathbf{H}_1\mathbf{X}_2$, $l = 1, 2, 3, \ldots, N$, contains elements that take on one of two values, namely $\frac{1}{k}$ in columns corresponding to treatments that are applied in the block that contains observation l, and 0 in columns corresponding to treatments that are *not* applied in the block that contains observation l. As an example, consider a BIBD for $t = 6$ treatments, in blocks of size $k = 4$. If treatments 1, 2, 3, and 6 are applied in a given block, the four rows of $\mathbf{H}_1\mathbf{X}_2$ associated with the units in that block are each:

$$\left(\frac{1}{4}, \ \frac{1}{4}, \ \frac{1}{4}, \ 0, \ 0, \ \frac{1}{4}\right).$$

Further, since $\mathbf{X}_{2|1} = (\mathbf{I} - \mathbf{H}_1)\mathbf{X}_2 = \mathbf{X}_2 - \mathbf{H}_1\mathbf{X}_2$, the corresponding rows of *this* matrix are:

$$\left(\frac{3}{4}, \ -\frac{1}{4}, \ -\frac{1}{4}, \ 0, \ 0, \ -\frac{1}{4}\right)$$

$$\left(-\frac{1}{4}, \ \frac{3}{4}, \ -\frac{1}{4}, \ 0, \ 0, \ -\frac{1}{4}\right)$$

$$\left(-\frac{1}{4}, \ -\frac{1}{4}, \ \frac{3}{4}, \ 0, \ 0, \ -\frac{1}{4}\right)$$

$$\left(-\frac{1}{4}, \ -\frac{1}{4}, \ -\frac{1}{4}, \ 0, \ 0, \ \frac{3}{4}\right).$$

In the general case, a given row of $\mathbf{X}_{2|1}$ contains three unique values:

- $1 - \frac{1}{k}$ in the column corresponding to the treatment applied to this unit,
- $-\frac{1}{k}$ in columns corresponding to the $k-1$ other treatments applied to units in this block, and
- 0 in columns corresponding to treatments not applied to any unit in this block.

Since all such rows have elements that sum to zero, it is clear that any linear combination of the rows of $\mathbf{X}_{2|1}$ has a zero sum, implying (once again) that contrasts are the only estimable linear combinations of the elements of $\boldsymbol{\tau}$. Further, it is clear that for any distinct pair of treatments j and j', $\tau_j - \tau_{j'}$ is estimable because the contrast can be expressed as a linear combination of just the rows of $(\mathbf{I} - \mathbf{H}_1)\mathbf{X}_2$ corresponding to a block in which these two treatments are both applied. Finally, since the pairwise contrast associated with *every* two treatments is estimable, *any* linear contrast in the elements of $\boldsymbol{\tau}$ is estimable, just as in the case of the CBD.

To understand the structure of the design information matrix $\mathcal{I} = \mathbf{X}'_{2|1}\mathbf{X}_{2|1}$ for this design, consider the characterization of $\mathbf{X}_{2|1}$ given above. The jth column (corresponding to treatment j) contains:

- r elements of value $1 - \frac{1}{k}$, corresponding to the r experimental runs in which treatment j is applied,

- $r(k-1)$ elements of value $-\frac{1}{k}$ corresponding to runs in which other treatments were applied, but which are grouped in blocks where treatment j was also applied, and

- $k(b-r)$ elements of value zero corresponding to runs in blocks where treatment j was not applied.

Hence, the sum of squared elements in any one column is:

$$r\left(1-\frac{1}{k}\right)^2 + r(k-1)\left(-\frac{1}{k}\right)^2 = \frac{r(k-1)}{k} = \frac{\lambda(t-1)}{k},$$

so this is the common diagonal element of $\mathcal{I}$. The inner product of any pair of distinct columns of $\mathbf{X}_{2|1}$ is comprised of:

- 2λ terms of value $-(1-\frac{1}{k})\frac{1}{k}$ corresponding to runs receiving one of the two treatments, from blocks in which they are both applied,

- $\lambda(k-2)$ terms of value $\frac{1}{k^2}$ corresponding to runs receiving neither of the two treatments, from blocks in which they are both applied (to other units), and

- $(b-\lambda)k$ terms of value 0 corresponding to runs from blocks in which at least one of the two treatments is not applied.

So, the complete inner product for two such columns is:

$$-2\lambda\left(1-\frac{1}{k}\right)\frac{1}{k} + \lambda(k-2)\frac{1}{k^2} = -\lambda\frac{1}{k},$$

the common off-diagonal element of $\mathcal{I}$. Taken together, these imply that

$$\mathcal{I} = \left(\frac{\lambda(t-1)}{k} + \frac{\lambda}{k}\right)\mathbf{I} - \frac{\lambda}{k}\mathbf{J} = \frac{\lambda}{k}(t\mathbf{I} - \mathbf{J}) = \frac{\lambda t}{k}\left(\mathbf{I} - \frac{1}{t}\mathbf{J}\right). \tag{7.5}$$

Since the reduced normal equations can be written as $\mathcal{I}\hat{\tau} = \mathbf{X}'_{2|1}\mathbf{y}$, we now consider the form of the right side of this system. Turning again to our description of the structure of $\mathbf{X}_{2|1}$, the inner product of the jth column (corresponding to treatment j) and $\mathbf{y}$ includes terms of:

- $(1-\frac{1}{k})$ times the total of responses associated with treatment j, and

- $-\frac{1}{k}$ times the total of responses associated with all other treatments, but only from blocks in which treatment j was assigned,

or, simplifying the sum of these, the total of responses associated with treatment j minus $\frac{1}{k}$ times the total of *all* responses from blocks in which treatment j was assigned. Letting $\mathbf{T}$ represent the t-element vector of treatment totals, and $\mathbf{B}$ represent the b-element vector of block totals, this means that the system of reduced normal equations can be written as

$$\frac{\lambda t}{k}\left(\mathbf{I} - \frac{1}{t}\mathbf{J}\right)\hat{\tau} = \mathbf{X}'_2(\mathbf{I} - \mathbf{H}_1)\mathbf{y} = \mathbf{T} - \frac{1}{k}(\mathbf{X}'_2\mathbf{X}_1)\mathbf{B}. \tag{7.6}$$

The elements of the vector on the right side of equation (7.6) are often called the "adjusted treatment totals,"

$$Q = \mathbf{T} - \frac{1}{k}\mathbf{X}_2'\mathbf{X}_1\mathbf{B}.$$

Dividing each side of the reduced normal equations by $\frac{\lambda t}{k}$ yields

$$\left(\mathbf{I} - \frac{1}{t}\mathbf{J}\right)\hat{\tau} = \frac{k}{\lambda t}Q \tag{7.7}$$

which in turn leads to the form of the estimate for any contrast in the treatment parameters:

$$\widehat{\mathbf{c}'\tau} = \frac{k}{\lambda t}\mathbf{c}'Q. \tag{7.8}$$

It is interesting to compare this equation to the form of the estimate for unblocked or orthogonally blocked experiments with r units assigned to each treatment:

$$\widehat{\mathbf{c}'\tau} = \frac{1}{r}\mathbf{c}'\mathbf{T}.$$

While these forms are similar, $\mathbf{T}$ and Q are not equal (nor even proportional), and $r \neq \lambda t/k$, reflecting the fact that treatments and blocks are not orthogonal in a BIBD.

Another result of this nonorthogonality of blocks and treatments is the somewhat more complicated form of an ANOVA decomposition. For CBDs, the partitioning of sums of squares is especially simple:

$$SSB \quad + \quad SST \quad + SSE = \quad \text{Total SS}$$

$$\sum_i t(\bar{y}_{i.} - \bar{y}_{..})^2 + \sum_j r(\bar{y}_{.j} - \bar{y}_{..})^2 + SSE = \sum_{i,j}(y_{ij} - \bar{y}_{..})^2$$

where SSE can be easily computed as a difference. The absence of orthogonality between blocks and treatments in a BIBD can be interpreted as meaning that *some* of the variation in the data can be attributed to either blocks *or* treatments. Computing SSB as above assigns all of this variation to blocks, and SST (which now quantifies variation associated for treatments after "adjustment" for blocks) can be written for a BIBD as

$$SST = \sum_j \frac{k}{\lambda t}(Q_j - \bar{Q}_.)^2 = \sum_j \frac{k}{\lambda t}Q_j^2 = \sum_j \frac{\lambda t}{k}\left(Q_j \Big/ \frac{\lambda t}{k}\right)^2 \tag{7.9}$$

since $\bar{Q}_. = 0$. Again, $\frac{\lambda t}{k}$ "plays the role of" r in the sum of squares, just as it did in the reduced normal equations, even though $r \neq \lambda t/k$. Given this adjustment, SSE can be computed as a difference between the mean-corrected total sum of squares and $(SSB + SST)$ for BIBDs, just as it is for CRDs.

7.3.1 Basic analysis: an example

We continue with the data of John (1961) described in subsection 7.2.2 and listed in Table 7.1. Block and treatment totals are included in the table, and the array of elements in $\mathbf{X}_2'\mathbf{X}_1$, the incidence matrix, follows the pattern of entries in the main body of the table with 1's corresponding to data locations and 0's corresponding to empty table cells. Using equation (7.6) above, this leads to:

$$
\frac{1}{3}[9\mathbf{I} - \mathbf{J}]\hat{\tau} =
\begin{pmatrix} 79 \\ 72 \\ 53 \\ 26 \\ 102 \\ 93 \\ 83 \\ 77 \\ 119 \end{pmatrix}
- \frac{1}{3}
\begin{pmatrix}
1 & 0 & 0 & 1 & 0 & 0 & 1 & 0 & 0 & 1 & 0 & 0 \\
1 & 0 & 0 & 0 & 1 & 0 & 0 & 1 & 0 & 0 & 1 & 0 \\
1 & 0 & 0 & 0 & 0 & 1 & 0 & 0 & 1 & 0 & 0 & 1 \\
0 & 1 & 0 & 1 & 0 & 0 & 0 & 0 & 1 & 0 & 1 & 0 \\
0 & 1 & 0 & 0 & 1 & 0 & 1 & 0 & 0 & 0 & 0 & 1 \\
0 & 1 & 0 & 0 & 0 & 1 & 0 & 1 & 0 & 1 & 0 & 0 \\
0 & 0 & 1 & 1 & 0 & 0 & 0 & 1 & 0 & 0 & 0 & 1 \\
0 & 0 & 1 & 0 & 1 & 0 & 0 & 0 & 1 & 1 & 0 & 0 \\
0 & 0 & 1 & 0 & 0 & 1 & 1 & 0 & 0 & 0 & 1 & 0
\end{pmatrix}
\begin{pmatrix} 47 \\ 55 \\ 68 \\ 47 \\ 62 \\ 69 \\ 77 \\ 65 \\ 40 \\ 63 \\ 52 \\ 59 \end{pmatrix}
$$

or, after algebra and reduction

$$
\hat{\tau} - \bar{\hat{\tau}}.\mathbf{1} = \frac{1}{3}\mathcal{Q} = \frac{1}{3}
\begin{pmatrix}
1.000 \\ -3.333 \\ -18.667 \\ -38.667 \\ 17.667 \\ 9.000 \\ 3.333 \\ -0.667 \\ 30.333
\end{pmatrix}
=
\begin{pmatrix}
0.333 \\ -1.111 \\ -6.222 \\ -12.889 \\ 5.889 \\ 3.000 \\ 1.111 \\ -0.222 \\ 10.111
\end{pmatrix}.
$$

So for example, differences between each of the first eight treatment parameters and the ninth (standard) are uniquely estimable as $\tau_1 - \tau_9 = 0.333 - 10.111 = -9.778$, et cetera. Following equation (7.9), SST can be computed as:

$$
\frac{k}{\lambda t}\sum_j \mathcal{Q}_j^2 = \frac{1}{3} \times 3180.4 = 1060.1333
$$

leading to

$$
MST = 1060.1333/8 = 132.5167.
$$

SSB is $\sum_i k(\bar{y}_{i.} - \bar{y}_{..})^2 = 426.2363$, and SSE can then be computed as a difference between the corrected total sum of squares, and sum of squares for blocks and (block-corrected) treatments:

$$1502.8889 - 426.2363 - 1060.1333 = 16.5193$$

for which the corresponding mean square is an estimate of error variance:

$$\hat{\sigma}^2 = 16.5193/16 = 1.0325.$$

While the derivation and use of these equations are helpful in demonstrating how inference for BIBDs and CBDs is both similar and different, data analysis for designs that are not orthogonally blocked (and often for simpler plans as well) is most often carried out using general computer programs. The format of input used by various programs differs, but the key is to be sure that the treatment sum of squares is being computed after correction for blocks, and not vice versa. A strategy that is generally safe (regardless of the program) is to assemble the ANOVA components using the least squares fits to two models, one containing only the nuisance parameters coded in $\mathbf{X}_1$, and the other containing all terms (i.e., model (7.3)). Call the first of these Model 1, and the second Model 1+2 (R7.1). Then the ANOVA decomposition of interest can be assembled as:

- SSB and associated degrees of freedom as the "model" component of the Model 1 fit,
- SSE and associated degrees of freedom as the "residual" component of the Model 1+2 fit,
- SST and associated degrees of freedom as the difference between "model" components for Model 1+2 and Model 1, and
- total SS as reported in either fit.

7.4 Influence of design on quality of inference

For an estimable linear combination of the elements of $\boldsymbol{\tau}$, that is, $\mathbf{c}'\boldsymbol{\tau}$ for which $\mathbf{c}'\mathbf{1} = 0$, the variance of $\widehat{\mathbf{c}'\boldsymbol{\tau}}$ is:

$$Var(\widehat{\mathbf{c}'\boldsymbol{\tau}}) = \sigma^2 \mathbf{c}'\mathcal{I}^-\mathbf{c} = \sigma^2 \frac{k}{\lambda t}\mathbf{c}'\left(\mathbf{I} - \frac{1}{t}\mathbf{J}\right)^-\mathbf{c} = \sigma^2\frac{k}{\lambda t}\mathbf{c}'\mathbf{c}.$$

Recall that the variance function for CRDs with each treatment applied to r experimental units is

$$Var(\widehat{\mathbf{c}'\boldsymbol{\tau}}) = \sigma^2\frac{1}{r}\mathbf{c}'\mathbf{c},$$

and that this variance function also applies to all orthogonally blocked designs for which each treatment is applied to r units. Since the variance formula for the BIBD differs by a factor of $\frac{kr}{\lambda t}$, the ratio of estimation variances for a

BIBD and CRD of the same number of experimental units is:

$$\frac{Var_{BIBD}}{Var_{CRD}} = \frac{kr}{\lambda t}\frac{\sigma^2_{BIBD}}{\sigma^2_{CRD}} = \frac{k(t-1)}{t(k-1)}\frac{\sigma^2_{BIBD}}{\sigma^2_{CRD}}.$$

Since $k < t$, the factor $\frac{k(t-1)}{t(k-1)}$ is larger than one, and becomes larger as the block size decreases relative to the number of treatments. This is a direct consequence of the comparative information lost to the absence, in some blocks, of pairs of units assigned to any two treatments. As with other blocked designs, the hope is that effective blocking leads to a smaller value of σ^2. The reduction in variance must be such that $\frac{\sigma^2_{BIBD}}{\sigma^2_{CRD}} < \frac{t(k-1)}{k(t-1)}$ if the BIBD is to result in smaller estimation variances for estimates. This requirement is likely to be met in many cases, where blocking is even modestly effective. The more important comparison, however, is between BIBDs and comparably sized CBDs. Since the latter are orthogonally blocked, the BIBD has superior estimation variance properties only if:

$$\frac{\sigma^2_{BIBD}}{\sigma^2_{CBD}} < \frac{t(k-1)}{k(t-1)}.$$

Here, the reduction in variance must come from reduction of the block size from t to k, and this change may be much less dramatic. Nonetheless, there are many applications in which very efficient blocking can be achieved only when k is at least somewhat less than t, and BIBD's are natural candidates in these settings.

As we have discussed, BIBDs are not orthogonally blocked designs, but their balanced structure makes their statistical efficiency easy to compare to the designs described in earlier chapters. In particular, for a t-treatment BIBD containing b blocks each of size k, and in which each treatment is replicated r times:

- The residual degrees of freedom are found by subtracting from $N = rt = bk$ the number of linearly independent fixed parameters in the block-treatment additive model, i.e., $N - t - b + 1$. In the following, we use "df" to stand for this quantity.

- The variance of an estimable function of treatment parameters is

$$Var(\widehat{\mathbf{c}'\boldsymbol{\tau}}) = \sigma^2 \frac{k}{\lambda t}\sum_{k=1}^{t} c_k^2. \tag{7.10}$$

- The Balanced Incomplete Block Design yields a smaller expected squared length for $(1-\alpha)100\%$ two-sided confidence intervals of estimable functions than a CRD with N/t units assigned to each treatment if

$$\sigma_{BIBD}/\sigma_{CRD} < \sqrt{\frac{t(k-1)}{k(t-1)}}[t_{1-\alpha/2}(N-t))/t_{1-\alpha/2}(df)], \tag{7.11}$$

and is superior in the same sense to a CBD in $b = N/t$ blocks if

$$\sigma_{BIBD}/\sigma_{CBD} < \sqrt{\frac{t(k-1)}{k(t-1)}}[t_{1-\alpha/2}(N-b-t+1)/t_{1-\alpha/2}(df)], \quad (7.12)$$

where σ_{CRD}, σ_{CBD}, and σ_{BIBD} are the standard deviations associated with ϵ in each of the three designs.

- For a given estimable function $\mathbf{c}'\boldsymbol{\tau}$ and signal-to-noise ratio $\psi = \mathbf{c}'\boldsymbol{\tau}/\sigma$, a desired $\Psi = \mathbf{c}'\boldsymbol{\tau}/\sqrt{Var(\widehat{\mathbf{c}'\boldsymbol{\tau}})}$ can be obtained with

$$m = \Psi^2 \frac{\mathbf{c}'\mathbf{c}}{\psi^2} \frac{k}{\lambda t} \qquad (7.13)$$

replicates of the BIBD.

- For testing $\mathrm{Hyp}_0 : \tau_1 = \tau_2 = \tau_3 = \ldots = \tau_t$, the noncentrality parameter associated with the F-test is:

$$\boldsymbol{\tau}'\boldsymbol{\mathcal{I}}\boldsymbol{\tau}/\sigma^2 = \frac{\lambda t}{k}\sum_k(\tau_k - \bar{\tau})^2/\sigma^2, \qquad (7.14)$$

and the power of this test for given values of $\boldsymbol{\tau}$ and σ^2 is

$$\mathrm{Prob}\{W > F_{1-\alpha}(t-1, df)\} \text{ where } W \sim F'\left(t-1, df, \frac{\lambda t}{k}\sum_k(\tau_k - \bar{\tau})^2/\sigma^2\right).$$
$$(7.15)$$

7.5 More general constructions

7.5.1 Extended complete block designs

The primary motivation for BIBDs given so far is the need to construct designs for which the natural or convenient block size is smaller than the number of treatments to be compared. However, the BIBD structure we have discussed is also useful for generating *extended complete block designs* introduced by John (1963), and further developed by others including Trail and Weeks (1973). These designs are similar to the *augmented complete block designs* discussed in Section 4.6, in that the number of units in each block is larger than the number of treatments. In extended complete block designs, each treatment is assigned to one of the first t units in each block, and a BIBD is used to determine the treatment assignments for the "extra" units.

More precisely, suppose that b blocks of $k > t$ units each are available for comparing t treatments, *and* that b, k, and t are such that a BIBD exists for t treatments in b blocks of size $k' = k - t$ with design parameters $r' = bk'/t$ and $\lambda' = r'(k'-1)/(t-1)$. An extended complete block design can then be constructed by applying the k' treatments included in a block of the BIBD to *two* randomly selected units, and each of the $t - k'$ other treatments to one randomly selected unit, in the corresponding block of the extended complete block design. The result is a design that has first- and second-order balance among treatments as defined in Section 7.1, *and* that affords bk' "pure error" degrees of freedom in within-block replication for estimation of σ^2 that does not depend on the assumption of additive block effects.

In thinking about the form of the design information matrix for the extended complete block design, it is helpful to use the expression $\mathcal{I} = \mathbf{X}_2'\mathbf{X}_2 - (\mathbf{H}_1\mathbf{X}_2)'(\mathbf{H}_1\mathbf{X}_2)$. For this design, $\mathbf{X}_2'\mathbf{X}_2 = (b + r')\mathbf{I}$. The form of $\mathbf{H}_1$ is as it would be for any design containing b blocks each of size k,

$$\mathbf{H}_1 = \frac{1}{k}\begin{pmatrix} \mathbf{J}_{k\times k} & \mathbf{0}_{k\times k} & \cdots & \mathbf{0}_{k\times k} \\ \mathbf{0}_{k\times k} & \mathbf{J}_{k\times k} & \cdots & \mathbf{0}_{k\times k} \\ \cdots & \cdots & \cdots & \cdots \\ \mathbf{0}_{k\times k} & \mathbf{0}_{k\times k} & \cdots & \mathbf{J}_{k\times k} \end{pmatrix}. \tag{7.16}$$

$\{\mathbf{H}_1\mathbf{X}_2\}_{ij}$ is equal to $\frac{1}{k}$ times the number of units assigned to treatment j within the block containing the ith run; in this case, the unique counts are 1 and 2. In the jth column of this matrix $r'k$ elements are $\frac{2}{k}$ (each row associated with a block in which treatment j occurs twice) and $(b - r')k$ elements are $\frac{1}{k}$ (each row associated with a block in which treatment j occurs once). Further, for any pair of columns, $j \neq j'$, corresponding elements are both $\frac{2}{k}$ in $\lambda'k$ rows, are mixed – i.e., $(\frac{1}{k}, \frac{2}{k})$ or $(\frac{2}{k}, \frac{1}{k})$ – in $2(r' - \lambda')k$ rows, and are both $\frac{1}{k}$ in $(b - 2r' + \lambda')k$ rows. With these results and some algebraic reduction, one expression for the design information matrix for the extended complete block design is:

$$\mathcal{I} = b\frac{k(k-3) + 2t}{k(t-1)}\left[\mathbf{I} - \frac{1}{t}\mathbf{J}\right].$$

7.5.2 Partially balanced incomplete block designs

In some cases, an incomplete block design is desirable due to operational or efficiency constraints, but the value of t, and the allowable values of k and b, do not permit construction of a BIBD. In these cases, a design from the larger class of *partially balanced incomplete block designs* (PBIBD) may be available. A PBIBD, like a BIBD, is a design for comparing t treatments in b blocks of $k < t$ units each, that requires each treatment to be applied to one unit in each of r blocks – the same first-order balance property required of BIBDs. However, the "partial" second-order balance requirements of a PBIBD are less stringent. Specifically, a PBIBD(2) (the "2" to be explained shortly) requires the following:

1. For treatment i, the remaining $t - 1$ treatments may be divided into two groups; call these A_i and B_i. For example, for eight treatments, A_3 and B_3 might be $\{1, 2, 4\}$ and $\{5, 6, 7, 8\}$. The treatments identified in A_i are called *first associates* of treatment i, and those identified in B_i are called *second associates* of treatment i.

2. For any treatment i, A_i contains t_1 elements, and B_i contains t_2 elements, $t_1 + t_2 = t - 1$.

3. Any two treatments that are first associates (e.g., treatment i and any treatment identified in A_i) appear together in λ_1 blocks. Any two treatments that are second associates appear together in λ_2 blocks.

4. For any two treatments that are first associates, say treatments i and j,

   • A_i and A_j have p_{11} elements in common,
   • A_i and B_j (or A_j and B_i) have p_{12} elements in common, and
   • B_i and B_j have p_{22} elements in common.

5. For any two treatments that are second associates, say treatments i and j,

   • A_i and A_j have q_{11} elements in common,
   • A_i and B_j (or A_j and B_i) have q_{12} elements in common, and
   • B_i and B_j have q_{22} elements in common.

Hence, the inter-related design parameters (values specifying the structure and properties of the design) are t, b, k, t_1, t_2, λ_1, λ_2, p_{11}, p_{12}, p_{22}, q_{11}, q_{12}, and q_{22}. Extensive tables of PBIBD(2) designs have been published by, for example, Clatworthy (1973).

For example, consider again the second example design given in the introduction to this chapter:

1	3	5	2	4	1	3	5	2	4
2	4	1	3	5	2	4	1	3	5
3	5	2	4	1	3	5	2	4	1

Each of the five treatments appears in six blocks, so the design has first-order balance. Each pair of treatments is applied together in blocks with frequencies:

Treatments	(1,2)	(1,3)	(1,4)	(1,5)	(2,3)	(2,4)	(2,5)	(3,4)	(3,5)	(4,5)
Frequency	2	1	1	2	2	1	1	2	1	2

So, for each treatment, the first and second associates are

Treatment	First Associates	Second Associates
1	2,5	3,4
2	1,3	4,5
3	2,4	1,5
4	3,5	1,2
5	1,4	2,3

with $\lambda_1 = 2$, and $\lambda_2 = 1$. We leave it to the reader to confirm that requirements 4 and 5 are satisfied, completing the verification that this design is a PBIBD(2).

Finally, the designation PBIBD(2) specifies that there are two associate classes for each treatment. The more general class of designs also includes subclasses PBIBD(m) for $m = 3, 4, 5, \ldots, t - 1$ for more associate classes (and so also even more design parameters). Where m is larger, the class of designs is more general (based on fewer restrictions) resulting in less simplicity in inference formulae, but more freedom for constructing designs with desired size characteristics.

7.6 Conclusion

Balanced incomplete block designs are attractive alternatives to CBDs when homogeneous blocks contain fewer units than the number of treatments under study ($k < t$). BIBD "balance" refers to the requirements that each treatment be assigned to the same number of units (r), and that each pair of treatments be assigned to units in the same number of blocks (λ). When k is not much smaller than t, the loss of efficiency of a BIBD due to incomplete block structure is minimal, and is often completely overcome by the smaller observation variance associated with smaller blocks. Apart from any difference in the error variances, variances of estimable functions under BIBDs are $\frac{k(t-1)}{t(k-1)}$ times their counterparts under CRDs or orthogonally blocked designs. *Extended complete block designs* contain blocks of $k > t$ in which the all t treatments are assigned to at least one unit in each block, and the treatments assigned to a second unit in each block are determined by a BIBD. *Partially balanced incomplete block designs* contain fewer units in each block than the number of treatments being compared, and require the same first-order balance requirement as BIBDs, but do not require all pairs of treatments to be assigned to units in the same number of blocks.

7.7 Exercises

1. A researcher wants to conduct a blocked experiment to compare five treatments, but operational constraints require that the blocks be of size 3.

   (a) She likes the idea of performing the experiment in five blocks since that would allow her to assign each treatment to three different units in the experiment. Can a BIBD for this experiment be constructed under these conditions? Prove your answer.

   (b) Construct a BIBD for this experiment using ten blocks (not five). Fully specify your design by making a table with 10 rows for blocks, with three entries in each row for the treatments to be included in that block.

   (c) Suppose we think of this experiment with reference to the model:

   $$y_{ij} = \alpha + \beta_i + \tau_j + \epsilon_{ij}.$$

where y is the response, β's are block effects, and τ's are treatment effects. Suppose that, in fact (although we don't know it as experimenters),

$$\tau_1 = \tau_2 = -1, \quad \tau_3 = 0, \quad \tau_4 = \tau_5 = +1, \quad \sigma = 2$$

Given this information, completely characterize the distribution of the F-statistic that would be used to test for equality of treatments. (That is, specify the distribution including numerical values of all parameters.)

(d) Continuing with part (c), suppose the researcher had been able to execute a randomized complete block design in ten blocks, rather than the BIBD we have been discussing. Note that this would have been a larger experiment, since each block would contain five units. Using the same model information provided in part (c), give a complete characterization of the distribution of the F-statistic that would be used to test for equality of treatments in this case.

2. A consumer products testing laboratory performed a study to compare four varieties of home radon detectors. Trials were performed in a laboratory chamber that was large enough for simultaneous testing of only three units. Units processed together in the chamber can safely be assumed to experience very similar exposures, but there may be some variation in achieved chamber conditions among the chamber operation "sessions." The design and resulting data (expressed in a unitless efficiency measure) are given in the following table:

| | Detector Type | | | |
Chamber Session	A	B	C	D
1	6.11	-	5.95	5.82
2	6.70	6.22	-	5.97
3	6.60	6.11	6.52	-
4	-	6.22	6.54	6.18

(a) Identify the treatments and units in this experiment.

(b) Use a computer package such as R to compute sums of squares for blocks, treatments after accounting for blocks, residuals, and corrected total. Do this by fitting two models, one containing terms for only chamber session effects, and one containing terms representing both chamber sessions and detector effects, and assemble the required information from the two fits.

(c) Perform an F-test for equality of detector types.

(d) Assuming that blocks constitute fixed effects, derive the least-squares estimates and standard errors of the six treatment differences:

$$\tau_1 - \tau_2, \quad \tau_1 - \tau_3, \quad \tau_1 - \tau_4, \quad \tau_2 - \tau_3, \quad \tau_2 - \tau_4, \quad \tau_3 - \tau_4$$

3. Continuing the previous radon detector problem, *now* suppose the experiment and resulting data set are twice as large:

Chamber Session	Detector Type			
	A	B	C	D
1	6.11	-	5.95	5.82
2	6.70	6.22	-	5.97
3	6.60	6.11	6.52	-
4	-	6.22	6.54	6.18
5	6.34	-	6.20	6.06
6	6.77	6.30	-	6.02
7	6.55	6.09	6.48	-
8	-	6.04	6.24	5.98

(a) Use a computer package such as R to compute sums of squares for blocks, treatments after accounting for blocks, residuals, and corrected total. Do this by fitting two models, one containing terms for only chamber session effects, and one containing terms representing both chamber sessions and detector effects, and assemble the required information from the two fits.

(b) Compute the adjusted data needed to make the six boxplots described in subsection 7.2.1. As each plot would be based on only four data values in this case, plot these as six parallel "dot plots," each comprised of four dots on a common vertical axis, rather than as boxplots. Does this plot suggest the existence of systematic differences among the four detector types?

4. Suppose that it might have been possible to perform the (simplified version of) the drug experiment discussed in subsection 7.1.1 as an extended complete block experiment in which each patient received two of the drugs in six-week periods as described, but could also receive all five of the drugs in five additional six-week periods. Continue to assume that the order of drug administration does not matter. For each of (1) the BIBD described in subsection 7.1.1, and (2) the extended complete block design described here, compute:

(a) the design information matrix for treatments

(b) the power of the level $\alpha = 0.05$ test for no treatment effects if in fact

$$\tau' = (-8, 2, 2, 2, 2) \quad \text{and} \quad \sigma = 10.$$

5. Consider the following incomplete block design in six treatments:

1	1	1	1	1	2	2	2	3	4
2	2	3	3	4	3	3	4	5	5
5	6	4	6	5	4	5	6	6	6

Verify that this is or is not a BIBD. If it is, give the values of r and λ. In any case, compute the design information matrix for the six treatments under comparison.

6. Consider the following partially balanced incomplete block design, a PBIBD(2) taken from Clatworthy (1973), for comparing eight treatments in eight blocks of size 4 units each:

1	5	2	6	3	7	4	8
2	6	7	3	8	4	1	5
3	7	8	4	1	5	6	2
4	8	1	5	6	2	7	3

(a) For each treatment, identify the first- and second-associates.

(b) Determine the values of all design parameters discussed in subsection 7.5.2:

$$t, \ b, \ k, \ t_1, \ t_2, \ \lambda_1, \ \lambda_2, \ p_{11}, \ p_{12}, \ p_{22}, \ q_{11}, \ q_{12}, \ \text{and} \ q_{22}.$$

7. Consider a BIBD in $b = 20$ blocks of size $k = 2$ units each for comparing $t = 5$ treatments. For this design:

(a) Apart from a factor of σ^2, what is the sampling variance of $\widehat{\tau_1 - \tau_2}$?

(b) What t-value would be used to construct a 95% two-sided confidence interval for $\tau_1 - \tau_2$?

(c) If $\tau_1 = \tau_2 = 0$, $\tau_3 = 1$, $\tau_4 = \tau_5 = 2$, and $\sigma^2 = 2$, what is the noncentrality parameter of the distribution of the test statistic for:

$$\text{Hyp}_0 : \tau_1 = \tau_2 = \tau_3 = \tau_4 = \tau_5?$$

8. Suppose a Latin square design is planned to compare $t = 5$ treatments, using 25 experimental units, each of which is part of one "row block" and one "column block." However, in the execution of the experiment, all observations to be taken from one of the row blocks are lost, so that the resulting experiment actually provides $N = 20$ data values. Note that in this case, the "row blocks" alone take the form of a CBD, while the "column blocks" partition units as a BIBD. (This is an example of a *Youden square design*, introduced by Youden (1940).)

(a) Using a computer, calculate the 5×5 design information matrix for treatments.

(b) Apart from a factor of σ^2, what is the sampling variance of $\widehat{\tau_1 - \tau_2}$?

(c) What t-value would be used to construct a 95% two-sided confidence interval for $\tau_1 - \tau_2$?

## CHAPTER 8

Random block effects

8.1 Introduction

In our discussion of blocked designs to this point, we have represented block effects as unknown, fixed model parameters. But there are also experimental situations in which it is reasonable to think of block effects as random, suggesting that a mixed effects model may be more appropriate (e.g., Dedidenko (2002), Hocking (2003)). If an industrial metallurgist performs an experiment using ingots of steel as units, and the ingots used in any particular block are chosen from those delivered by a supplier on a particular day, this might *not* lead to a random-blocks assumption since the metallurgist might not (at least without extensive study) be willing to assume that day-to-day differences from the supplier are reasonably considered to be random draws from some conceptual population of batches. However, if the metallurgist's company already owned a very large inventory of ingots, and each batch was assembled by selecting ingots stored together at a randomly chosen location in a warehouse (and so likely produced together, et cetera), an analysis based on an assumption of random block effects might be more sensible.

At first glance, it might seem that the distinction between random blocks and fixed blocks has little practical importance because block effects are usually not of experimental interest. In fact, many experimenters prefer to treat block effects as fixed in the analysis, regardless of arguments that might be made about them being reasonably thought of as random. But where the random blocks assumption is reasonable, it can lead to additional analysis options. In some designs, such as the split-plot designs described in Chapter 10, a full analysis of experimental treatments cannot be made *unless* block effects can be treated as random.

8.2 Inter- and intra-block analysis

Suppose that an N-run experiment is divided into b blocks of equal size k (so $N = bk$), and that a partitioned model of form:

$$\mathbf{y} = \mathbf{X}_1\boldsymbol{\beta} + \mathbf{X}_2\boldsymbol{\tau} + \boldsymbol{\epsilon}, \quad E(\boldsymbol{\epsilon}) = \mathbf{0}, \quad Var(\boldsymbol{\epsilon}) = \sigma^2\mathbf{I}$$

is appropriate, where

$$\mathbf{X}_1 = \begin{pmatrix} \mathbf{1}_k & \mathbf{0}_k & \cdots & \mathbf{0}_k \\ \mathbf{0}_k & \mathbf{1}_k & \cdots & \mathbf{0}_k \\ \cdots & \cdots & \cdots & \cdots \\ \mathbf{0}_k & \mathbf{0}_k & \cdots & \mathbf{1}_k \end{pmatrix},$$

β is a b-element random vector, $E(\beta) = \mu_\beta \mathbf{1}$, and $Var(\beta) = \sigma_\beta^2 \mathbf{I}$. If the elements of β are also independent of those in ϵ, it follows from the model that:

$$E(\mathbf{y}) = \mathbf{1}\mu_\beta + \mathbf{X}_2\boldsymbol{\tau}$$

$$Var(\mathbf{y}) = \sigma_\beta^2 \mathbf{X}_1 \mathbf{X}_1' + \sigma^2 \mathbf{I}.$$

While each observation has the same variance, $\sigma_\beta^2 + \sigma^2$, pairs of observations associated with the same block have covariance σ_β^2. It follows that the best linear unbiased estimator of $\boldsymbol{\tau}$ is the *generalized least-squares estimate*, any solution to the normal equations:

$$\mathbf{X}_2'(\sigma_\beta^2 \mathbf{X}_1 \mathbf{X}_1' + \sigma^2 \mathbf{I})^{-1}\mathbf{X}_2\hat{\boldsymbol{\tau}} = \mathbf{X}_2'(\sigma_\beta^2 \mathbf{X}_1 \mathbf{X}_1' + \sigma^2 \mathbf{I})^{-1}\mathbf{y},$$

or equivalently,

$$\mathbf{X}_2'(\rho^2 \mathbf{X}_1 \mathbf{X}_1' + \mathbf{I})^{-1}\mathbf{X}_2\hat{\boldsymbol{\tau}} = \mathbf{X}_2'(\rho^2 \mathbf{X}_1 \mathbf{X}_1' + \mathbf{I})^{-1}\mathbf{y},$$

where $\rho = \sigma_\beta/\sigma$. In most practical cases, ρ is not known and so $\hat{\boldsymbol{\tau}}$ cannot be determined. *Iteratively reweighted least-squares* procedures (Jennrich and Moore (1975), Gentle (2007)) that cycle between estimation of $\boldsymbol{\tau}$ using the normal equations with an estimated value of ρ^2 in place of the true variance ratio, and estimation of ρ treating an estimated value of $\boldsymbol{\tau}$ as the true parameter vector, are sometimes used in such cases.

In designed experiments, it is often possible to manage the relationship between $\mathbf{X}_2$ and $\mathbf{X}_1$ so that other options are available. Let $\mathbf{U}$ be any N-row matrix such that $\mathbf{X}_1'\mathbf{U} = \mathbf{0}$. Then define two linear transformations of the data vector, $\mathbf{y}_1 = \mathbf{U}'\mathbf{y}$ and $\mathbf{y}_2 = \mathbf{X}_1'\mathbf{y}$. Note in particular that $\mathbf{y}_2$ is the b-element vector of block totals. Statistical models for $\mathbf{y}_1$ and $\mathbf{y}_2$, following from the original data model, are:

$$\mathbf{y}_1 = \mathbf{U}'\mathbf{X}_2\boldsymbol{\tau} + \mathbf{U}'\mathbf{X}_1\beta + \mathbf{U}'\epsilon$$

$$= \mathbf{U}'\mathbf{X}_2\boldsymbol{\tau} + \epsilon_1$$

where $E(\epsilon_1) = \mathbf{0}$ and $Var(\epsilon_1) = \sigma^2 \mathbf{U}'\mathbf{U}$, and

$$\mathbf{y}_2 = \mathbf{X}_1'\mathbf{X}_2\boldsymbol{\tau} + \mathbf{X}_1'\mathbf{X}_1\beta + \mathbf{X}_1'\epsilon$$

$$= \mathbf{X}_1'\mathbf{X}_2\boldsymbol{\tau} + k\beta + \mathbf{X}_1'\epsilon$$

$$= k\mu_\beta\mathbf{1} + \mathbf{X}_1'\mathbf{X}_2\boldsymbol{\tau} + \epsilon_2$$

where $E(\epsilon_2) = \mathbf{0}$ and $Var(\epsilon_2) = (k^2\sigma_\beta^2 + k\sigma^2)\mathbf{I}$. Hence the transformed data can be represented as two linear models, each containing only a single random element. Further,

$$Cov(\mathbf{y}_1, \mathbf{y}_2) = \mathbf{U}'(\sigma^2\mathbf{I} + \sigma_\beta^2\mathbf{X}_1\mathbf{X}_1')\mathbf{X}_1 = \sigma^2\mathbf{U}'\mathbf{X}_1 + \sigma_\beta^2\mathbf{U}'\mathbf{X}_1\mathbf{X}_1'\mathbf{X}_1 = \mathbf{0}$$

so if all random elements are normally distributed, analyses based on the two transformed models are statistically independent. With this structure, it is desirable that the treatments be assigned (i.e., $\mathbf{X}_2$ have structure relative to $\mathbf{X}_1$ and $\mathbf{U}$), so that the interesting linear functions of $\boldsymbol{\tau}$ are $\mathbf{c}'\boldsymbol{\tau}$, such that

- $\mathbf{c}' = \mathbf{l}_1'(\mathbf{U}'\mathbf{X}_2)$, so that $\mathbf{c}'\boldsymbol{\tau}$ is estimable based on the analysis of $\mathbf{y}_1$, or
- $\mathbf{c}' = \mathbf{l}_2'(\mathbf{I} - \frac{1}{b}\mathbf{J})(\mathbf{X}_1'\mathbf{X}_2)$, so that $\mathbf{c}'\boldsymbol{\tau}$ is estimable based on the analysis of $\mathbf{y}_2$.

We can add some specificity to these ideas by noting that one matrix meeting the requirements specified for $\mathbf{U}$ is $\mathbf{I} - \mathbf{X}_1(\mathbf{X}_1'\mathbf{X}_1)^{-1}\mathbf{X}_1' = \mathbf{I} - \mathbf{H}_1$, the projection matrix associated with the compliment of the column space of $\mathbf{X}_1$. From this, models for $\mathbf{y}_1$ and $\mathbf{y}_2$ can be rewritten in more familiar terms as:

$$\mathbf{y}_1 = (\mathbf{I} - \mathbf{H}_1)\mathbf{X}_2\boldsymbol{\tau} + \epsilon_1 \qquad Var(\epsilon_1) = \sigma^2(\mathbf{I} - \mathbf{H}_1) \qquad (8.1)$$

$$\mathbf{y}_2 = k\mu_\beta\mathbf{1} + \mathbf{X}_1'\mathbf{X}_2\boldsymbol{\tau} + \epsilon_2 \qquad Var(\epsilon_2) = k(k\sigma_\beta^2 + \sigma^2)\mathbf{I}. \qquad (8.2)$$

Analysis based on the first model leads to normal equations of form:

$$\mathbf{X}_2'(\mathbf{I} - \mathbf{H}_1)[\sigma^2(\mathbf{I} - \mathbf{H}_1)]^-(\mathbf{I} - \mathbf{H}_1)\mathbf{X}_2\hat{\boldsymbol{\tau}}_{intra} = \mathbf{X}_2'(\mathbf{I} - \mathbf{H}_1)[\sigma^2(\mathbf{I} - \mathbf{H}_1)]^-\mathbf{y}_1,$$

or eliminating σ^{-2} from each side,

$$\mathbf{X}_2'(\mathbf{I} - \mathbf{H}_1)(\mathbf{I} - \mathbf{H}_1)^-(\mathbf{I} - \mathbf{H}_1)\mathbf{X}_2\hat{\boldsymbol{\tau}}_{intra} = \mathbf{X}_2'(\mathbf{I} - \mathbf{H}_1)(\mathbf{I} - \mathbf{H}_1)^-\mathbf{y}_1.$$

Here, note that the identity matrix is a generalized inverse of $\mathbf{I} - \mathbf{H}_1$, because

$$(\mathbf{I} - \mathbf{H}_1)\mathbf{I}(\mathbf{I} - \mathbf{H}_1) = (\mathbf{I} - \mathbf{H}_1)$$

so we may rewrite the reduced normal equations as:

$$\mathbf{X}_2'(\mathbf{I} - \mathbf{H}_1)\mathbf{X}_2\hat{\boldsymbol{\tau}}_{intra} = \mathbf{X}_2'(\mathbf{I} - \mathbf{H}_1)\mathbf{y}_1 = \mathbf{X}_2'(\mathbf{I} - \mathbf{H}_1)\mathbf{y}, \qquad (8.3)$$

exactly the same reduced normal equations we have seen for the fixed-block scenario. In this context, $\hat{\boldsymbol{\tau}}_{intra}$ is called the *intra-block* estimate of $\boldsymbol{\tau}$, because it relies on the data only through linear combinations that are contrasts within each block (since $\mathbf{U}'\mathbf{X}_1 = \mathbf{0}$). Analysis based on the second model is a regression of the vector of block totals ($\mathbf{y}_2$) on a model matrix of form $(k\mathbf{1}|\mathbf{X}_1'\mathbf{X}_2)$ and parameter vector of form $(\mu_\beta, \boldsymbol{\tau}')$, and resulting in the *inter-block* estimate $\hat{\boldsymbol{\tau}}_{inter}$ (via reduced normal equations corrected for μ_β):

$$\mathbf{X}_2'\mathbf{X}_1\left(\mathbf{I} - \frac{1}{b}\mathbf{J}\right)\mathbf{X}_1'\mathbf{X}_2\hat{\boldsymbol{\tau}}_{inter} = \mathbf{X}_2'\mathbf{X}_1\left(\mathbf{I} - \frac{1}{b}\mathbf{J}\right)\mathbf{y}_2.$$

This inter-block estimate represents the additional, sometimes called "recovered," information about $\boldsymbol{\tau}$ based only on block totals, that comes as a direct result of the random-blocks assumption. Analyses of this form are sometimes used with extended complete block designs and balanced incomplete block designs, and are usually used with split-plot designs.

Because this approach to analysis is based on two different statistical models, we can think of the design as having two different design information matrices. The intra-block analysis information matrix is as we have previously defined it for fixed-block analysis:

$$\mathcal{I}_{intra} = \mathbf{X}_2'(\mathbf{I} - \mathbf{H}_1)\mathbf{X}_2.$$

The inter-block reduced normal equations feature a "left side" matrix of form

$$\mathbf{X}_2'\mathbf{X}_1 \left(\mathbf{I} - \frac{1}{b}\mathbf{J}\right) \mathbf{X}_1'\mathbf{X}_2$$

which might be regarded as the design information matrix for this analysis. However, recall that the full (Fisher) information matrix and noncentrality parameters each depend on both the design information matrix *and* a variance, specifically

$$Var(\widehat{\mathbf{c}'\boldsymbol{\tau}}) = \mathbf{c}'\mathcal{I}^-\mathbf{c}\sigma^2 \quad \text{and} \quad \lambda = \boldsymbol{\theta}'\mathcal{I}\boldsymbol{\theta}/\sigma^2$$

respectively. Hence any common factor can be added to both the design information matrix and the variance without influencing the result. We apply the factor $\frac{1}{k}$ here, and define the inter-block information matrix to be

$$\mathcal{I}_{inter} = \frac{1}{k}\mathbf{X}_2'\mathbf{X}_1 \left(\mathbf{I} - \frac{1}{b}\mathbf{J}\right) \mathbf{X}_1'\mathbf{X}_2$$

relative to the variance element

$$\frac{1}{k}(k^2\sigma_\beta^2 + \sigma^2) = k\sigma_\beta^2 + \sigma^2.$$

The latter variance is estimated unbiasedly by $MSE_{inter} = \frac{1}{k}MSE_{BT}$, where MSE_{BT} is the mean square error for the regression model of block totals (8.2). This adjustment has two practical advantages. First, it yields inter- and intra-block variance elements ($k\sigma_\beta^2 + \sigma^2$ and σ^2) that are directly comparable, because the coefficient of σ^2 is 1 in each case. Second, MSE_{inter} is a "natural" component of unified ANOVA decompositions that are used in the analysis of split-plot experiments (one of the most important applications of combined inter- and intra-block analysis, to be discussed in Chapter 10).

8.3 Complete block designs (CBDs) and augmented CBDs

Recall the experiment of Kocaoz et al. (2005) described in subsection 4.1.1, involving the comparison of $t = 4$ coatings on the tensile strength of steel reinforcement bars. The experiment was organized following a randomized complete block design in which four experimental units (steel bars) were treated together (one with each coating) and tested together in a process involving a single batch of cementitious grout. Each block contained one unit assigned to each treatment $(k = t)$, and $b = 8$ blocks were used. The design information matrix associated with the usual intra-block analysis is

$$\mathcal{I}_{intra} = \mathbf{X}_2'(\mathbf{I} - \mathbf{H}_1)\mathbf{X}_2 = b\left(\mathbf{I} - \frac{1}{t}\mathbf{J}\right) = 8\left(\mathbf{I}_{4\times4} - \frac{1}{4}\mathbf{J}_{4\times4}\right).$$

If, throughout the course of the experiment, coatings were consistently applied to each batch of bars, batches of grout were prepared in a consistent manner, and there was no reason to believe that any systematic differences could be associated with any of the experimental material used or time periods within which groups of bars were tested, it might be reasonable to think of block-to-block differences as being random, i.e., a second source of unexplainable random "noise" akin to, but not necessarily of the same general magnitude as, that associated with individual strength determinations.

If we attempt to construct an inter-block model for the block totals in this experiment, we find that the block totals all have the same expectation:

$$E(\mathbf{y}_2) = \left(k\mu_\beta + \sum_{i=1}^{4}\tau_i\right)\mathbf{1}$$

and so no informative inter-block estimator is available. Another way to see this is to observe that $\mathbf{X}_1'\mathbf{X}_2$, the matrix of "regressors" associated with $\boldsymbol{\tau}$ in the model for block totals, is $\mathbf{J}_{8\times4}$, so that the matrices in both sides of the reduced normal equations:

$$\left[\mathbf{X}_2'\mathbf{X}_1\left(\mathbf{I} - \frac{1}{b}\mathbf{J}\right)\mathbf{X}_1'\mathbf{X}_2\right]\hat{\boldsymbol{\tau}}_{inter} = \left[\mathbf{X}_2'\mathbf{X}_1\left(\mathbf{I} - \frac{1}{b}\mathbf{J}\right)\right]\mathbf{y}_2$$

have only zero elements. A little thought shows that this would *always* be true for complete block designs in which each treatment is applied to one unit in each block. More generally, it is also true of any blocked design in which any given treatment is applied to the same number of units in each block, even if this number is not the same for each treatment, since each column of $\mathbf{X}_1'\mathbf{X}_2$ would still be some multiple of $\mathbf{1}$. Hence, a minimal requirement for recovery of some inter-block information is that *the pattern of treatment assignments not be the same in each block*.

Suppose the experiment of Kocaoz et al. had been enlarged so that five bars were included in each of the eight blocks, and that this was done in a balanced fashion so that the "extra" bar received coating 1 in blocks 1 and 2, coating 2 in blocks 3 and 4, et cetera. Note that in this *augmented* CBD, N would be $5 \times 8 = 40$ rather than 32 as in the actual experiment described in Chapter 4. In this case, $\mathbf{X}_1'\mathbf{X}_1 = 5\mathbf{I}_8$, and $\mathbf{H}_1$ is a block-diagonal matrix, with eight blocks of $\frac{1}{5}\mathbf{J}_{5\times5}$ and other elements equal to zero. The design information matrix associated with the usual intra-block analysis is

$$\mathcal{I}_{intra} = \mathbf{X}_2'(\mathbf{I} - \mathbf{H}_1)\mathbf{X}_2 = \frac{48}{5}\left(\mathbf{I}_{4\times4} - \frac{1}{4}\mathbf{J}_{4\times4}\right).$$

The inter-block model matrix

$$(5 \times 1 | \mathbf{X}_1' \mathbf{X}_2) = \left(\begin{array}{c|cccc} 5 & 2 & 1 & 1 & 1 \\ 5 & 2 & 1 & 1 & 1 \\ 5 & 1 & 2 & 1 & 1 \\ 5 & 1 & 2 & 1 & 1 \\ 5 & 1 & 1 & 2 & 1 \\ 5 & 1 & 1 & 2 & 1 \\ 5 & 1 & 1 & 1 & 2 \\ 5 & 1 & 1 & 1 & 2 \end{array} \right)$$

is not of full column rank because the sum of columns 2–5 equals column 1, but inter-block information about $\boldsymbol{\tau}$ can be recovered under an assumption of random block effects. The design information matrix associated with this analysis is

$$\mathcal{I}_{inter} = \frac{1}{5} \mathbf{X}_2' \mathbf{X}_1 \left(\mathbf{I} - \frac{1}{8} \mathbf{J} \right) \mathbf{X}_1' \mathbf{X}_2 = \frac{2}{5} \left(\mathbf{I}_{4 \times 4} - \frac{1}{4} \mathbf{J}_{4 \times 4} \right).$$

8.4 Balanced incomplete block designs (BIBDs)

Yates (1940) first demonstrated how inter-block analysis can increase the information available in BIBDs with random blocks. Characterization of this additional information takes much the same form as in the case of "balanced" augmented CBDs, due to the first- and second-order balance properties of BIBDs. However, inter-block estimates can sometimes be of more practical value in BIBDs, especially when there are many blocks (i.e., the "sample size" for inter-block analysis is large) and blocks are small relative to the number of treatments (so that there is non-negligible loss of efficiency in the intra-block analysis). Of course, this depends on the size of ρ as well; the value of an inter-block analysis will always be limited, regardless of the design, if σ_β is large relative to σ.

Recall that BIBDs are characterized by five related design "parameters":

- b = the number of blocks,
- t = the number of treatments,
- k = the number of units in each block (less than t),
- r = the number of units allocated to each treatment in the design, and
- λ = the number of blocks in which any two treatments are both applied to units.

The intra-block analysis for BIBDs was developed in Chapter 7, where we found the reduced normal equations to be:

$$\frac{\lambda}{k} [t\mathbf{I} - \mathbf{J}] \hat{\boldsymbol{\tau}}_{intra} = \mathbf{T} - \frac{1}{k} (\mathbf{X}_2' \mathbf{X}_1) \mathbf{B} \tag{8.4}$$

where $\mathbf{T}$ is the t-element vector of treatment totals and $\mathbf{B}$ is the b-element vector of block totals, and the associated design information matrix is:

$$\mathcal{I}_{intra} = \frac{\lambda t}{k}\left(\mathbf{I} - \frac{1}{t}\mathbf{J}\right). \tag{8.5}$$

Inter-block analysis again depends fundamentally on the matrix $\mathbf{X}_1'\mathbf{X}_2$, which for general BIBDs has

- elements of 0 and 1 (since treatments are applied either once or not at all in each block),
- r 1's and $b - r$ 0's in each column, and
- λ rows in which both entries in any two columns are 1.

Because $\mathbf{I} - \mathbf{H}_1 = \mathbf{I}_{b \times b} - \frac{1}{b}\mathbf{J}_{b \times b}$, and since

$$(\mathbf{X}_1'\mathbf{X}_2)'(\mathbf{X}_1'\mathbf{X}_2) = (r - \lambda)\mathbf{I}_{t \times t} + \lambda \mathbf{J}_{t \times t}$$

and

$$\frac{1}{b}(\mathbf{X}_1'\mathbf{X}_2)'\mathbf{J}(\mathbf{X}_1'\mathbf{X}_2) = \frac{r^2}{b}\mathbf{J}_{t \times t}$$

the reduced normal equations for inter-block analysis are

$$(r - \lambda)\left[\mathbf{I} - \frac{1}{t}\mathbf{J}\right]\hat{\tau}_{inter} = \mathbf{X}_2'\mathbf{X}_1(\mathbf{B} - \bar{B}\mathbf{1}) \tag{8.6}$$

where $\bar{B}$ is the average block total. The corresponding design information matrix for the inter-block analysis is

$$\mathcal{I}_{inter} = \frac{1}{k}\mathbf{X}_2'\mathbf{X}_1\left(\mathbf{I} - \frac{1}{b}\mathbf{J}\right)(\mathbf{X}_1'\mathbf{X}_2) = \frac{r - \lambda}{k}\left(\mathbf{I} - \frac{1}{t}\mathbf{J}\right). \tag{8.7}$$

8.5 Combined estimator

Because intra- and inter-block estimators are uncorrelated, a weighted average of the two, with weights proportional to the inverse of their respective variances, is their optimal (minimum variance) linear combination. To see this, write a general combined estimate of any estimable function $\mathbf{c}'\tau$ as

$$\widehat{\widehat{\mathbf{c}'\tau}} = w_1\widehat{\mathbf{c}'\tau}_{intra} + w_2\widehat{\mathbf{c}'\tau}_{inter},$$

constrain $w_1 + w_2 = 1$ to ensure unbiasedness, and minimize

$$Var(\widehat{\widehat{\mathbf{c}'\tau}}) = w_1^2\, Var(\widehat{\mathbf{c}'\tau}_{intra}) + w_2^2\, Var(\widehat{\mathbf{c}'\tau}_{inter})$$

with respect to choice of w_1 and w_2 using the Method of Lagrangian Multipliers (subsection 3.4.1). For the BIBD, since $\mathcal{I}_{intra} = \frac{\lambda t}{k}(\mathbf{I} - \frac{1}{t}\mathbf{J})$ and $\mathcal{I}_{inter} = \frac{r - \lambda}{k}(\mathbf{I} - \frac{1}{t}\mathbf{J})$ the variances of the intra- and inter-block estimates of any estimable function are proportional to

$$Var(\widehat{\mathbf{c}'\tau}_{intra}) \propto \sigma^2\frac{k}{\lambda t}$$

$$Var(\widehat{\mathbf{c}'\tau}_{inter}) \propto (k\sigma_\beta^2 + \sigma^2)\frac{k}{r - \lambda}$$

which leads to:

$$w_1 = \frac{(k\sigma_\beta^2 + \sigma^2)/(r - \lambda)}{(k\sigma_\beta^2 + \sigma^2)/(r - \lambda) + \sigma^2/(\lambda t)}, \qquad w_2 = \frac{\sigma^2/(\lambda t)}{(k\sigma_\beta^2 + \sigma^2)/(r - \lambda) + \sigma^2/(\lambda t)}.$$

In nearly all applications, the values of the variance components are not known and so the weights cannot be computed exactly. σ^2 can be estimated unbiasedly by MSE_{intra}, the mean square error for the fixed-block analysis based on model (8.1). $k\sigma_\beta^2 + \sigma^2$ can be estimated by MSE_{inter}, the mean square error for the regression analysis of block totals based on model (8.2), divided by the block size k. Substituting these estimates for their parameter counterparts leads to expressions for estimated weights:

$$\hat{w}_1 = \frac{MSE_{inter}/(r - \lambda)}{MSE_{inter}/(r - \lambda) + MSE_{intra}/\lambda t},$$

$$\hat{w}_2 = \frac{MSE_{intra}/\lambda t}{MSE_{inter}/(r - \lambda) + MSE_{intra}/\lambda t}.$$

Due to the imprecision in variance component estimates, the actual variance of the combined estimate based on estimated weights can be larger than that of the intra-block estimate alone, especially when σ_β/σ is large or b/N is small, since either condition suggests that the inter-block analysis brings relatively little additional information to the inference.

8.5.1 Example: dishwashing detergents reprise

Return to the experiment reported by John (1961) described in subsection 7.2.2, performed to compare the effectiveness of nine dishwashing detergent formulations. Reduced normal equations for the intra-block analysis were developed and solved in Chapter 7. The inter-block analysis can be performed as a regression of the 12 block totals with a model matrix of form $(k\mathbf{1}|\mathbf{X}_1'\mathbf{X}_2)$, i.e., a fit of the model:

$$
\begin{pmatrix} 47 \\ 55 \\ 68 \\ 47 \\ 62 \\ 69 \\ 77 \\ 65 \\ 40 \\ 63 \\ 52 \\ 59 \end{pmatrix}
=
\left(\begin{array}{c|ccccccccc}
3 & 1 & 1 & 1 & 0 & 0 & 0 & 0 & 0 & 0 \\
3 & 0 & 0 & 0 & 1 & 1 & 1 & 0 & 0 & 0 \\
3 & 0 & 0 & 0 & 0 & 0 & 0 & 1 & 1 & 1 \\
3 & 1 & 0 & 0 & 1 & 0 & 0 & 1 & 0 & 0 \\
3 & 0 & 1 & 0 & 0 & 1 & 0 & 0 & 1 & 0 \\
3 & 0 & 0 & 1 & 0 & 0 & 1 & 0 & 0 & 1 \\
3 & 1 & 0 & 0 & 0 & 1 & 0 & 0 & 0 & 1 \\
3 & 0 & 1 & 0 & 0 & 0 & 1 & 1 & 0 & 0 \\
3 & 0 & 0 & 1 & 1 & 0 & 0 & 0 & 1 & 0 \\
3 & 1 & 0 & 0 & 0 & 0 & 1 & 0 & 1 & 0 \\
3 & 0 & 1 & 0 & 1 & 0 & 0 & 0 & 0 & 1 \\
3 & 0 & 0 & 1 & 0 & 1 & 0 & 1 & 0 & 0
\end{array}\right)
\begin{pmatrix} \mu_\beta \\ \tau \end{pmatrix}
+ \epsilon_2.
$$

Table 8.1 Intra- and Inter-Block Estimates of $\tau_i - \tau_9$, $i = 1, 2, 3, \ldots, 8$, for the Data of John (1961)

Contrast	$\tau_1 - \tau_9$	$\tau_2 - \tau_9$	$\tau_3 - \tau_9$	$\tau_4 - \tau_9$	$\tau_5 - \tau_9$	$\tau_6 - \tau_9$	$\tau_7 - \tau_9$	$\tau_8 - \tau_9$
Intra-Block Estimate	-9.778	-11.222	-16.333	-23.000	-5.000	-7.111	-9.000	-10.333
Inter-Block Estimate	-10.667	-13.333	-17.000	-24.000	-4.333	-4.667	-9.000	-11.000

Note that the model matrix is rank-deficient; the sum of columns 2 through 10 is equal to the first column, reflecting the fact that the only estimable linear combinations of τ are contrasts, as in the intra-block analysis. Because detergent 9 was regarded as the control or standard treatment, differences between each of the other treatment parameters and τ_9 are likely of interest to the experimenters. Table 8.1 displays intra-block estimates for these quantities described in the analysis of subsection 7.3.1, as well as the inter-block estimates computed from fitting the above model using a linear regression program. From the analysis of Chapter 7, we have $MSE_{intra} = 1.0325$, while MSE_{inter} is 2.9630 (computed as $\frac{1}{k}$ times MSE from the regression fit of block totals, $\frac{1}{3} \times 8.8887$). These lead to estimates $\hat{w}_1 = 0.896$ and $\hat{w}_2 = 0.104$ for a combined estimator, indicating that the bulk of information about these contrasts comes from the intra-block analysis.

8.6 Why can information be "recovered"?

The reader may find it less than intuitively obvious *why* an assumption of random block effects *should* permit the recovery of more information about τ. Generally, for a given amount of data, an increase in information that leads to decreases in expected standard errors and more powerful tests requires an increase in the strength of assumptions the analyst is willing to make. For example, in simple linear regression, the standard deviation of the estimated regression coefficient in the first-order term is no greater under an assumed first-order model (a relatively stronger assumption) than under an assumed second-order model (a relatively weaker assumption), and this difference can be substantial for some experimental designs. In the present case, we are not talking about different algebraic forms of the model, but whether the β_i's should be regarded as random or fixed-but-unknown. Is the random-β assumption actually "stronger," in this sense, than a fixed-β assumption?

Although it may not be as clear in this case as it is in the example of the regression model mentioned above, the answer is "yes." Saying that block effects are "fixed" is really saying nothing at all about them; the unknown elements of β could equally well be any set of b real numbers. The intra-block analysis is not affected by what these values might be because it relies only on linear combinations of data that are contrasts within each block, so that the elements of β "cancel out" in the expectations of these contrasts.

An assumption of random block effects does not specify "hard" structure on the data, such as deterministic relationships among elements of β. However, it *does* imply a "softer" probabilistic relationship among them that goes beyond the fixed-block assumption. For example, in a CBD $(t = k)$ if $t\mu_\beta + \sum_{i=1}^{t} \tau_i$ *were* a quantity of interest to us, in addition to the usual intra-block analysis, an assumption of random blocks would imply that the collection of block totals is an i.i.d. sample of size b with mean $t\mu_\beta + \sum_{i=1}^{t} \tau_i$; the inter-block analysis would be a one-sample inference on this common mean. But if (for example) the actual value of β in this case were $(0, 0, 0, \ldots, 0, 10000)'$, one of these block totals would likely be an outlier, and the intra-block analysis would likely give an invalid result. The same is true of the more realistic inter-block inferences about estimable functions of τ in augmented CBDs or BIBDs. A random-blocks assumption is certainly "stronger" than a fixed-blocks assumption, because it effectively rules out values of β containing any extreme outliers relative to the rest. Like any other statistical assumption, this can lead to more informative inference if it is justified, but to invalid inference if it is not.

8.7 CBD reprise

In Section 8.3, we noted that an assumption of random block effects does not yield additional information with a CBD; the inter-block estimator is noninformative because the data totals from each block have a common expectation. This would seem to indicate that a random blocks assumption is of no benefit with CBDs. There is, however, an additional advantage to assuming random blocks with an CBD. Suppose that we do assume random block effects, but *add treatment-by-block* interactions to the model:

$$y_{ij} = \alpha + \beta_i + \tau_j + \theta_{ij} + \epsilon_{ij}$$
$$\epsilon_{ij} \text{ i.i.d. with } E(\epsilon_{ij}) = 0 \text{ and } Var(\epsilon_{ij}) = \sigma^2$$
$$\beta_i \text{ i.i.d. with } E(\beta_i) = 0 \text{ and } Var(\beta_i) = \sigma_\beta^2$$
$$\theta_{ij} \text{ i.i.d. with } E(\theta_{ij}) = 0 \text{ and } Var(\theta_{ij}) = \sigma_\theta^2.$$

Because each observation contains independent, random components associated with both ϵ and θ, these sources of variability are essentially inseparable in the experiment, and for purposes of data analysis they can be viewed as combined:

$$y_{ij} = \alpha + \beta_i + \tau_j + \epsilon_{ij}^*$$
$$\epsilon_{ij}^* \text{ i.i.d. with } E(\epsilon_{ij}^*) = 0 \text{ and } Var(\epsilon_{ij}^*) = \sigma^2 + \sigma_\theta^2$$
$$\beta_i \text{ i.i.d. with } E(\beta_i) = 0 \text{ and } Var(\beta_i) = \sigma_\beta^2.$$

This model is functionally the same as the random blocks, additive model for a CBD. That is, we can justify the usual intra-block analysis if we assume *either*:

1. blocks and treatments do not interact (whether block effects are treated as fixed or random), *or*
2. block effects are random (whether there are block-by-treatment interactions or not).

8.8 Conclusion

When block effects are regarded as fixed, standard intra-block least-squares estimates of estimable $c'\tau$ are unbiased because the weights associated with each data value sum to zero within each block, resulting in cancellation of additive block effects. When block effects are regarded as random, additional information can sometimes be "recovered" through differences between block totals. Such inter-block estimates are generally less precise than intra-block estimates because the random block effects add to the "noise" component of between-block differences, but they are unbiased under the random-blocks model because the expected difference between any two random block effects is zero. Inter-block estimation cannot be used with designs in which each treatment is assigned to the same number of units in each block because the block total responses all have the same expectation. But for designs in which treatment assignment patterns are different from block-to-block, such as BIBDs, inter-block estimators can provide additional information about the estimable functions of interest. Because inter- and intra-block estimators are uncorrelated, the optimal (minimum variance) combined estimator is a weighted average of the two if the variances associated with blocks and units are known, and can be approximated based on estimated weights when they are not known.

8.9 Exercises

1. Consider the following two data sets resulting from two executions of the indicated small BIBD:

	Experiment A				Experiment B		
		Treatment				Treatment	
Block	1	2	3	Block	1	2	3
1	10	21	-	1	15	26	-
2	8	-	29	2	6	-	27
3	-	22	31	3	-	31	40

Without doing any more than a few simple mental arithmetic operations:

(a) Comment on the difference or similarity between the intra-block estimates of $c'\tau$ for these two experiments.

(b) Which experiment provides a more informative inter-block analysis, and why?

(c) Why could the combined estimator based on estimated weights $(\hat{w}_1, \hat{w}_2)$ not be calculated for either of these data sets?

2. For each experiment listed below, write a brief paragraph arguing why blocks might or might not be considered a random source of variation based on what you have been told about each study.

(a) Rainfall and Grassland Experiment, Fay et al. (2000), Section 1.1.

(b) Web Page Design Experiment, Murphy et al. (2006), subsection 5.1.1. Make arguments separately for the row-blocks and column-blocks of the Latin Square.

(c) Drug Sequence Experiment, Kraiczi et al. (2000), subsection 7.1.1.

3. Consider an experimental setting in which there are t treatments to be investigated in blocks of two units. Treatment 1 is a control condition. The two units in each block will be treated with treatment 1 and one of the other treatments; specifically, blocks numbered 1 through r contain units receiving treatment 1 and treatment 2, units in blocks numbered $r + 1$ through $2r$ receive treatments 1 and 3, ..., and units in blocks numbered $(t - 2)r + 1$ through $b = (t - 1)r$ receive treatments 1 and t.

(a) Write a random-blocks model for the data to be collected in this experiment, including a full explanation of all indices, and a characterization of all random quantities. Use the notation you develop here to work parts (b) and (c).

(b) Derive a closed form expression for the intra-block estimate of $\tau_2 - \tau_3$, and a closed form expression for its standard deviation.

(c) Derive a closed form expression for the inter-block estimate of $\tau_2 - \tau_3$, and a closed form expression for its standard deviation.

4. Suppose you have executed a BIBD under circumstances that lead you to believe that the effect of blocks can be regarded as random. You compute intra- and inter-block estimates of the treatment comparisons of interest, and calculate residual mean squares from each analysis in preparation for computing a combined estimate. However, you are surprised by the fact that *MSE* for the intra-block analysis is larger than that for the inter-block analysis. What (other than calculation error) might cause this to happen? Does this have implications for whether you should "trust" either the intra- or inter-block estimates, and whether the combined estimate should be

computed? (Hint: "Surprises" in statistical analysis often come as the result of erroneous assumptions; think about the "standard assumptions" that would be made here, and whether some specific kinds of violations of them might lead to this result.)

5. Consider the following design in five treatments, ten row-blocks and four column-blocks, and resulting data:

Row Blocks	Column Blocks			
	1	2	3	4
1	trt 1, $y = 32$	trt 2, $y = 29$	trt 3, $y = 31$	trt 4, $y = 24$
2	trt 2, $y = 30$	trt 3, $y = 31$	trt 4, $y = 25$	trt 5, $y = 34$
3	trt 3, $y = 23$	trt 4, $y = 21$	trt 5, $y = 29$	trt 1, $y = 24$
4	trt 4, $y = 26$	trt 5, $y = 32$	trt 1, $y = 32$	trt 2, $y = 30$
5	trt 5, $y = 35$	trt 1, $y = 31$	trt 2, $y = 33$	trt 3, $y = 30$
6	trt 1, $y = 34$	trt 2, $y = 28$	trt 3, $y = 32$	trt 4, $y = 24$
7	trt 2, $y = 22$	trt 3, $y = 22$	trt 4, $y = 17$	trt 5, $y = 23$
8	trt 3, $y = 22$	trt 4, $y = 23$	trt 5, $y = 28$	trt 1, $y = 24$
9	trt 4, $y = 33$	trt 5, $y = 41$	trt 1, $y = 39$	trt 2, $y = 36$
10	trt 5, $y = 37$	trt 1, $y = 32$	trt 2, $y = 34$	trt 3, $y = 29$

Assume that row-blocks, column-blocks, and treatments can be regarded as additive effects (i.e., that no two of them interact).

(a) Using a computer, calculate the least-squares estimates of all 10 treatment parameter differences $\tau_i - \tau_j$ under a model in which row-blocks and column-blocks are each assumed to have fixed effects. Compute the common margin of error for these differences corresponding to 95% confidence (i.e., half the length of the usual 95% confidence interval).

(b) Calculate the inter-block estimates of the 10 treatment parameter differences under a model in which row-blocks are assumed to be random but column-blocks are assumed to be fixed. Compute the common margin of error for these differences corresponding to 95% confidence (again, based only on inter-block information).

6. Consider again exercise 3 at the end of Chapter 7, the 8-block version of the radon detector problem.

(a) Assuming that chamber sessions are represented by fixed effects, compute the least-squares estimate of $\tau_1 - \tau_2$, and compute MSE for the full model (blocks and treatments). Use statistical software to work this and all parts of this exercise.

(b) Assuming that chamber sessions are represented by random effects, construct a separate inter-block estimate of $\tau_1 - \tau_2$ using only the chamber session totals, and calculate the MSE for this model.

(c) Use the estimates and *MSE*s of the two models you have fit to construct a single combined estimate of $\tau_1 - \tau_2$ appropriate for the random-sessions model.

(d) Suppose your estimates of the unit and block variances are exactly correct, rather than estimates. Pretending this is true, what is the standard deviation of each of the following?

- The intra-block (i.e., fixed-block) estimate of $\tau_1 - \tau_2$.
- The inter-block estimate based on block totals.
- The combined estimate.

7. Consider an experiment designed to estimate the difference in mean response between two treatments, executed in nine blocks of two units each. Suppose units are assigned to treatments as follows:

- In each of blocks 1–3, one unit is assigned to each treatment.
- In each of blocks 4–6, both units are assigned to treatment 1.
- In each of blocks 7–9, both units are assigned to treatment 2.

Letting y_{ij} represent the jth response recorded in the ith block, and for convenience letting $j = \#$ of the applied treatment in blocks 1-3:

(a) Derive the intra-block estimator of $\tau_1 - \tau_2$ and compute the number of degrees of freedom on which its standard error would be based.

(b) Assuming block effects can be regarded as random, derive the inter-block estimator of $\tau_1 - \tau_2$, and compute the number of degrees of freedom on which its standard error would be based.

8. Consider a BIBD to compare $t = 4$ treatments in 12 blocks of size $k = 3$. For analysis purposes, suppose the model:

$$y_{ij} = \beta_i + \tau_j + \epsilon_{ij}$$

is adequate, where $i = 1, 2, 3, \ldots, 12$ indexes blocks and $j = 1, 2, 3, 4$ indexes treatments. (Note that not all combinations of i and j are used.) If the β_i are i.i.d. random variables with variance 1, and ϵ_{ij} are i.i.d. random variables, also with variance 1:

(a) What is the standard deviation of the intra-block estimate of $\tau_1 - \tau_2$?

(b) What is the standard deviation of the inter-block estimate of $\tau_1 - \tau_2$?

(c) What is the standard deviation of the (optimal) combined estimate of $\tau_1 - \tau_2$?

CHAPTER 9

Factorial treatment structure

9.1 Introduction

To this point, the experimental designs discussed in this book have been presented in the context of "unstructured" treatments, identified only as a collection of distinct operations, additives, procedures, recipes, et cetera, that can be applied to experimental units in the interest of comparing the resulting responses. In some cases, we have discussed how one or more of the treatments might be regarded as a control, indicating that it serves primarily as a base-line or reference condition. But otherwise, our mathematical description and handling of treatments have included no assumptions about relationships between any pair of them – they have played essentially exchangeable roles in our development.

In contrast to this, many experiments are designed to compare treatments defined by selecting a *level* related to each of a collection of *factors*. For example, a mechanical engineer might have interest in understanding the properties of a certain kind of metal part that can be formed by either of two different forging processes, and coated with any one of three types of surface plating. As a result, there are six different treatments of interest (types of parts that can be produced), each defined by specifying one level of the forging factor, and one level of the plating factor. We describe any experiment in which the treatments have factorial structure as a *factorial experiment*. In general, if f factors are used to define a treatment, and the ith of these factors has l_i levels, the number of treatments that can be defined is $t = \prod_{i=1}^{f} l_i$. For any treatment, exactly one level of each factor is specified, and we shall restrict our attention to situations in which all possible combinations of factor levels are meaningful. Fisher (1971) was an early advocate of factorial experimentation, and contributed many of the basic ideas that are still important in the design and analysis of such studies.

In factorial experiments, the most interesting or important experimental questions are generally framed in a way that do *not* treat relationships between treatments symmetrically. For example, in the hypothetical experiment involving metal parts, the difference between average response for the three treatments involving one forging process, and the average response for the three treatments involving the other, might be relatively interesting because it is a natural "overall" indication of the physical influence of one factor. But the difference between average response for:

(forging 1, plating 1), (forging 1, plating 2), (forging 2, plating 1)

Table 9.1 Factors and Levels from the Concrete Experiment of Soudki et al. (2001)

Factor	Levels (unitless)
Water/cement ratio	0.4, 0.5, 0.6
Total aggregate/cement ratio	3, 4, 5, 6
Coarse aggregate/total aggregate ratio	0.55, 0.60, 0.65, 0.70

and the average response for:

(forging 1, plating 3), (forging 2, plating 2), (forging 2, plating 3)

would usually not be so interesting since it is not so easily interpreted relative to the physical factorial structure of the treatments. In this chapter we will consider the structure of *full* factorial experiments, i.e., those in which all possible combinations of factor levels are represented.

9.1.1 Example: strength of concrete

Soudki, El-Salakawy, and Elkum (2001) described an experiment carried out to test the effects of a number of factors on the compressive strength of concrete in hot climates. Batches of concrete were prepared with differing relative quantities of water, total aggregate, coarse aggregate, and cement. Since *relative* concentration of each component in the mixture is the meaningful characterization, the four quantities can be represented as three ratios that serve as the factors in this experiment. The factors and their respective levels are listed in Table 9.1, representing $3 \times 4 \times 4 = 48$ experimental treatments. As a part of the experiment, samples of concrete were cast from batches made according to each of the 48 formulae, emersed in water heated to 52 degrees Celsius for 28 days, and then tested for compressive strength at room temperature. The average strength of samples associated with each treatment is summarized in Table 9.2.

9.2 An overparameterized model

Despite the definition of treatments via factors, it is certainly possible to frame the analysis of data from a factorial experiment within the context of unstructured linear models. Let $(i, j, \ldots, k)$ be a set of f indices representing the levels of each factor, that is, $i = 1, 2, 3, \ldots, l_1; \ j = 1, 2, 3, \ldots, l_2; \ \ldots; \ k = 1, 2, 3, \ldots, l_f$. Then, a cell means model and an effects model for an unblocked experiment in which each treatment is replicated r times can be written as:

$$y_{ij\ldots kt} = \mu_{ij\ldots k} + \epsilon_{ij\ldots kt} \quad \text{and} \quad y_{ij\ldots kt} = \alpha + \tau_{ij\ldots k} + \epsilon_{ij\ldots kt}$$

$$t = 1, 2, 3, \ldots, r$$

$$\epsilon_{ij\ldots kt} \text{ i.i.d. with } E(\epsilon_{ij\ldots kt}) = 0 \text{ and } Var(\epsilon_{ij\ldots kt}) = \sigma^2 \tag{9.1}$$

Table 9.2 Average Response Values
(Megapascals, MPa) for the Concrete
Experiment of Soudki et al. (2001)

TA/C	W/C	\multicolumn{4}{c}{CA/TA}
		0.55	0.60	0.65	0.70
3	0.4	33.7	34.2	33.2	31.9
3	0.5	21.6	33.0	23.7	27.3
3	0.6	18.6	17.3	19.9	18.2
4	0.4	32.5	32.7	33.6	30.1
4	0.5	25.1	29.3	28.2	23.9
4	0.6	20.0	21.8	21.3*	23.9
5	0.4	13.5	24.9	25.6	18.0
5	0.5	27.3	20.8	28.7	22.5
5	0.6	20.2	19.4	19.6	19.6
6	0.4	9.6	6.2	7.7	10.8
6	0.5	27.0	15.0	23.6	22.5
6	0.6	19.6	18.3	21.6	13.6

* Recorded as 2.13 in Soudki et al.

respectively, where the subscript t identifies a unique unit and response associated with any specified treatment. These models are essentially the same as those presented in Chapter 3; the only difference is that, rather than using a single subscript to identify a treatment, a collection of f subscripts is used to identify the treatment through the selected factor levels. In most applications, however, a mathematically equivalent *factorial* model is more useful. Suppose for specificity that $f = 3$ factors are being used to define treatments. Then, a factorial model for a CRD might be written as:

$$y_{ijkt} = \mu + \dot\alpha_i + \dot\beta_j + \dot\gamma_k + (\dot{\alpha\beta})_{ij} + (\dot{\alpha\gamma})_{ik} + (\dot{\beta\gamma})_{jk} + (\dot{\alpha\beta\gamma})_{ijk} + \epsilon_{ijkt}. \quad (9.2)$$

Here we are following a common notational convention in which "α", "β", and "γ" are used to represent effects associated with the first, second, and third experimental factors; "μ", like "α" in preceding chapters, is a component common to all response expectations representing the summary of all experiment-specific effects. (The over-dots are used here to distinguish this parameterization from a different one to be introduced in Section 9.3.) In this parameterization, the collection of l_1 parameters $\dot\alpha_1$, $\dot\alpha_2$, $\dot\alpha_3$, ... $\dot\alpha_{l_1}$ describe the *main effect* associated with the first factor, and likewise for the $\dot\beta$ and $\dot\gamma$ parameters and the second and third factors, respectively. $(\dot{\alpha\beta})_{ij}$ is the *two-factor interaction* associated with levels i and j of factors 1 and 2, respectively, and represents the nonadditive component of the joint contribution of these two factors, i.e., that which cannot be expressed by the sum of associated main effects, $\dot\alpha_i + \dot\beta_j$, to the mean response. (The notation $(\dot{\alpha\beta})_{ij}$ is intended to denote a single parameter; the parentheses are used to make it clear that this is not a product of main effects.) Similarly, synergistic effects of pairs of levels of

Table 9.3 Cell Means for Three-Factor Example

Level of Factor:		Level of Factor 3 (k)			
1 (i)	2 (j)	1	2	3	4
1	1	12	22	17	7
1	2	30	40	35	25
1	3	50	60	55	45
2	1	20	30	25	15
2	2	42	52	47	37
2	3	60	70	65	55

factors 1 and 3, and factors 2 and 3, are denoted by indexed parameters $(\dot{\alpha\gamma})$ and $(\dot{\beta\gamma})$, respectively. Finally, $(\alpha\beta\gamma)$ denotes a three-factor interaction, the component of a cell mean that cannot be reproduced by a sum of effects each involving fewer than all three factors. The definition of these *factorial effects* is extended in the obvious way for situations involving more than three factors; for example, where $f = 5$, the interactions of *highest order* are five-factor interactions, and there are five groups of four-factor interactions, each representing the synergistic effect of the specified levels of four of the five factors.

As noted above, statistical models for factorial treatment structures *can* be written just as in the case of unstructured treatments. But in many applications, a large proportion of the variation among responses for different treatments is associated with factorial effects of relatively low order, i.e., main effects and interactions involving relatively few factors. Factorial model parameterization such as (9.2) facilitate the examination of data for this characteristic, and the description of the treatment structure when it occurs. For example, suppose that, in our three-factor example, factors 1, 2, and 3 have $l_1 = 2$, $l_2 = 3$, and $l_3 = 4$ levels, and that the cell means are actually as presented in Table 9.3. (For illustrative purposes, these hypothetical values have less complex structure than what would be expected in many real applications, but they represent the kind of physical simplicity that is often present in real factorial studies.)

Examination of the table reveals that most differences among cell means can be described simply with reference to the individual experimental factors. The most dominant pattern of variation is associated with factor 2; within the neighboring groups of three rows representing a single level of factor 1, moving from row to row produces an increase of approximately 20 when j is changed from 1 to 2, and another increase of approximately 20 when j is changed from 2 to 3. Major systematic variation is also apparent in comparing the sections of three neighboring rows associated with the first factor; changing i from 1 to 2 is associated with an increase of approximately 10 in the response, regardless of the specific values of j and k chosen. Finally, within any row in the table, moving from left to right, i.e., changing from k of 1 to 2 to 3 to 4 for any specified (i, j) pair results in an *exact* (not approximate in this case) increase of 10, followed by a decrease of five and a second decrease of 10.

In fact, extending the observations made in the preceding paragraph, the tabulated cell means can actually be written as:

$$\mu_{ijk} = \mu + \dot{\alpha}_i + \dot{\beta}_j + \dot{\gamma}_k + (\dot{\alpha\beta})_{ij}$$

where

$$\mu = 10$$

$$\dot{\alpha}_i = \begin{cases} 0, & i = 1 \\ 10, & i = 2 \end{cases} \quad \dot{\beta}_j = \begin{cases} 0, & j = 1 \\ 20, & j = 2 \\ 40, & j = 3 \end{cases} \quad \dot{\gamma}_k = \begin{cases} 0, & k = 1 \\ 10, & k = 2 \\ 5, & k = 3 \\ -5, & k = 4 \end{cases}$$

$$(\dot{\alpha\beta})_{ij} = \begin{cases} 2, & (i,j) = (1,1), (2,2) \\ 0, & \text{otherwise} \end{cases}.$$

The other factorial effects – two-factor interactions associated with factors 1 and 3, or 2 and 3, and three-factor interactions requiring specification of levels of all factors – are not needed to explain the patterns observed in the table. The fact that changes in index k result in exactly the same changes in cell means (i.e., the relationships among elements within each row of the table) is reflected in the fact that no interactions involving the third factor are needed. The fact that marginal changes in i and j result in *nearly* the same changes in treatment means in each section of the table is reflected in relatively large main effects for factors 1 and 2, and a relatively small two-factor interaction that represents the minor deviations from that pattern. In fact, if the entries in the first and fifth rows of the table were reduced by 2 (i.e., if they ended in "0" or "5" as the other entries in their respective columns), the word "approximately" would not be needed in the last paragraph, and the small two-factor interaction parameters involving factors 1 and 2 would not be needed (i.e., they would be exactly zero) in the factorial representation of the cell means. In this case, the differences among treatments are primarily due to the *additive* action of the factors, and components of treatment differences attributable to the additional joint action of factors are minimal and limited to factors 1 and 2.

9.2.1 Graphical logic

A major benefit of full factorial experimentation is that it is *fully efficient* for investigating the effects associated with each factor. This means that a factorial experiment with $N = r \times l_1 \times l_2 \times \ldots \times l_f$ runs provides as much information about $\{\dot{\alpha}_1, \dot{\alpha}_2, \ldots, \dot{\alpha}_{l_1}\}$ as an experiment of N runs in which only factor 1 is varied. But it *also* provides as much information about $\{\dot{\beta}_1, \dot{\beta}_2, \ldots, \dot{\beta}_{l_2}\}$ as an experiment of N runs in which only factor 2 is varied, and so forth for all f factors. That is, an N-run full factorial experiment in f factors provides the same information about main effects as f single-factor experiments, each of size N. But *further*, the factorial experiment provides information on how the factors affect the response in combination through interactions, something that cannot be learned from one-factor-at-a-time studies.

Taking this view, suppose we *did* momentarily regard an unreplicated ($r = 1$) factorial experiment as an experiment designed to estimate the effects associated with changing factor 1 across its l_1 levels. In this case, we should think of it as a blocked experiment because, even if we temporarily say we are not interested in the effects associated with the levels of the other factors, those effects are not constant across runs. If factors 2 through f are regarded as variables that simply identify blocks, we might think of the experiment as a CBD with $t = l_1$ treatments (the levels of factor 1), and $b = l_2 \times l_3 \times \ldots \times l_f$ complete blocks (within which the levels of factors 2 through f are fixed). As in Chapter 4, we might then propose plotting block-corrected values for each level of factor 1, i.e., the b values $\bar{y}_{ijk\ldots} - \bar{y}_{.jk\ldots}$ for each $i = 1, 2, \ldots, l_1$. Parallel boxplots might then be used, each representing the b values associated with one level of factor 1 after correcting for, or "canceling out," effects associated with factors 2 through f. Differences of location between parallel plots reflect the main effect associated with factor 1, the differences in response associated with its levels. The individual quantities used in each plot are actually the main effect estimates for factor 1 that would be computed from each "slice" of the factorial study defined by the selected levels of the other factors. Differences of spread between parallel plots indicate the possibility of interaction between factor 1 and other factors in the experiment (i.e., the *variation in the effect of factor 1* as the other factors change), but if this pattern is observed, more detailed examination of the data is required to determine which factors are actually involved in the interactions. A collection of parallel boxplots made for each of the f factors gives an overall idea of which have the most dramatic individual effects, and which are most likely interacting with other factors. These "main-effect boxplots" are shown in Figure 9.1 for the total aggregate to cement ratio (TA/C) factor for the data of Soudki et al. given in Table 9.2 (R9.1).

9.2.2 Matrix development for the overparameterized model

Model (9.2) is severely *overparameterized*, i.e., there are far more parameters than experimental treatments. For our example with $f = 3$, the full model, including all main effects and interactions of each order, for expressing the expected response of $2 \times 3 \times 4 = 24$ unique experimental conditions contains 60 parameters – μ, 9 main effects, 26 two-factor interactions, and 24 three-factor interactions. The degree of overparameterization becomes even more severe for larger values of f; in fact, it is easy to see that the number of interactions of highest order (alone) is always equal to the number of experimental conditions defined by the factors. One result of this is that the set of solutions to the reduced normal equations is especially ambiguous, and without external constraints offers little insight to the structure of the cell means.

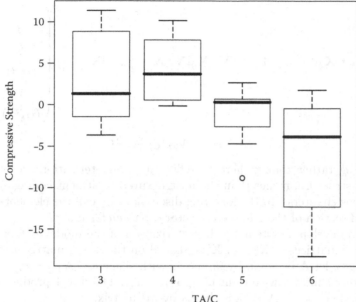

Figure 9.1 Main-effect boxplots for TA/C, data from Soudki et al. (2001)(R9.1).

Nonetheless, we certainly could proceed with characterization of model (9.2) by developing a model matrix of indicator variables to select the correct collection of main effects and interactions in the expectation of each response. In the three-factor example we are describing, there would be groups of columns in $\mathbf{X}_2$ corresponding to each main effect group and each interaction group:

- 2 columns of indicator variables corresponding to $(\dot{\alpha}_1, \dot{\alpha}_2)' = \dot{\boldsymbol{\alpha}}$
- 3 columns of indicator variables corresponding to $(\dot{\beta}_1, \dot{\beta}_2, \dot{\beta}_3)' = \dot{\boldsymbol{\beta}}$
- 4 columns of indicator variables corresponding to $(\dot{\gamma}_1, \dot{\gamma}_2, \dot{\gamma}_3, \dot{\gamma}_4)' = \dot{\boldsymbol{\gamma}}$
- 6 columns of indicator variables corresponding to $((\dot{\alpha}\beta)_{11} \ldots (\dot{\alpha}\beta)_{23})' = (\dot{\alpha}\beta)$
- 8 columns of indicator variables corresponding to $((\dot{\alpha}\gamma)_{11} \ldots (\dot{\alpha}\gamma)_{24})' = (\dot{\alpha}\gamma)$
- 12 columns of indicator variables corresponding to $((\dot{\beta}\gamma)_{11} \ldots (\dot{\beta}\gamma)_{34})' = (\dot{\beta}\gamma)$
- 24 columns of indicator variables corresponding to $((\dot{\alpha}\beta\gamma)_{111} \ldots (\alpha\dot{\beta}\gamma)_{234})' = (\alpha\dot{\beta}\gamma)$

A matrix model for the entire experiment could then be written as

$$y = 1\mu + X_2\dot{\phi} + \epsilon = 1\mu + (X_{\dot{\alpha}}|X_{\dot{\beta}}|X_{\dot{\gamma}}|X_{(\dot{\alpha\beta})}|\ldots|X_{(\dot{\alpha\beta\gamma})}) \begin{pmatrix} \dot{\alpha} \\ \dot{\beta} \\ \dot{\gamma} \\ (\dot{\alpha\beta}) \\ \ldots \\ (\dot{\alpha\beta\gamma}) \end{pmatrix} + \epsilon$$

$$E(\epsilon) = 0, \quad Var(\epsilon) = \sigma^2 I. \tag{9.3}$$

(We use $\dot{\phi}$ rather than τ here to reflect a parameterization motivated by *factorial* structure, rather than the unstructured treatment coding employed in previous chapters.) In the following discussion, we call the elements of each indicated section of the parameter vector a *parameter group*.

We can develop a systematic characterization of the model matrix and the various submatrices of X_2 and $X_2'X_2$ based on the use of matrix *direct products*. Briefly, let A be any n-by-m matrix with elements $\{A\}_{ij} = a_{ij}$ and let B be any s-by-t matrix with elements $\{B\}_{ij} = b_{ij}$. The direct product of these matrices, denoted by $A \times B$, is the ns-by-mt matrix:

$$A \times B = \begin{pmatrix} Ab_{11} & Ab_{12} & Ab_{13} & \ldots & Ab_{1t} \\ Ab_{21} & Ab_{22} & Ab_{23} & \ldots & Ab_{2t} \\ Ab_{31} & Ab_{32} & Ab_{33} & \ldots & Ab_{3t} \\ \ldots & \ldots & \ldots & \ldots & \ldots \\ Ab_{s1} & Ab_{s2} & Ab_{s3} & \ldots & Ab_{st} \end{pmatrix}.$$

Note that in general, $A \times B \neq B \times A$, even though the two direct products are of the same dimension. Two useful (and easily verified) properties of direct products are:

- $(A \times B)' = A' \times B'$
- $(A \times B)(F \times G) = (AF) \times (BG)$

where the indicated matrices have conformable dimensions for (regular) matrix products. A more complete discussion of direct products and their use in linear statistical models is given by Graybill (1983).

Now suppose that the elements of the response vector are ordered as:

$$y = (y_{1111} \cdots y_{111r}, \ y_{1121} \cdots y_{112r}, \ y_{1211} \cdots y_{121r}, \ldots, \ y_{2341} \cdots y_{234r})'$$

i.e., with the index for the first factor changing most slowly, and that for the third factor changing most quickly, in lexicographical order, and replicates within experimental treatments (represented by the 4th index) are grouped together. Given this ordering, the submatrices of X_2 in model (9.3) can be written as:

$$\mathbf{X}_{\dot\alpha} = \mathbf{1}_r \times \mathbf{1}_4 \times \mathbf{1}_3 \times \mathbf{I}_{2\times2} \quad \mathbf{X}_{\dot\beta} = \mathbf{1}_r \times \mathbf{1}_4 \times \mathbf{I}_{3\times3} \times \mathbf{1}_2$$
$$\mathbf{X}_{\dot\gamma} = \mathbf{1}_r \times \mathbf{I}_{4\times4} \times \mathbf{1}_3 \times \mathbf{1}_2$$
$$\mathbf{X}_{(\dot\alpha\beta)} = \mathbf{1}_r \times \mathbf{1}_4 \times \mathbf{I}_{3\times3} \times \mathbf{I}_{2\times2} \quad \mathbf{X}_{(\dot\alpha\gamma)} = \mathbf{1}_r \times \mathbf{I}_{4\times4} \times \mathbf{1}_3 \times \mathbf{I}_{2\times2}$$
$$\mathbf{X}_{(\dot\beta\gamma)} = \mathbf{1}_r \times \mathbf{I}_{4\times4} \times \mathbf{I}_{3\times3} \times \mathbf{1}_2$$
$$\mathbf{X}_{(\dot\alpha\beta\gamma)} = \mathbf{1}_r \times \mathbf{I}_{4\times4} \times \mathbf{I}_{3\times3} \times \mathbf{I}_{2\times2}$$

where $\mathbf{1}$ is a column vector of 1's and $\mathbf{I}$ is an identity matrix, each dimensioned as indicated. Given these representations, it is easy to verify that in $\mathbf{X}_2'\mathbf{X}_2$, the *diagonal block* corresponding to any parameter group is a multiple of the identity matrix. For our example:

$$\begin{aligned}\mathbf{X}_{\dot\alpha}'\mathbf{X}_{\dot\alpha} &= (\mathbf{1}_r \times \mathbf{1}_4 \times \mathbf{1}_3 \times \mathbf{I}_{2\times2})'(\mathbf{1}_r \times \mathbf{1}_4 \times \mathbf{1}_3 \times \mathbf{I}_{2\times2})\\ &= (\mathbf{1}_r' \times \mathbf{1}_4' \times \mathbf{1}_3' \times \mathbf{I}_{2\times2}')(\mathbf{1}_r \times \mathbf{1}_4 \times \mathbf{1}_3 \times \mathbf{I}_{2\times2})\\ &= \mathbf{1}_r'\mathbf{1}_r \times \mathbf{1}_4'\mathbf{1}_4 \times \mathbf{1}_3'\mathbf{1}_3 \times \mathbf{I}_{2\times2}'\mathbf{I}_{2\times2} = 12r\mathbf{I}_{2\times2}.\end{aligned}$$

Similar calculations show that

$$\mathbf{X}_{\dot\beta}'\mathbf{X}_{\dot\beta} = 8r\mathbf{I}_{3\times3} \quad \mathbf{X}_{\dot\gamma}'\mathbf{X}_{\dot\gamma} = 6r\mathbf{I}_{4\times4}$$

$$\mathbf{X}_{(\dot\alpha\beta)}'\mathbf{X}_{(\dot\alpha\beta)} = 4r\mathbf{I}_{6\times6} \quad \mathbf{X}_{(\dot\alpha\gamma)}'\mathbf{X}_{(\dot\alpha\gamma)} = 3r\mathbf{I}_{8\times8} \quad \mathbf{X}_{(\dot\beta\gamma)}'\mathbf{X}_{(\dot\beta\gamma)} = 2r\mathbf{I}_{12\times12}$$

$$\mathbf{X}_{(\dot\alpha\beta\gamma)}'\mathbf{X}_{(\dot\alpha\beta\gamma)} = r\mathbf{I}_{24\times24}.$$

Off-diagonal blocks of $\mathbf{X}_2'\mathbf{X}_2$ for which the parameter groups associated with matrix rows and columns do not reference common factors are multiples of a matrix of 1's; for example

$$\mathbf{X}_{\dot\alpha}'\mathbf{X}_{\dot\beta} = (\mathbf{1}_r \times \mathbf{1}_4 \times \mathbf{1}_3 \times \mathbf{I}_{2\times2})'(\mathbf{1}_r \times \mathbf{1}_4 \times \mathbf{I}_{3\times3} \times \mathbf{1}_2) = 4r\mathbf{J}_{2\times3}.$$

Similarly,

$$\mathbf{X}_{\dot\alpha}'\mathbf{X}_{(\dot\beta\gamma)} = r\mathbf{J}_{2\times12}.$$

Finally, off-diagonal blocks for which the parameter groups associated with rows and columns do reference common factors have both zero and common nonzero elements; for example

$$\mathbf{X}_{\dot\alpha}'\mathbf{X}_{(\dot\alpha\beta)} = (\mathbf{1}_r \times \mathbf{1}_4 \times \mathbf{1}_3 \times \mathbf{I}_{2\times2})'(\mathbf{1}_r \times \mathbf{1}_4 \times \mathbf{I}_{3\times3} \times \mathbf{I}_{2\times2}) = 4r\begin{pmatrix} 1 & 1 & 1 & 0 & 0 & 0 \\ 0 & 0 & 0 & 1 & 1 & 1 \end{pmatrix}.$$

Some thought about the structure of $\mathbf{X}_{\dot\alpha}$ and $\mathbf{X}_{(\dot\alpha\beta)}$ reveals that the nonzero elements of this matrix appear at the intersection of matrix rows and columns that reference the same level of factor 1 (that symbolized by α), that is:

$$\begin{array}{c} & (\dot\alpha\beta)_{11} & (\dot\alpha\beta)_{12} & (\dot\alpha\beta)_{13} & (\dot\alpha\beta)_{21} & (\dot\alpha\beta)_{22} & (\dot\alpha\beta)_{23} \\ \dot\alpha_1 & 4r & 4r & 4r & 0 & 0 & 0 \\ \dot\alpha_2 & 0 & 0 & 0 & 4r & 4r & 4r \end{array}$$

Similarly, $\mathbf{X}'_{(\alpha\beta)}\mathbf{X}_{(\beta\gamma)}$ is a 6×12 matrix in which elements corresponding to $((\alpha\beta)_{ij}, (\beta\gamma)_{jk})$ are r, and all other elements are 0.

Given this structure, an overwhelming variety of solutions to the least-squares problem exist. Some are of relatively simple form while others are quite complicated, reflecting the particular generalized inverse selected for $\mathbf{X}'\mathbf{X}$. In any case, the overparameterization of the problem makes the unconstrained reduced normal equations fairly difficult to fully characterize.

9.3 An equivalent full-rank model

Advantages of the model parameterization presented in Section 9.2 include an intuitive interpretation and a simple representation of each factor level within each main effect and combination of factor levels within each interaction. However, because this model is overparameterized, it leads to technical complications from the viewpoint of the general linear model, associated primarily with characterization of solutions to the reduced normal equations. In this section we present an alternative full-rank parameterization that eliminates many of these complications, at the cost of some interpretative simplicity.

The alternative representation is facilitated by thinking of the modeling problem in a context of regression with continuous prediction variables (even though that does not actually need to be the case in the application being studied). Consider models that may contain independent variables other than binary-coded indicator variables, and write:

$$E(y) = \mu + \mathbf{x}\phi + \epsilon$$

or in matrix form

$$E(\mathbf{y}) = \mathbf{1}\mu + \mathbf{X}_2\phi + \epsilon$$

where the row vector $\mathbf{x}$ and model matrix $\mathbf{X}_2$ can contain elements other than 0 and 1, and ϕ represents a set of treatment-related parameters that are related to, but not the same as, those denoted by $\dot{\phi}$ in Section 9.2. In order to construct a convenient model, we need to define a set of $l_i - 1$ "regressors" for the ith factor, $i = 1, 2, 3, \ldots, f$. For example, the levels of a three-level factor can be "coded" using x_1 and x_2 defined as:

Factor Level	x_1	x_2
1	$-\sqrt{\frac{3}{2}}$	$-\frac{1}{\sqrt{2}}$
2	0	$+\sqrt{2}$
3	$+\sqrt{\frac{3}{2}}$	$-\frac{1}{\sqrt{2}}$

For our purposes, the key features of these values are:

- The sum of values for each of x_1 and x_2 is zero.
- The sum of squared values for each of x_1 and x_2 is 3 $(= l)$.
- The sum of values of products $x_1 x_2$ is zero.

It is convenient to specify such a coding by defining a l-by-$(l-1)$ matrix $\mathbf{F}$ containing the elements of the code as listed in the table above; for our example, that matrix is:

$$\mathbf{F} = \begin{pmatrix} -\sqrt{\dfrac{3}{2}} & -\dfrac{1}{\sqrt{2}} \\ 0 & +\sqrt{2} \\ +\sqrt{\dfrac{3}{2}} & -\dfrac{1}{\sqrt{2}} \end{pmatrix}.$$

Similar codings can be easily constructed for factors with any number of levels. For example, factors with $l = 2$ and 4 levels could be represented by:

$$\mathbf{F} = \begin{pmatrix} -1 \\ +1 \end{pmatrix} \quad \text{and} \quad \mathbf{F} = \begin{pmatrix} -\dfrac{3}{\sqrt{5}} & +1 & -\dfrac{1}{\sqrt{5}} \\ -\dfrac{1}{\sqrt{5}} & -1 & +\dfrac{3}{\sqrt{5}} \\ +\dfrac{1}{\sqrt{5}} & -1 & -\dfrac{3}{\sqrt{5}} \\ +\dfrac{3}{\sqrt{5}} & +1 & +\dfrac{1}{\sqrt{5}} \end{pmatrix},$$

respectively. Suppose we denote the matrix of this type selected for the first factor as $\mathbf{F}^\alpha$, denote the ith row of $\mathbf{F}^\alpha$ by $\mathbf{f}_i^\alpha$, and use similar notation for matrices associated with the other factors. With this, we may write a model for data from a three-factor experiment as:

$$y_{ijkt} = \mu + \mathbf{f}_i^\alpha \boldsymbol{\alpha} + \mathbf{f}_j^\beta \boldsymbol{\beta} + \mathbf{f}_k^\gamma \boldsymbol{\gamma} + (\mathbf{f}_i^\alpha \times \mathbf{f}_j^\beta)(\boldsymbol{\alpha\beta}) + \ldots + (\mathbf{f}_i^\alpha \times \mathbf{f}_j^\beta \times \mathbf{f}_k^\gamma)(\boldsymbol{\alpha\beta\gamma}) + \epsilon_{ijkt},$$
$$(9.4)$$

where $\boldsymbol{\alpha} = (\alpha_1, \alpha_2, \alpha_3, \ldots, \alpha_{l_1-1})'$ and likewise for the other groups of factorial parameters. Note that while we are using parameter notation similar to that in the overparameterized model, the symbols (now without over-dots) do not represent the same quantities. However, the general role of the parameters in this model is similar to that of the parameters in model (9.2); parameters denoted by single Greek letters are still associated with single factors and can still be called "main effects," those denoted by pairs of Greek letters are still associated with pairs of factors and can still be called "two-factor interactions," and so forth. In this model, the collection of all α's *collectively* describes the variation associated with the levels of factor 1, rather than each clearly describing the deviation of one level from the overall average as in the overparameterized model. For example, the additive component of the mean responses that depends only on the three-level factor 2 is

- $\mathbf{f}_1^\beta \boldsymbol{\beta} = (-\sqrt{\frac{3}{2}}, -\frac{1}{\sqrt{2}})\boldsymbol{\beta} = -\sqrt{\frac{3}{2}}\beta_1 - \frac{1}{\sqrt{2}}\beta_2$ at level 1
- $\mathbf{f}_2^\beta \boldsymbol{\beta} = (0, \sqrt{2})\boldsymbol{\beta} = +\sqrt{2}\beta_2$ at level 2, and
- $\mathbf{f}_3^\beta \boldsymbol{\beta} = (\sqrt{\frac{3}{2}}, -\frac{1}{\sqrt{2}})\boldsymbol{\beta} = +\sqrt{\frac{3}{2}}\beta_1 - \frac{1}{\sqrt{2}}\beta_2$ at level 3.

Averaging over the levels of all other factors, the difference between expected responses at level 1 and level 3 of this factor is $\sqrt{6}\beta_1$, and the difference between expected response at level 2 and the average of expected responses at levels 1 and 3 combined is $3\beta_2/\sqrt{2}$. In fact, any *contrast* in $E(\bar{y}_{.1..})$, $E(\bar{y}_{.2..})$ and $E(\bar{y}_{.3..})$ can be expressed as a linear combination of β_1 and β_2. Similarly, $2\alpha_1 = E(\bar{y}_{2...}) - E(\bar{y}_{1...})$, and any contrast in $E(\bar{y}_{..1.})$, $E(\bar{y}_{..2.})$, $E(\bar{y}_{..3.})$, and $E(\bar{y}_{..4.})$ can be represented as a linear combination of γ_1, γ_2, and γ_3. Likewise (for example), any contrast in the six expected responses specified by the levels of factors 1 and 2, after averaging over the levels of factor 3, restricted so that the sum of contrast weights over either the levels of factor 1 or factor 2 alone is zero, can be represented as a linear combination of the two parameters in $(\alpha\beta)$. As a result, the test for differences between levels of factor 1, averaged over the levels of factors 2 and 3, is based on the null hypothesis:

$$\text{Hyp}_0 : \boldsymbol{\alpha} = \mathbf{0}.$$

Similarly, the test for the three factor interaction is based on the null hypothesis:

$$\text{Hyp}_0 : (\boldsymbol{\alpha\beta\gamma}) = \mathbf{0}.$$

A second difference between the models is that (9.4) contains far fewer parameters. Specifically, in the full-rank model:

- Main effects for l-level factors are defined by $l - 1$ parameters.
- Two-factor interactions for l_1- and l_2-level factors are defined by $(l_1 - 1)(l_2 - 1)$ parameters.
- ...
- f-factor interactions are defined by $\prod_{i=1}^{f}(l_i - 1)$ parameters.

In fact, the (overall) number of model parameters used to describe the mean structure, including μ, is exactly the same as the number of treatments. Further, there is a one-to-one linear relationship between the t cell means $\mu_{ij...k}$ of (9.1) and the t elements of the parameter vector in the full-rank model $(\mu, \boldsymbol{\phi}')'$.

9.3.1 Matrix development for the full-rank model

In matrix notation, the full-rank model may be written as:

$$\mathbf{y} = \mathbf{X}\begin{pmatrix}\mu \\ \boldsymbol{\phi}\end{pmatrix} + \boldsymbol{\epsilon} = \mathbf{1}\mu + \mathbf{X}_2\boldsymbol{\phi} + \boldsymbol{\epsilon}$$

$$E(\boldsymbol{\epsilon}) = \mathbf{0}, \qquad Var(\boldsymbol{\epsilon}) = \sigma^2\mathbf{I},$$

where $\mathbf{X}$ contains $r \times \prod_{i=1}^{f} l_i$ rows and $\prod_{i=1}^{f} l_i$ columns. In fact, writing the expectation of both sides of the model with $r = 1$ yields the one-to-one relationship between cell means and model parameters referenced at the end of the last paragraph:

$$E(\mathbf{y}) = \mathbf{X} \begin{pmatrix} \mu \\ \phi \end{pmatrix}$$

since in this case $\mathbf{X}$ is a square matrix of full rank. In characterizing $\mathbf{X}$, it is convenient to define square matrices for each factor comprised of the associated matrix $\mathbf{F}$ with an appended column of 1's, e.g.,

$$\mathbf{G}^\alpha = (\mathbf{1}|\mathbf{F}^\alpha).$$

A key property of these matrices is that, apart from a factor of $l^{1/2}$, they are orthogonal, e.g.,

$$\mathbf{G}^{\alpha'}\mathbf{G}^\alpha = l_1\mathbf{I}.$$

In the more general case $(r \geq 1)$, the complete model matrix may be written as:

$$\mathbf{X} = \mathbf{1}_r \times \ldots \times \mathbf{G}^\gamma \times \mathbf{G}^\beta \times \mathbf{G}^\alpha.$$

The analytical simplicity associated with this model comes from the structure of the model matrix and implications for $\mathbf{X}'\mathbf{X}$:

$$\mathbf{X}'\mathbf{X} = (\mathbf{1}_r \times \ldots \times \mathbf{G}^\gamma \times \mathbf{G}^\beta \times \mathbf{G}^\alpha)'(\mathbf{1}_r \times \ldots \times \mathbf{G}^\gamma \times \mathbf{G}^\beta \times \mathbf{G}^\alpha)$$
$$= \mathbf{1}_r'\mathbf{1}_r \times \ldots \times \mathbf{G}^{\gamma'}\mathbf{G}^\gamma \times \mathbf{G}^{\beta'}\mathbf{G}^\beta \times \mathbf{G}^{\alpha'}\mathbf{G}^\alpha$$
$$= N \times \ldots \times \mathbf{I} \times \mathbf{I} \times \mathbf{I} = N \times \mathbf{I}$$

due to the orthogonal structure of each $\mathbf{G}$. Note that this convenient notation results in an ordering of columns in $\mathbf{X}$ that does not place all main effect together, et cetera. Columns associated with α can be reconstructed as $\mathbf{1}_r \times \mathbf{1}_{l_f} \times \mathbf{1}_{l_{f-1}} \times \ldots \times \mathbf{1}_{l_2} \times \mathbf{F}^\alpha$, while those associated with $(\alpha\gamma)$ can be reconstructed as $\mathbf{1}_r \times \mathbf{1}_{l_f} \times \ldots \times \mathbf{F}^\gamma \times \mathbf{1}_{l_2} \times \mathbf{F}^\alpha$, et cetera. Because all pairs of columns in $\mathbf{X}$ are orthogonal, matrix forms associated with estimation are especially simple. In this case, $\mathbf{X}_{2|1} = \mathbf{X}_2$, and so the reduced normal equations are:

$$\mathbf{X}_2'\mathbf{X}_2\hat{\phi} = \mathbf{X}_2'\mathbf{y}.$$

But since $\mathbf{X}_2'\mathbf{X}_2 = N\mathbf{I}$, this leads immediately to the unique least-squares estimator:

$$\hat{\phi} = \frac{1}{N}\mathbf{X}_2'\mathbf{y}.$$

The experiment offers equal information about each parameter in ϕ, as reflected by the fact that:

$$\mathcal{I} = \mathbf{X}_2'\mathbf{X}_2 = N\mathbf{I}_{(t-1)\times(t-1)}.$$

9.4 Estimation

As noted earlier, factorial structure generally has implications for the interesting contrasts in expected treatment responses. For example, questions associated with the "overall" effect of factor 1 may be addressed by linear contrasts of form

$$\theta = \mathbf{c}' \begin{pmatrix} E(\bar{y}_{1...}) \\ E(\bar{y}_{2...}) \\ ... \\ E(\bar{y}_{l_1...}) \end{pmatrix}.$$

Under the full-rank model, this may be written as $\mathbf{c}'\mathbf{F}^\alpha\alpha$, and its (unique) least-squares estimator is:

$$\mathbf{c}'\widehat{\mathbf{F}^\alpha\alpha} = \mathbf{c}'\mathbf{F}^\alpha\hat{\alpha} = \mathbf{c}'\mathbf{F}^\alpha\frac{1}{N}(\mathbf{1}' \times \mathbf{1}' \times \mathbf{1}' \times \ldots \times \mathbf{F}^{\alpha'})\mathbf{y}$$

$$= \frac{1}{N}(\mathbf{1}' \times \mathbf{1}' \times \mathbf{1}' \times \ldots \times \mathbf{c}'\mathbf{F}^\alpha\mathbf{F}^{\alpha'})\mathbf{y}$$

$$= \frac{1}{N}(\mathbf{1}' \times \mathbf{1}' \times \mathbf{1}' \times \ldots \times \mathbf{c}'(l_1\mathbf{I} - \mathbf{J}))\mathbf{y}$$

$$= \frac{1}{N}(\mathbf{1}' \times \mathbf{1}' \times \mathbf{1}' \times \ldots \times l_1\mathbf{c}')\mathbf{y}$$

$$= \frac{l_1}{N}\sum_{i=1}^{l_1} c_i y_{i...} = \sum_{i=1}^{l_1} c_i \bar{y}_{i...}$$

where the last matrix notation step is possible because $\mathbf{c}'\mathbf{1} = 0$. That is, the estimate of the linear contrast of expected average responses is the same linear contrast of the observed average responses. The variance of the estimate is simple, and can easily be represented in general form since the design information matrix for any subset of ϕ is $\mathcal{I} = N\mathbf{I}$ of appropriate dimension. In particular, the variance of the estimate described above is:

$$Var(\hat{\theta}) = \sigma^2\mathbf{c}'\mathbf{F}^\alpha\mathcal{I}^{-1}\mathbf{F}^{\alpha'}\mathbf{c}$$

$$= \frac{\sigma^2}{N}\mathbf{c}'\mathbf{F}^\alpha\mathbf{F}^{\alpha'}\mathbf{c}$$

$$= \frac{\sigma^2}{N}\mathbf{c}'(l_1\mathbf{I} - \mathbf{J})\mathbf{c}$$

$$= \frac{l_1}{N}\sigma^2\mathbf{c}'\mathbf{c}.$$

Similarly, a comparison among levels of the first two factors, averaging over the third:

$$(\mathbf{c}_1' \times \mathbf{c}_2') \begin{pmatrix} E(\bar{y}_{11..}) \\ E(\bar{y}_{12..}) \\ ... \\ E(\bar{y}_{l_1l_2..}) \end{pmatrix}$$

where the elements of each of $\mathbf{c}_1'$ and $\mathbf{c}_2'$ sum to zero, is estimated by

$$(\mathbf{c}_1' \times \mathbf{c}_2')(\mathbf{F}^\beta \times \mathbf{F}^\alpha)\widehat{(\alpha\beta)} = (\mathbf{c}_1' \times \mathbf{c}_2') \begin{pmatrix} \bar{y}_{11..} \\ \bar{y}_{12..} \\ \cdots \\ \bar{y}_{l_1 l_2 ..} \end{pmatrix}$$

and has variance

$$\frac{l_1 l_2}{N}\sigma^2(\mathbf{c}_1' \times \mathbf{c}_2')(\mathbf{c}_1 \times \mathbf{c}_2) = \frac{l_1 l_2}{N}\sigma^2(\mathbf{c}_1'\mathbf{c}_1)(\mathbf{c}_2'\mathbf{c}_2).$$

An example of the latter, for $l_1 = 2$ and $l_2 = 3$, is:

$$\mathbf{c}_1' = (-1, +1), \quad \mathbf{c}_2' = (-1, 0, +1)$$

expressing the component of the factor-1-by-factor-2 interaction

$$E(\bar{y}_{11..}) - E(\bar{y}_{13..}) - E(\bar{y}_{21..}) + E(\bar{y}_{23..}).$$

The least-squares estimator is the corresponding contrast of data averages:

$$\bar{y}_{11..} - \bar{y}_{13..} - \bar{y}_{21..} + \bar{y}_{23..}$$

with variance

$$\frac{2 \times 3}{N}\sigma^2(2 \times 2) = \frac{24}{N}\sigma^2.$$

9.5 Partitioning of variability and hypothesis testing

The ANOVA decomposition is also especially simple in the case of model (9.4), and facilitates the examination of variability that can be associated with each parameter group (e.g., the variation attributable to the effects of individual factors through main effects, attributable to synergisms between pairs of factors through two-factor interactions, et cetera). The general from of the treatment sum of squares can be written as:

$$SST = \mathbf{y}'\mathbf{H}_{2|1}\mathbf{y}$$

$$= \frac{1}{N}\mathbf{y}'\mathbf{X}_2'\mathbf{X}_2\mathbf{y}$$

$$= N\hat{\phi}'\hat{\phi}.$$

This expression can be further reduced to individual sums of squares of estimates from each parameter group (i.e., each segment of $\hat{\phi}$) as, in our three-factor example:

$$SST = N(\hat{\alpha}'\hat{\alpha} + \hat{\beta}'\hat{\beta} + \hat{\gamma}'\hat{\gamma} + \widehat{(\alpha\beta)}'\widehat{(\alpha\beta)} + \widehat{(\alpha\gamma)}'\widehat{(\alpha\gamma)} + \widehat{(\beta\gamma)}'\widehat{(\beta\gamma)}$$
$$+ \widehat{(\alpha\beta\gamma)}'\widehat{(\alpha\beta\gamma)}).$$

If ϵ has a multivariate normal distribution the seven terms in this sum are independent sums of squares because they are orthogonal contrasts in the data. For example, $\hat{\alpha} = \frac{1}{N}\mathbf{X}_\alpha'\mathbf{y}$ and $\widehat{(\alpha\beta)} = \frac{1}{N}\mathbf{X}_{\alpha\beta}'\mathbf{y}$ are independent because

Table 9.4 ANOVA Format for Three-Factor Example, $l_1 = 2$, $l_2 = 3$, $l_3 = 4$

Source	Degrees of Freedom	Sums of Squares
main effects:		
factor 1, α	1	$12r \sum_i (\bar{y}_{i...} - \bar{y}_{....})^2$
factor 2, β	2	$8r \sum_j (\bar{y}_{.j..} - \bar{y}_{....})^2$
factor 3, γ	3	$6r \sum_k (\bar{y}_{..k.} - \bar{y}_{....})^2$
two-factor interactions:		
factors 1 and 2, $(\alpha\beta)$	2	$4r \sum_{ij} (\bar{y}_{ij..} - \bar{y}_{i...} - \bar{y}_{.j..} + \bar{y}_{....})^2$
factors 1 and 3, $(\alpha\gamma)$	3	$3r \sum_{ik} (\bar{y}_{i.k.} - \bar{y}_{i...} - \bar{y}_{..k.} + \bar{y}_{....})^2$
factors 2 and 3, $(\beta\gamma)$	6	$2r \sum_{jk} (\bar{y}_{.jk.} - \bar{y}_{.j..} - \bar{y}_{..k.} + \bar{y}_{....})^2$
three-factor interaction	6	difference
residual (or "error")	$24(r-1)$	$\sum_{ijkt}(y_{ijkt} - \bar{y}_{ijk.})^2$
corrected total	$24r - 1$	$\sum_{ijkt}(y_{ijkt} - \bar{y}_{...})^2$

$\mathbf{X}'_\alpha \mathbf{X}_{\alpha\beta} = \mathbf{0}$. Each has degrees of freedom equal to the number of elements in the parameter group, because the corresponding submatrix of $\mathbf{X}$ is of full rank, e.g., $(l_2 - 1)(l_3 - 1)$ degrees of freedom in the case of $N\widehat{(\beta\gamma)}'\widehat{(\beta\gamma)}$. Finally, these representations can be easily shown to be equivalent to the familiar scalar notation formulae for ANOVA sums of squares in the case of balanced data; for example:

$$N\hat{\alpha}'\hat{\alpha} = N\left(\frac{1}{N}\mathbf{X}_\alpha \mathbf{y}\right)'\left(\frac{1}{N}\mathbf{X}_\alpha \mathbf{y}\right) = \frac{1}{N}\mathbf{y}'\mathbf{X}_\alpha \mathbf{X}'_\alpha \mathbf{y}$$

$$= \frac{1}{N}(y_{1...}, y_{2...}, \ldots, y_{l_1...})'\mathbf{F}^\alpha \mathbf{F}^{\alpha'}\begin{pmatrix} y_{1...} \\ y_{2...} \\ \cdots \\ y_{l_1...} \end{pmatrix}.$$

Denote the vector of data totals specific to the levels of factor 1 by $\mathbf{y}_1$, and note that

$$\mathbf{F}^\alpha \mathbf{F}^{\alpha'} = \mathbf{G}^\alpha \mathbf{G}^{\alpha'} - \mathbf{11}' = l_1 \mathbf{I} - \mathbf{J},$$

so the above can be written as

$$N\hat{\alpha}'\hat{\alpha} = \frac{1}{N}\mathbf{y}'_1(l_1\mathbf{I} - \mathbf{J})\mathbf{y}_1 = \frac{l_1}{N}\sum_{i=1}^{l_1} y_{i...}^2 - \frac{1}{N}y_{....}^2 = \frac{N}{l_1}\sum_{i=1}^{l_1}(\bar{y}_{i...} - \bar{y}_{....})^2.$$

Independent sums of squares can be written for each of the $2^f - 1$ parameter groups, although *tests* for these groups are not independent because each relies on the same denominator mean square error (*MSE*). An ANOVA decomposition for our three-factor example is presented in Table 9.4. Here, for example, a test of $\text{Hyp}_0 : (\beta\gamma) = \mathbf{0}$, i.e., that there is no interaction between factors 2 and 3, can be carried out by comparing:

$$\frac{N\widehat{(\beta\gamma)}'\widehat{(\beta\gamma)}/6}{MSE} = \frac{2r\sum_{jk}(\bar{y}_{.jk.} - \bar{y}_{.j..} - \bar{y}_{..k.} + \bar{y}_{....})^2/6}{\sum_{ijkt}(\bar{y}_{ijkt} - \bar{y}_{ijk.})^2/24(r-1)}$$

to an appropriate quantile from a central $F(6, 24(r-1))$ distribution. Since the design information matrix for $(\beta\gamma)$ is $24r\mathbf{I}_{6\times6}$, the noncentrality parameter associated with this test is

$$24r(\beta\gamma)'(\beta\gamma)/\sigma^2.$$

So for nonzero $(\beta\gamma)$ the power of the test performed at level 0.01 is

$$\text{Prob}\{W > F_{0.99}(6, 24(r-1))\} \quad \text{where} \quad W \sim F'(6, 24(r-1), 24r(\beta\gamma)'(\beta\gamma)/\sigma^2).$$

9.6 Factorial experiments as CRDs, CBDs, LSDs, and BIBDs

The three fundamental experimental design plans discussed in Chapters 3, 4, and 5 can be employed in factorial settings by simply ignoring the factorial structure. Hence a three-factor treatment structure with $l_1 = 2$, $l_2 = 3$, and $l_3 = 4$ can be examined via a CRD with $24r$ unblocked units, a CBD in r blocks of 24 units each, or a Latin square design (LSD) in 576 units organized in 24 rows and 24 columns in which each treatment is applied $r = 24$ times. For each plan, treatment-to-unit randomization can be carried out exactly as described in previous chapters, again by simply ignoring the factorial nature of treatments.

For each of these designs, inference can be based on a partitioned model separating the nuisance parameters (overall mean, and block effects where applicable) from parameters used to quantify differences between treatments (main effects and interactions). Blocking terms do not affect the form of the reduced normal equations because blocks and treatments are orthogonal in these plans; the reduced normal equations for factorial treatment effects in each case are the same as those that "correct" only for the overall mean. As usual, ANOVA decompositions include components associated with blocks, and for CBDs and LSDs these are calculated ignoring treatment assignments – that is, each block is represented in the block sum of squares by:

$$\text{block size} \times (\text{ block average } - \text{ overall average })^2.$$

As with experiments organized around unstructured treatments, association of this variability with blocks reduces the residual variation against which significance of the factorial treatment "signals" are assessed, but also reduces the number of degrees of freedom available to estimate this residual variability. So, for example, a two-factor experiment with $l_1 = 2$ and $l_2 = 3$ (and so $t = 6$ treatments), with six units assigned to each treatment (and so a total sample size of 36), would yield 30 degrees of freedom for the MSE in a CRD, 25 degrees of freedom for a CBD, and 20 degrees of freedom for a LSD, just as with unstructured treatments. And as usual, *MSE* is used as the denominator for tests of overall treatment differences or of the presence of specific groups of factorial effects (α, β, and $(\alpha\beta)$), and as the basis of standard errors for estimates of these parameters or of other contrasts in the cell means.

Experiments for factorial structures can also be implemented using BIBD plans, again by ignoring the factors and using randomization of treatments to units as in the unstructured case. So for example, letting (i, j) denote the treatment specified by the ith level of a two-level factor and jth level of a three-level factor, a BIBD in 10 blocks of size 3 could be arranged as:

(1,1)	(1,2)	(2,2)	(1,2)	(1,3)	(2,1)
(1,1)	(1,2)	(2,3)	(1,2)	(1,3)	(2,2)
(1,1)	(1,3)	(2,1)	(1,2)	(2,1)	(2,3)
(1,1)	(1,3)	(2,3)	(1,3)	(2,2)	(2,3)
(1,1)	(2,1)	(2,2)	(2,1)	(2,2)	(2,3)

As with BIBDs used with unstructured treatments, all treatment contrasts are estimable, with a sacrifice in efficiency resulting in an increase in the estimation standard deviation by a factor of $\sqrt{\frac{k(t-1)}{t(k-1)}}$ (or $\sqrt{1.25}$ in this case) compared to a CRD with the same number of units assigned to each treatment and the same value of σ^2.

9.7 Model reduction

One substantial and practical difficulty that often arises with factorial experiments is the potentially very large number of parameters included in the model. However, as described in the hypothetical example of Section 9.2, it often turns out that the interactions of relatively high order are not especially important for accurate representation of the response mean structure. That is, it is often the fortunate case that the parameters that are most easily explained physically (main effects and interactions of low order) are associated with the majority of variation in the data, while the factorial parameters associated with the most complex patterns of potential variation (interactions of high order) are zero or near zero.

There are substantial statistical advantages to reducing the number of parameters in the model used in analyzing data from a factorial experiment. Suppose we rewrite the full-rank model in further partitioned form as:

$$E(\mathbf{y}) = \mathbf{1}\mu + \mathbf{X}_2\boldsymbol{\phi} = \mathbf{1}\mu + \mathbf{W}_1\boldsymbol{\phi}_1 + \mathbf{W}_2\boldsymbol{\phi}_2$$

where the columns of $\mathbf{W}_1$ and $\mathbf{W}_2$ form a partition of those in $\mathbf{X}_2$, and $\boldsymbol{\phi}_1$ and $\boldsymbol{\phi}_2$ form the corresponding partition of $\boldsymbol{\phi}$. Further, say that we have made the decision to assume that $\boldsymbol{\phi}_2 = \mathbf{0}$. Adopting the reduced model, the least-squares estimate of $\boldsymbol{\phi}_1$ is:

$$\hat{\boldsymbol{\phi}}_1 = \frac{1}{N}\mathbf{W}_1'\mathbf{y}$$

and, since the columns of $\mathbf{W}_1$ are taken from $\mathbf{X}_2$, the estimates included in this vector are exactly the same functions of $\mathbf{y}$ as they would be under the full model. Now, any contrast in cell means is estimated by $\mathbf{l}'\mathbf{X}_2\hat{\phi}$ under the full model, or $\mathbf{l}'\mathbf{W}_2\hat{\phi}_1$ under the reduced model. The variance of this estimate is:

$$\mathbf{l}'\mathbf{X}_2\,Var(\hat{\phi})\mathbf{X}_2'\mathbf{l} = \frac{\sigma^2}{N}\mathbf{l}'\mathbf{X}_2\mathbf{X}_2'\mathbf{l}$$

under the full model, or

$$\mathbf{l}'\mathbf{W}_1\,Var(\hat{\phi}_1)\mathbf{W}_1'\mathbf{l} = \frac{\sigma^2}{N}\mathbf{l}'\mathbf{W}_1\mathbf{W}_1'\mathbf{l}$$

under the reduced model. But since

$$\mathbf{X}_2\mathbf{X}_2' = \mathbf{W}_1\mathbf{W}_1' + \mathbf{W}_2\mathbf{W}_2' \qquad \mathbf{l}'\mathbf{X}_2\mathbf{X}_2'\mathbf{l} = \mathbf{l}'\mathbf{W}_1\mathbf{W}_1'\mathbf{l} + \mathbf{l}'\mathbf{W}_2\mathbf{W}_2'\mathbf{l}$$

the estimation variance based on the reduced model can be no more than that based on the full model, and depending on the specific vector $\mathbf{l}$ of interest and partitioning of ϕ, may be much less.

In any particular setting, the questions that must be answered are whether the full model can be reduced, and if so, which terms can be eliminated. One approach is to construct a series of up to $2^f - 1$ F-statistics, one appropriate for testing the null hypothesis of "no effect" for each parameter group. These tests are generally performed on the parameter groups of highest order first, and are often performed with "side conditions" requiring that when a three-factor interaction group is included in the model (for example), all two-factor interaction groups and main effect groups involving the same factors must also be retained. This particular restriction is often called a *hierarchical model* requirement, and is discussed in more detail for two-level factorial experiments in Chapter 11.

Because the number of unique treatments increases exponentially with the number of factors, and multiplicatively with the number of levels used with any factor, a seemingly simple and even modest factorial structure can lead to suggested experimental designs requiring more units than can be easily accommodated in an experiment of practical size. However, especially in the early stages of an experimental program, a primary goal is often to understand which of several factors or combinations of factors are responsible for most of the variation in the response or responses of interest. One compromise between what is statistically desirable and operationally feasible is to execute an unreplicated experiment – that is, to include only one experimental unit representing each treatment, resulting in a data set for which $N = t$. The statistical disadvantages of this are obvious, and the most important of these is the lack of an "honest" estimate of σ^2 against which the estimates of treatment effects can be compared. If one is willing to *assume* that the highest-order interaction terms are actually absent or negligible, the corresponding component of the

ANOVA decomposition can be used as the basis of a $\prod_{i=1}^{f}(l_i - 1)$ degree-of-freedom estimate of σ^2. In this case, one model-trimming strategy can be organized as follows:

1. Initialize $i = f$, and compute a "denominator sum of squares," SSD, as the sum of squares for the highest-order (f) interaction group, and a "denominator degrees of freedom," df_D, as the degrees of freedom for the f-order interaction.

2. Compute a "denominator mean square" as $MSD = SSD/df_D$.

3. Test each parameter group of the order $i-1$ by constructing an F-statistic comparing the parameter group mean square to MSD. Use a fairly liberal level for this test, e.g., $\alpha = 0.25$, to avoid accidentally removing small-but-real effects from the model.

4. For each parameter group of order i for which the null hypothesis of "no effect" is not rejected, add the sum of squares to SSD and the degrees of freedom to df_D. (At this point, SSD and df_D include sums of squares and degrees of freedom for all effects from order $i - 1$ through f that are assumed or appear to be negligible or absent). Decrease the value of i by 1, and return to step 2.

5. When no terms of a given order appear to be negligible, stop the process, and tentatively adopt a "working model" that contains all parameter groups of this or lower order, and all higher-order parameter groups for which the null hypothesis was rejected.

In order to preserve hierarchical model structure, step 3 can be modified to test only ith-order interaction groups for which the associated factors are not a subset of higher-order interactions for which the null hypothesis was rejected at an earlier iteration. While approaches of this kind can be useful at the factor-screening stage of an experimental program, follow-up studies focusing on the apparently active factors, designed so as to include true replication, should also be planned to provide more dependable inferences.

9.8 Conclusion

In many important settings, experimental treatments are defined by selection of levels corresponding to a number of factors. Strictly speaking, this does not change the broad approach outlined for experimental design and analysis presented in previous chapters, for example:

- Factorial experiments can be executed as CRDs, CBDs, or LSDs, in which least-squares estimates are unaffected by blocking, and the variation associated with blocks can be easily computed and removed from unexplained variability in an ANOVA decomposition.

- Factorial experiments can be executed as BIBDs in which blocks have the same impact on analysis as is described in Chapter 7.

However, factorial treatment structure generally *does* have implications for what treatment contrasts are of most scientific interest. In particular, differences in expected response that can be associated with changes in the levels of an individual factor or small groups of factors are described by the main effects and low-order interactions in factorial parameterizations. In many applications, interactions of higher order have less influence on responses, and where this is true and can be verified in data analysis, it can lead to a reduction in the number of parameters in the model required to summarize the systematic differences in observed responses. Since factorial studies often involve many treatments, methods that can be used to assist in "trimming" unneeded high-order interactions are useful. Reduced models provide more precise estimates of treatment contrasts, and more degrees of freedom for estimating error variance.

9.9 Exercises

1. The rainfall example of Fay et al. (2000) featured in Chapter 1 described four experimental treatments defined as a 2^2 factorial structure. But the experimental design also featured a fifth experimental condition used as a "control." Re-read the description of that study, and discuss how the data from this experiment might be most usefully analyzed, given the scientific questions of apparent interest.

2. A cell biologist designs an experiment to study the joint effect of exposure to ionizing radiation and a particular toxic chemical to the survival of marrow stem cells taken from a specific strain of mouse. He designs a factorial experiment to study all combinations of three radiation doses (including no radiation) and four concentrations of the chemical (including zero concentration). Thirty-six cell cultures are prepared, each consisting of a standard number of cells in medium in a petri dish, and three cell cultures are exposed to each treatment. After exposure to the treatment, the data value associated with each culture is a cell count that reflects the viability of the culture after exposure.

   (a) Suppose the experiment can be executed as a completely randomized design, i.e., with each of the 36 cell cultures prepared independently, and assigned randomly to treatments with only the restriction that each treatment be applied to three cultures. In an analysis of variance that might be used following this experiment, what are the degrees of freedom that would be associated with:

   - the main effect for radiation
   - the main effect for chemical concentration
   - the radiation by chemical concentration interaction
   - residual (error) variation

(b) Suppose the experiment can be executed as a randomized complete block design, using three batches of 12 cell cultures each. Assuming that blocks and treatments do not interact, what are the analysis of variance degrees of freedom, under this design, for:

- the main effect for radiation
- the main effect for chemical concentration
- the radiation by chemical concentration interaction
- blocks
- residual (error) variation

(c) Suppose that using the randomized complete block version of the experiment results in 5% less unit-to-unit variability (represented by σ^2 in our model notation) than what would be realized in the completely randomized design, because batches of 12 cell cultures can be made somewhat more uniformly than batches of 36. Is this sufficient information to determine which design would result in a more powerful test for the radiation main effect if, in fact, $E(\bar{y}_{3..}) = E(\bar{y}_{2..}) + 1 = E(\bar{y}_{1..}) + 2$? If so, determine which design would be preferred from this point of view. If not, what minimal additional information would be needed to make this comparison?

(d) Continuing with the information supplied in part (c), is this sufficient information to determine which design would result in smaller expected squared lengths of 95% confidence intervals on treatment contrasts? If so, determine which design would be preferred from this point of view. If not, what minimal additional information would be needed to make this comparison?

3. Suppose now that the experiment described in exercise 2 is to be executed as a CRD, but that 40 petri dishes are prepared, and different numbers of petri dishes are to be allocated to some treatments. Specifically,

- two cell cultures are allocated to each treatment including both nonzero radiation and chemical exposure
- four cell cultures are allocated to each treatment with either no chemical exposure, or no radiation, but not both, and
- eight cell cultures are allocated to the "control" treatment with no radiation or chemical exposure.

Derive the noncentrality parameter associated with the test for a radiation main effect under this design. If $\sigma = 3$ and $E(\bar{y}_{3..}) - 1 = E(\bar{y}_{2..}) = E(\bar{y}_{1..})$, what is the power of this test performed at level 0.05 under this design?

4. Using the data from the concrete experiment of Soudki et al. (subsection 9.1.1), compute a three-way analysis of variance, and test for the presence of each group of main effects and two-factor interactions, *assuming* the three-factor interaction can be removed from the model.

5. Continuing to use the data from the example of subsection 9.1.1, make two sets of parallel boxplots from the data as described in subsection 9.2.1 to depict the effects of the total aggregate/cement ratio and the coarse aggregate/total aggregate ratio. Together with the similar graph presented in subsection 9.2.1 for the water/cement ratio, is the appearance of these graphs consistent with the results of the ANOVA? What (if anything) could you not infer from looking at the plots, that is apparent in the ANOVA?

Consider an industrial experiment carried out to improve the properties of a certain kind of manufactured sandpaper. The sandpaper is made of several layers of material. These layers are glued together – glue is inserted between the layers, and pressure is applied to both sides for a fixed length of time. The measurement of interest in this experiment is the amount of force that is required to pull the layers of the finished sandpaper apart (that is, to destroy the product). The controlled experimental factors in this experiment are the kind of glue used (two kinds), and the amount of pressure used (three levels) in the assembly of the sandpaper, resulting in six unique experimental treatments. A specific measurement of force will be denoted throughout as y_{ijl}, where $i = 1, 2$ denotes the level of the glue factor and $j = 1, 2, 3$ denotes the level of the pressure factor. The following quantities may be used as needed in working exercises 6 and 7. In each case, the sum is over all combinations of values of the indicated indices that appear together in the experiment. (Some of these values may not be useful or meaningful in both of the experiments described.) Note that most of these quantities are *not* actually ANOVA sums of squares, but that ANOVA sums of squares can be easily computed from them.

$$\sum_{ijl}(y_{ijl} - \bar{y}_{...})^2 = 360$$

$$\sum_{ij}(\bar{y}_{ij.} - \bar{y}_{...})^2 = 60$$

$$\sum_{ij}(\bar{y}_{ij.} - \bar{y}_{i..} - \bar{y}_{.j.} + \bar{y}_{...})^2 = 10$$

$$\sum_{i}(\bar{y}_{i..} - \bar{y}_{...})^2 = 10$$

$$\sum_{j}(\bar{y}_{.j.} - \bar{y}_{...})^2 = 10$$

$$\sum_{l}(\bar{y}_{..l} - \bar{y}_{...})^2 = 5$$

$$\sum_{il}(\bar{y}_{i.l} - \bar{y}_{...})^2 = 60$$

$$\sum_{jl}(\bar{y}_{.jl} - \bar{y}_{...})^2 = 70$$

6. Suppose the experiment is executed using a completely randomized design. Each piece of sandpaper is individually manufactured, independently from all others used in the experiment. For each combination of glue and pressure, four pieces are made, and l indexes these. For example, y_{234} is the measurement on the 4th piece of sandpaper from those made with glue 2

and pressure level 3. For an F-test for the main effect of pressure, what are the values of:

(a) The numerator sum of squares

(b) The numerator degrees of freedom

(c) The denominator sum of squares

(d) The denominator degrees of freedom

7. Suppose the experiment is executed as a randomized complete block design. Sets of six pieces of sandpaper are made together, one using each of the glue-pressure combinations. Four such sets are made, and l indexes these sets. So for example, y_{234} is the measurement taken from the single sample made with glue 2 and pressure level 3, in set (or block) number 4. Assume that effects associated with blocks are fixed and additive (i.e., blocks do not interact with treatments). For an F-test for the main effect of pressure, what are the values of:

(a) The numerator sum of squares

(b) The numerator degrees of freedom

(c) The denominator sum of squares

(d) The denominator degrees of freedom

8. An experimenter wants to understand the effect of a three-level factor ("factor 1") on a response. She also has some interest in the effects of "factor 2" and "factor 3," each of which has three levels, and can be used jointly with factor 1 to define treatments, but these factors are not the focus of her current research. She is considering three different experimental designs, each requiring $N = 54$ units:

• Design A: A 3^1 design in factor 1, with each treatment applied to $r = 18$ units.

• Design B: A 3^2 design in factors 1 and 2, with each treatment applied to $r = 6$ units.

• Design C: A 3^3 design in factors 1, 2, and 3, with each treatment applied to $r = 2$ units.

Each design can be executed without blocking.

(a) For each design, what is the expected squared length of a 95% confidence interval for α_1 (full-rank model) if $\sigma^2 = 1$?

(b) For each design, what is the power of the level 0.05 test for Hyp$_0$: $\boldsymbol{\alpha} = \mathbf{0}$ if, in fact, $\alpha_1 = \alpha_2 = 0.1$ and $\sigma^2 = 1$?

CHAPTER 10

Split-plot designs

10.1 Introduction

Suppose an engineer wishes to perform an experiment to examine how a chemical reaction takes place in a small reactor. In particular, he would like to compare product made using all combinations of three different reaction temperatures and four different stirring rates, and so naturally thinks of the study as a factorial experiment with two factors with three and four levels, respectively. If his budget allows enough material for 24 runs, $r = 2$ data values can be collected under each of the $t = 12$ treatment conditions. He will perform the study using a single reactor, and so must develop a schedule for sequential execution of all 24 experimental runs. If the experiment were executed as a completely randomized design (CRD), good practice would suggest that the entire schedule be constructed randomly to avoid any uncontrolled and unknown systematic changes in reactor operating conditions that might occur over time.

Suppose, however, that operational circumstances suggest that four runs of the reactor are the most that can reasonably be accomplished in a day, and possible day-to-day variations in the experimental environment suggest that the runs made during a given day should be regarded as a block. One possible approach would be to consider whether a balanced incomplete block design (BIBD) for $t = 12$ treatments in $b = 6$ blocks of size $k = 4$ might be constructed. (It is simple to show that such a BIBD does not exist; think about how large the experiment would have to be in order to accommodate BIBD structure.) Suppose, further, that the levels of the second factor, stirring rate, can be changed very quickly, simply by turning a knob and waiting a few minutes for the mixture in the reactor to change. However, raw material is preheated in a relatively large vessel outside of the reactor, and the uniform temperature of this material cannot be changed quickly. In fact, the only way that four runs can be executed in one day is if all four treatments are run at the same temperature, i.e., that changes among the levels of the relatively easy-to-change factor, stirring rate, be the only changes made from run to run within a block.

If the effects associated with blocks are treated as a fixed source of variation, these constraints present an obvious problem. Since temperature is completely confounded with blocks, there can be no information available from the data on the main effect of temperature once correction has been made for blocks/days. However, if day-to-day differences can be regarded as random, an inter-block analysis can be used to recover information about the fixed main effects associated with temperature. The quality of this information will likely not be as high as that provided by the intra-block inference about the main effects associated with stirring rate, since the former requires characterizing patterns attributable to differences in temperature through noise that includes random components associated with both day-to-day variation, and unit-to-unit variability and measurement errors attributed to each experimental run.

In this setting, blocks are sometimes called *plots*, and the experimental design is often called a *split-plot design* (SPD). This terminology originated with agricultural experiments in which, for example, the joint effects of crop variety and fertilizer type on yield were important, but only one of the two factors could be easily varied in a small spatial area. Other studies called *repeated measures* experiments make use of test subjects that can only be treated at a single level of one factor, but for which a series of data values can be collected corresponding to all levels of another factor in a collection of "repeated" tests. For example, in an evaluation of the effect of sleep deprivation, each subject (the plot/block) may be assigned to one of several levels of sleep deprivation (the *among-subjects factor*), and then asked to perform a series of tests to evaluate ability to concentrate under different levels of distracting noise (the *within-subjects factor*). While split-plot studies and repeated measures studies are generally not *physically* similar, they share a common statistical structure in that one or more treatment factors are *confounded* with blocks (plots or subjects), while all levels of one or more other factors are represented within each block. We will use the "split-plot" label for both types of experiments.

10.1.1 Example: strength of fabrics

Rong, Leon, and Bhat (2005) describe an experiment performed to characterize the effects of some process variables on the strength of cotton-based nonwoven fabrics. Fabrics were produced under 24 different conditions defined by combinations of three experimental factors summarized in Table 10.1. One response variable was peak load (measured in kg), a measure of tensile strength of the resulting fabric; average values for each of the 24 experimental conditions are given in Table 10.2.

The authors indicate that levels of calendering temperature could not be changed as often as the other two factors, presumably due to operational characteristics of the process. Hence six treatments corresponding to all combinations of binder fibers and binder content were processed while the process was operating at each temperature. As a result, differences among fabrics

Table 10.1 Factors and Levels in the Experiment of Rong et al. (2005)

Factor	Levels
Calendering temperature (°C)	90, 100, 110, 120
Binder fibers	unicomponent, bicomponent
Binder content (%)	15, 30, 50

Table 10.2 Average Peak Load (Kilograms, kg) for Unicomponent/Bicomponent Fibers in the Experiment of Rong et al. (2005)

Temperature	Binder Content		
	15%	30%	50%
90	0.10/0.24	0.12/0.42	0.13/0.54
100	0.14/0.30	0.15/0.54	0.15/1.15
110	0.17/0.36	0.30/0.65	0.36/1.21
120	0.17/0.29	0.30/0.56	0.31/0.85

that vary in binder fiber and binder content can be assessed as differences among data values generated in close temporal proximity, while differences attributable to temperature must be assessed by comparing "blocks" of six binder fiber/binder content tests. Hence, while experimental material did not naturally come in batches sufficient to make only six specimens, the authors recognize that samples made consecutively (and so within a common temperature level) are likely to be more "alike" than those separated by more time.

10.1.2 Example: English tutoring

Denton, Anthony, Parker, and Hasbrouck (2004) performed a study to compare the effectiveness of two tutoring programs used to teach English reading to Spanish-dominant bilingual elementary school students. Students in grades 2–5 were recruited from multiple schools. Because many conditions can have a major impact on educational success, students from the same school and grade were matched in pairs, one student in each pair to be assigned to one of the two tutoring programs, and the other to serve as a control (i.e., standard curriculum without the additional tutoring program). Each student was given a reading mastery test at the beginning of the study (that is, before tutoring began, sometimes called a *pretest*), and again at the end of the study. Hence a level of the factor *tutoring program* is applied to a student (or more precisely, to a block of two matches students, one of which is randomly selected to receive the tutoring), while both levels of the factor *time* are represented within each student.

Note that in the strict sense, this study cannot be called a true experiment because the order of application of the time levels cannot be randomized. Other repeated-measures studies *do* admit within-subject randomization, for example, where subjects are each asked to perform a series of related tasks coinciding with a factor level, and the tasks can be performed in any order.

10.2 SPD(R,B)

The critical distinction that makes split-plot experiments different from other factorial experiments is the presence of two (or even more) definitions of the *experimental unit*. One way to see this is to think of the individual factors as being applied to units separately, rather than all factors (and therefore the entire treatment) being applied together. The entities we are calling *blocks* are essentially the units to which levels of the among-blocks factor or factors are applied. So, for example, in the hypothetical chemical process experiment, "temperature" is set once for all data generated on a given day (or block, plot, or "whole-plot unit"). This means that blocks-within-temperature level variation is the "noise" against which differences between the levels of this factor must be judged. On the other hand, levels of the within-blocks factor or factors are applied to the experimental material associated with a single data value, e.g., the four individual reactor runs that are performed in a day using the four different stirring rates. In fact, it is sometimes helpful to think of a split-plot experiment as being carried out in multiple *strata*, or even as a combination of distinct, nested experiments. In the case of our chemical plant example, the "whole-plot" experiment is a CRD in which $3r$ days (at this stratum, experimental units) are randomly assigned to three different treatments (levels of the among-block factor). The "split-plot" experiment is a complete block design (CBD) in which the four runs (now the units) made in each day (now the block) are randomly assigned to the four levels of the within-block factor. We refer to the design of a split-plot experiment organized in this way as a SPD(R,B), indicating that the portion of the experiment executed at the "top" stratum is randomized, while that executed at the "bottom" stratum is blocked.

10.2.1 A model

The portion of the model that expresses treatment structure is no different in this case than in any other factorial model. What must be added here is a term to account for random block-to-block differences. Consider a two-factor experiment; treatment structure can be expressed in overparameterized form as:

$$\dot{\alpha}_i + \dot{\beta}_j + (\dot{\alpha\beta})_{ij}$$
$$i = 1, 2, 3, \ldots, l_1, \quad j = 1, 2, 3, \ldots, l_2.$$

If only one level of the factor represented by $\dot{\alpha}$, but all levels of the factor represented by $\dot{\beta}$, are applied to the units within a block, we might write

$$y_{itj} = \dot{\alpha}_i + \dot{\beta}_j + (\dot{\alpha\beta})_{ij} + \zeta_{it(i)} + \epsilon_{it(i)j}$$
$$i = 1, 2, \ldots, l_1; \quad j = 1, 2, \ldots, l_2; \quad t = 1, 2, \ldots, r.$$

Here, y_{itj} represents the datum resulting from application of the jth level of factor 2, taken from the tth block in which the first factor is applied at level i. Hence the number of blocks used in the experiment is $b = rl_1$, and the total number of data values generated is $N = bl_2 = rl_1 l_2$. The two random elements in this model are $\zeta_{it(i)}$ and $\epsilon_{it(i)j}$, with

$$E(\zeta_{it(i)}) = \mu_\zeta, \quad Var(\zeta_{it(i)}) = \sigma_\zeta^2$$
$$E(\epsilon_{it(i)j}) = 0, \quad Var(\epsilon_{it(i)j}) = \sigma^2$$

representing block variation and unit-within-block variation, respectively. These two random terms are generally also assumed to be independent.

For our purposes, the full-rank parameterization described in Section 9.3 is helpful. Order the elements of $\mathbf{y}$ lexicographically from top to bottom by (i, t, j), that is, with all l_2 values associated with a block together, and all subvectors representing blocks associated with the same level of the first factor together. Then following the notation introduced in Chapter 9, and given that $\mathbf{F}^\alpha$ and $\mathbf{F}^\beta$ have been defined, we can write a model for all data from the experiment as:

$$\mathbf{y} = \mathbf{X}_1\boldsymbol{\zeta} + \mathbf{X}_\alpha\boldsymbol{\alpha} + \mathbf{X}_\beta\boldsymbol{\beta} + \mathbf{X}_{(\alpha\beta)}(\alpha\beta) + \boldsymbol{\epsilon}$$
$$E(\boldsymbol{\zeta}) = \mu_\zeta \mathbf{1} \quad Var(\boldsymbol{\zeta}) = \sigma_\zeta^2\mathbf{I}$$
$$E(\boldsymbol{\epsilon}) = \mathbf{0} \quad Var(\boldsymbol{\epsilon}) = \sigma^2\mathbf{I}$$

where $\boldsymbol{\alpha}$, $\boldsymbol{\beta}$, and $(\alpha\beta)$ are vectors of length $l_1 - 1$, $l_2 - 1$, and $(l_1 - 1)(l_2 - 1)$, respectively, $\boldsymbol{\zeta}$ is the $l_1 r$-element vector of random block effects, and the model matrices can be written as direct products of simpler matrices:

$$\mathbf{X}_\alpha = \mathbf{1}_{l_2} \times \mathbf{1}_r \times \mathbf{F}^\alpha \quad \mathbf{X}_\beta = \mathbf{F}^\beta \times \mathbf{1}_r \times \mathbf{1}_{l_1}$$
$$\mathbf{X}_{(\alpha\beta)} = \mathbf{F}^\beta \times \mathbf{1}_r \times \mathbf{F}^\alpha \quad \mathbf{X}_1 = \mathbf{1}_{l_2} \times \mathbf{I}_{r\times r} \times \mathbf{I}_{l_1\times l_1}.$$

10.2.2 Analysis

The intra-block model is based on $\mathbf{y}_1$, the projection of the data into the complement of the space spanned by the columns of $\mathbf{X}_1$ (Section 8.2). Let

$$\mathbf{H}_1 = \mathbf{X}_1(\mathbf{X}_1'\mathbf{X}_1)^{-1}\mathbf{X}_1' = \frac{1}{l_2}(\mathbf{J}_{l_2\times l_2} \times \mathbf{I}_{r\times r} \times \mathbf{I}_{l_1\times l_1})$$

and define:

$$\mathbf{y}_1 = (\mathbf{I} - \mathbf{H}_1)\mathbf{y}$$
$$= (\mathbf{I}-\mathbf{H}_1)\mathbf{X}_\alpha\boldsymbol{\alpha}+(\mathbf{I}-\mathbf{H}_1)\mathbf{X}_\beta\boldsymbol{\beta}+(\mathbf{I}-\mathbf{H}_1)\mathbf{X}_{(\alpha\beta)}(\alpha\beta)+(\mathbf{I}-\mathbf{H}_1)\mathbf{X}_1\boldsymbol{\zeta}+(\mathbf{I}-\mathbf{H}_1)\boldsymbol{\epsilon}.$$

Since $\mathbf{H}_1$ projects vectors into the space spanned by the columns of $\mathbf{X}_1$, $(\mathbf{I} - \mathbf{H}_1)\mathbf{X}_1 = \mathbf{0}$. And since the columns of $\mathbf{X}_\alpha$ can be formed as sums of those in $\mathbf{X}_1$, $(\mathbf{I} - \mathbf{H}_1)\mathbf{X}_\alpha\boldsymbol{\alpha} = \mathbf{0}$ also. Hence

$$\mathbf{y}_1 = (\mathbf{I} - \mathbf{H}_1)\mathbf{X}_\beta\boldsymbol{\beta} + (\mathbf{I} - \mathbf{H}_1)\mathbf{X}_{(\alpha\beta)}(\alpha\beta) + (\mathbf{I} - \mathbf{H}_1)\boldsymbol{\epsilon}.$$

This simplifies further as:

$$\mathbf{y}_1 = \left(\mathbf{X}_\beta - \frac{1}{l_2}(\mathbf{J}_{l_2\times l_2} \times \mathbf{I}_{r\times r} \times \mathbf{I}_{l_1\times l_1})(\mathbf{F}^\beta \times \mathbf{1}_r \times \mathbf{1}_{l_1})\right)\boldsymbol{\beta}$$
$$+ \left(\mathbf{X}_{(\alpha\beta)} - \frac{1}{l_2}(\mathbf{J}_{l_2\times l_2} \times \mathbf{I}_{r\times r} \times \mathbf{I}_{l_1\times l_1})(\mathbf{F}^\beta \times \mathbf{1}_r \times \mathbf{F}^\alpha)\right)(\alpha\beta)$$
$$+ (\mathbf{I} - \mathbf{H}_1)\boldsymbol{\epsilon}$$
$$= \mathbf{X}_\beta\boldsymbol{\beta} + \mathbf{X}_{(\alpha\beta)}(\alpha\beta) + (\mathbf{I} - \mathbf{H}_1)\boldsymbol{\epsilon}.$$

The best linear unbiased estimators associated with this model form are weighted least-squares estimators, because:

$$E[(\mathbf{I} - \mathbf{H}_1)\epsilon] = \mathbf{0}, \qquad Var[(\mathbf{I} - \mathbf{H}_1)\epsilon] = \sigma^2(\mathbf{I} - \mathbf{H}_1).$$

Letting $\mathbf{X}_{\beta,(\alpha\beta)} = (\mathbf{X}_\beta|\mathbf{X}_{(\alpha\beta)})$, the appropriate weighted normal equations in this case are

$$\mathbf{X}'_{\beta,(\alpha\beta)}(\mathbf{I} - \mathbf{H}_1)\mathbf{X}_{\beta,(\alpha\beta)}\begin{pmatrix}\widehat{\beta}\\ \widehat{(\alpha\beta)}\end{pmatrix} = \mathbf{X}'_{\beta,(\alpha\beta)}(\mathbf{I} - \mathbf{H}_1)\mathbf{y}_1 = \mathbf{X}'_{\beta,(\alpha\beta)}(\mathbf{I} - \mathbf{H}_1)\mathbf{y}.$$

$$\tag{10.1}$$

But note that these are exactly the same as the reduced normal equations for the fixed-block model, omitting the main effect associated with the first factor because it is confounded with block-to-block differences. Inferences concerning the main effect for the within-block factor, and the interaction involving the among- and within-block factors, are based on this model.

The design information matrix associated with this intra-block analysis is

$$\mathcal{I}_{intra} = \mathbf{X}'_{\beta,(\alpha\beta)}(\mathbf{I} - \mathbf{H}_1)\mathbf{X}_{\beta,(\alpha\beta)} = N\mathbf{I}$$

as suggested by the reduced normal equations (10.1). Note that this matrix has rows and columns associated with the elements of β and $(\alpha\beta)$. We *could* add l_1 rows and columns, containing only zero elements, corresponding to the parameters α, but there is no practical reason to do this in the split-plot context. The variance of the estimate of any linear combination of β and $(\alpha\beta)$ is then:

$$Var((\beta'|(\alpha\beta)')\mathbf{c}) = \mathbf{c}'\mathcal{I}_{intra}^{-}\mathbf{c}\sigma^2 = \frac{\mathbf{c}'\mathbf{c}}{N}\sigma^2.$$

The inter-block analysis is based on the transformed data vector:

$$\mathbf{y}_2 = \mathbf{X}'_1\mathbf{y} = \mathbf{X}'_1\mathbf{X}_\alpha\alpha + \mathbf{X}'_1\mathbf{X}_\beta\beta + \mathbf{X}'_1\mathbf{X}_{(\alpha\beta)}(\alpha\beta) + \mathbf{X}'_1\mathbf{X}_1\zeta + \mathbf{X}'_1\epsilon.$$

Due to the relationship between blocks and the levels of the first treatment factor, the inner products of the columns of $\mathbf{X}_1$ with those of the other partitioned model matrices take simple forms. Specifically:

$$\mathbf{X}'_1\mathbf{1} = l_2\mathbf{1}$$
$$\mathbf{X}'_1\mathbf{X}_\alpha = (\mathbf{1}'_{l_2} \times \mathbf{I}_{r\times r} \times \mathbf{I}_{l_1\times l_1})(\mathbf{1}_{l_2} \times \mathbf{1}_r \times \mathbf{F}^\alpha) = l_2\mathbf{1}_r \times \mathbf{F}^\alpha$$
$$\mathbf{X}'_1\mathbf{X}_\beta = (\mathbf{1}'_{l_2} \times \mathbf{I}_{r\times r} \times \mathbf{I}_{l_1\times l_1})(\mathbf{F}^\beta \times \mathbf{1}_r \times \mathbf{1}_{l_1}) = \mathbf{0}$$
$$\mathbf{X}'_1\mathbf{X}_{(\alpha\beta)} = (\mathbf{1}'_{l_2} \times \mathbf{I}_{r\times r} \times \mathbf{I}_{l_1\times l_1})(\mathbf{F}^\beta \times \mathbf{1}_r \times \mathbf{F}^\alpha) = \mathbf{0}.$$

As a result, the inter-block model simplifies to:

$$\mathbf{y}_2 = l_2(\mathbf{1}_r \times \mathbf{F}^\alpha)\alpha + \mathbf{X}'_1\mathbf{X}_1\zeta + \mathbf{X}'_1\epsilon = l_2(\mathbf{1}_r \times \mathbf{F}^\alpha)\alpha + \mathbf{X}'_1(\mathbf{X}_1\zeta + \epsilon).$$

Finally, letting $\epsilon^* = \mathbf{X}'_1(\mathbf{X}_1(\zeta - \mu_\zeta\mathbf{1}) + \epsilon)$, we can write an inter-block model as:

$$\mathbf{y}_2 = l_2\mu_\zeta\mathbf{1} + l_2(\mathbf{1}_r \times \mathbf{F}^\alpha)\alpha + \epsilon^*$$
$$E(\epsilon^*) = \mathbf{0} \qquad Var(\epsilon^*) = (l_2^2\sigma_\zeta^2 + l_2\sigma^2)\mathbf{I}.$$

$$\tag{10.2}$$

Model (10.2) provides the basis for comparative inferences that can be made about the main effect for the among-block factor.

The design information matrix associated with this inter-block analysis is

$$\mathcal{I}_{inter} = l_2(\mathbf{1}_r \times \mathbf{F}^\alpha)'(\mathbf{1}_r \times \mathbf{F}^\alpha) = N\mathbf{I}.$$

Again, we define this matrix with only (l_1-1) rows and columns corresponding to the elements of $\boldsymbol{\alpha}$ since there is no inter-block information available about $\boldsymbol{\beta}$ or $(\boldsymbol{\alpha\beta})$. The variance of the estimate of any linear combination of the elements of $\boldsymbol{\alpha}$ is

$$Var(\mathbf{c}'\hat{\boldsymbol{\alpha}}) = \mathbf{c}'\mathcal{I}_{inter}^{-}\mathbf{c}[l_2\sigma_\zeta^2 + \sigma^2] = \frac{\mathbf{c}'\mathbf{c}}{N}[l_2\sigma_\zeta^2 + \sigma^2].$$

The dual nature of split-plot experiments means that experimental "noise" is different for some comparisons than for others. Block-within-factor variation is the noise against which the among-blocks factor associated with $\boldsymbol{\alpha}$ must be judged. Because a change in the level of the within-block factor implies a change in both β_j and $(\alpha\beta)_{ij}$, the noise against which within-block factor main effects *and* the interactions involving within-block factors must be judged is associated with units-within-blocks.

The split-plot ANOVA decomposition is often written in a unified form, in which degrees of freedom and sums of squares for all factorial effects are computed by the same formulae that would be applied in a variance decomposition associated with a CRD and CBD, but with two residual or error sums of squares computed corresponding to the noise associated with the two strata of the experiment. Table 10.3 displays a general ANOVA format for the SPD(R,B) with one among-blocks factor and one within-blocks factor.

Although the "two-experiments story" is usually most easily told beginning with the whole-plot portion of the study, the ANOVA decomposition may be more easily understood by thinking about the split-plot portion first. The second section of the ANOVA table of Table 10.3 shows a "corrected total" sum of squares and degrees of freedom as they would be computed for a CBD in $l_1 r$ blocks, with $l_1(l_2-1)$ degrees of freedom used to describe treatment variation,

Table 10.3 ANOVA Format for a Two-Factor SPD(R,B)

Stratum	Source	Degrees of Freedom	Sum of Squares
whole-plot	α	$l_1 - 1$	$\sum_i r l_2 (\bar{y}_{i..} - \bar{y}_{...})^2$
	resid	$l_1(r-1)$	$\sum_{it} l_2 (\bar{y}_{it.} - \bar{y}_{i..})^2$
	corrected total	$l_1 r - 1$	$\sum_{it} l_2 (\bar{y}_{it.} - \bar{y}_{...})^2$
split-plot	blocks	$l_1 r - 1$	$\sum_{it} l_2 (\bar{y}_{it.} - \bar{y}_{...})^2$
	β	$l_2 - 1$	$\sum_j l_1 r (\bar{y}_{..j} - \bar{y}_{...})^2$
	$(\alpha\beta)$	$(l_1 - 1)(l_2 - 1)$	$\sum_{ij} r(\bar{y}_{i.j} - \bar{y}_{i..} - \bar{y}_{..j} + \bar{y}_{...})^2$
	resid	$l_1(r-1)(l_2-1)$	difference
	corrected total	$l_1 r l_2 - 1$	$\sum_{itj} (y_{itj} - \bar{y}_{...})^2$

which in turn is partitioned into $l_2 - 1$ and $(l_1 - 1)(l_2 - 1)$ degree-of-freedom components associated with those treatment contrasts represented by β and $(\alpha\beta)$. Sums of squares and degrees of freedom associated with blocks, treatments, and residual each sum to their corrected total counterparts, allowing the residual values to be computed as differences. The F-statistics appropriate for testing $\text{Hyp}_0 : \beta = 0$ and $\text{Hyp}_0 : (\alpha\beta) = 0$ are formed as the ratios of the corresponding mean squares to the $l_1(r - 1)(l_2 - 1)$ degree-of-freedom split-plot residual mean square.

The experimental units in the whole-plot portion of the experiment are the same physical entities as the blocks in the split-plot portion, and so the ANOVA component for whole-plot corrected total is the same as that for split-plot blocks. This is the basis for the first section of Table 10.3, where the corrected total sum of squares and degrees of freedom are partitioned as they would be in a CRD with l_1 treatment groups of equal size. The whole-plot residual sum of squares represents variability among whole-plot units (or split-plot blocks) within groups assigned to the same level of the factor associated with α. The F-statistic appropriate for testing $\text{Hyp}_0 : \alpha = 0$ is formed as the ratio of the $l_1 - 1$ degree-of-freedom mean square associated with this factor, and the $l_1(r - 1)$ degree-of-freedom residual mean square from this section of the table.

Because we are working with the full-rank factorial parameterization of treatment effects defined in Chapter 9, all linear combinations of the elements of α, β, and $(\alpha\beta)$ are estimable. (Due to the way the $\mathbf{F}$ matrices were constructed, factorial effects all correspond to contrasts in the treatments, and so for example, $\mathbf{c}'\alpha$ is a treatment contrast for *any* $(l_1 - 1)$-element vector $\mathbf{c}$.) Based on

$$\mathbf{y}_2 = l_2 \mu_\zeta \mathbf{1} + l_2(\mathbf{1}_r \times \mathbf{F}^\alpha)\alpha + \epsilon^*$$

the unique least-squares estimate of α is:

$$\hat{\alpha} = l_2^{-2}((\mathbf{1}_r \times \mathbf{F}^\alpha)'(\mathbf{1}_r \times \mathbf{F}^\alpha))^{-1} l_2(\mathbf{1}_r \times \mathbf{F}^\alpha)'\mathbf{y}_2 = \frac{1}{l_1}\mathbf{F}^{\alpha'} \begin{pmatrix} \bar{y}_{1..} \\ \bar{y}_{2..} \\ \ldots \\ \bar{y}_{l_1..} \end{pmatrix},$$

so

$$\widehat{\mathbf{c}'\alpha} = \frac{1}{l_1}\mathbf{c}'\mathbf{F}^{\alpha'} \begin{pmatrix} \bar{y}_{1..} \\ \bar{y}_{2..} \\ \ldots \\ \bar{y}_{l_1..} \end{pmatrix},$$

and

$$Var(\widehat{\mathbf{c}'\alpha}) = \frac{1}{l_1^2}\mathbf{c}'\mathbf{F}^{\alpha'}\mathbf{F}^\alpha\mathbf{c}\left(\frac{1}{r}\sigma_\zeta^2 + \frac{1}{rl_2}\sigma^2\right) = \frac{\mathbf{c}'\mathbf{c}}{N}\left(l_2\sigma_\zeta^2 + \sigma^2\right).$$

The quantity enclosed in parentheses in the last expression is the expectation of the whole-plot residual mean square, and so can be estimated by this mean square in t-based confidence intervals:

$$\widehat{c'\alpha} \pm t_{1-\alpha/2}(l_1(r-1))\sqrt{\frac{c'c}{N}MSE_{wp}}.$$

The intra-block (or split-plot) analysis leads to point estimates of similar form:

$$\hat{\beta} = \frac{1}{l_2}F^{\beta'}\begin{pmatrix}\bar{y}_{..1} \\ \bar{y}_{..2} \\ \dots \\ \bar{y}_{..l_2}\end{pmatrix}$$

$$\widehat{(\alpha\beta)} = \frac{1}{l_1 l_2}(F^{\beta} \times F^{\alpha})'\begin{pmatrix}\bar{y}_{1.1} \\ \bar{y}_{1.2} \\ \dots \\ \bar{y}_{l_1.l_2}\end{pmatrix}.$$

But in this case, the same whole-plot blocks are represented in *all* elements of the vector of averages. Because the rows of $F^{\beta'}$ and $(F^{\beta} \times F^{\alpha})'$ have zero sums, the correlations among these averages cancel out in the variance formulae, leaving:

$$Var(\widehat{c'\beta}) = \frac{c'c}{N}\sigma^2,$$

$$Var(c'\widehat{(\alpha\beta)}) = \frac{c'c}{N}\sigma^2.$$

In this case, confidence intervals can be based on the residual mean square from the split-plot portion of the ANOVA, for which the expectation is σ^2, e.g.,

$$\widehat{c'\beta} \pm t_{1-\alpha/2}(l_1(r-1)(l_2-1))\sqrt{\frac{c'c}{N}MSE_{sp}}.$$

10.3 SPD(B,B)

A two-factor split-plot experiment can also be organized so that both the among-blocks and within-blocks components of the study are executed as CBDs. In our hypothetical chemical plant experiment, suppose that instead of randomly allocating $l_1 r$ days to l_1 levels of the first factor (under the constraint that r days are selected for each level), the experiment is run in r *replicates*, each replicate executed during a different week. Within a given replicate/week, l_1 days are randomly allocated to the l_1 levels of the hard-to-change, among-blocks factor, temperature. As a result, the top stratum of the design is a CBD in which the l_1 treatments (levels of factor 1) are applied once each to the units (days) of each replicate/block (week). The bottom stratum of the design is a CBD as before, with the l_2 treatments (levels of factor 2) applied once each to the units (individual runs) of each block (day). The potential advantage of operating in this way is the improvement in power and precision that might be attained in comparing levels of the whole-plot factor. This is possible because factor 1 contrasts can now be calculated from the data collected within weeks, rather than across all days used in the experiment. We denote a split-plot design organized in this way as a SPD(B,B), indicating that the portion of the experiment executed in each stratum is organized in blocks. As with any CBD, we must be willing to assume that the influence of weeks and factor 1 levels are additive.

10.3.1 A model

For this experimental strategy, we may write an overparameterized model for the data as:

$$y_{tij} = \rho_t + \dot{\alpha}_i + \zeta_{ti} + \dot{\beta}_j + (\dot{\alpha\beta})_{ij} + \epsilon_{tij}$$

$$i = 1, 2, \ldots, l_1; \quad j = 1, 2, \ldots, l_2; \quad t = 1, 2, \ldots, r,$$

where

$$\zeta_{ti} \sim \text{i.i.d.}, \quad E(\zeta_{ti}) = \mu_\zeta, \quad Var(\zeta_{ti}) = \sigma_\zeta^2$$

$$\epsilon_{tij} \sim \text{i.i.d.}, \quad E(\epsilon_{tij}) = 0, \quad Var(\epsilon_{tij}) = \sigma^2.$$

In this model, the influence of the new blocking structure (weeks) is represented by ρ_t, $t = 1, 2, 3, \ldots, r$. This requires that we re-allocate $r - 1$ degrees of freedom that would have been included in the whole-plot residual in a SPD(R,B). But if the random effects associated with the days within a week are small relative to week-to-week variation, and effects due to weeks and to the levels of factor 1 can be assumed to be additive, the $(r-1)(l_1-1)$-degree-of-freedom source of variation remaining among days after accounting for the effects of weeks and factor 1 levels may be smaller than would be the case with a SPD(R,B).

Following the conventions of the last section, a matrix formulation of the full rank model parameterization for the full experiment may be written as

$$\mathbf{y} = \mathbf{W}\rho + \mathbf{X}_1\zeta + \mathbf{X}_\alpha\alpha + \mathbf{X}_\beta\beta + \mathbf{X}_{(\alpha\beta)}(\alpha\beta) + \epsilon$$

where ρ is an r-element vector of fixed-effect block parameters,

$$\mathbf{W} = \mathbf{1}_{l_2} \times \mathbf{I}_{r \times r} \times \mathbf{1}_{l_1}$$

and all other matrices and parameter vectors are as defined in subsection 10.2.1. A little thought should make it clear that intra-block estimators of $\mathbf{c}'\beta$ and $\mathbf{c}'(\alpha\beta)$ are not affected by the addition of whole-plot blocks to the model because the columns of $\mathbf{W}$ (representing *groups* of whole-plot units or split-plot blocks) can be formed as linear combinations of the columns of $\mathbf{X}_1$. As a result, the data transformation $\mathbf{y}_1 = (\mathbf{I} - \mathbf{H}_1)\mathbf{y}$ has already "eliminated" effects associated with the columns of $\mathbf{W}$. The model on which inter-block estimates is based must be modified to reflect the new stratum of blocking with the addition of:

$$\mathbf{X}_1'\mathbf{W}\boldsymbol{\rho} = (\mathbf{1}_{l_2}' \times \mathbf{I}_{r \times r} \times \mathbf{I}_{l_1 \times l_1})(\mathbf{1}_{l_2} \times \mathbf{I}_{r \times r} \times \mathbf{1}_{l_1})\boldsymbol{\rho} = (l_2 \times \mathbf{I}_{r \times r} \times \mathbf{1}_{l_1})\boldsymbol{\rho}$$

leading to:

$$\mathbf{y}_2 = l_2(\mathbf{I}_r \times \mathbf{1}_{l_1})\boldsymbol{\rho} + l_2(\mathbf{1}_r \times \mathbf{F}^\alpha)\boldsymbol{\alpha} + \boldsymbol{\epsilon}^*.$$

However, *this* is just another way of writing a model for a CBD with r blocks, each containing l_1 units distributed among the l_1 levels of the factor associated with $\boldsymbol{\alpha}$. Because CRDs and CBDs that each assign r units to each of l_1 treatments are Condition E-equivalent, the inter-block estimates are also not affected by the addition of whole-plot blocks.

10.3.2 Analysis

While estimates of treatment-related parameters have the same form whether orthogonal blocking is used at the whole-plot stratum or not, the additional structure must be taken into consideration in the ANOVA decomposition. Because there is no pure replication when the whole-plot section is cast as a CBD, the whole-plot-block by α-main-effect interaction (assumed to be absent) becomes the source of information about residual variation, i.e., the denominator mean square for testing $\text{Hyp}_0 : \boldsymbol{\alpha} = \mathbf{0}$ and the variance estimate for calibrating confidence intervals for $\mathbf{c}'\boldsymbol{\alpha}$. The format for an ANOVA decomposition for a SPD(B,B), with r whole-plot blocks, one whole-plot factor, and one split-plot factor, is given in Table 10.4.

Estimates of linear combinations of $\boldsymbol{\alpha}$, $\boldsymbol{\beta}$, and $(\boldsymbol{\alpha\beta})$, and their standard errors, are computed just as in the case outlined for SPD(R,B), with the whole-plot residual mean square used in inference about $\boldsymbol{\alpha}$ and the split-plot residual mean square used in inference about $\boldsymbol{\beta}$ and $(\boldsymbol{\alpha\beta})$. The only adjustment necessary to the formulae in this case is the reduction of degrees of freedom for the whole-plot residual component, reflecting the fitting of nuisance parameters corresponding to whole-plot blocks.

Table 10.4 ANOVA Format for a Two-Factor SPD(B,B)

Stratum	Source	Degrees of Freedom	Sum of Squares
whole-plot	blocks	$r - 1$	$\sum_r l_1 l_2 (\bar{y}_{t..} - \bar{y}_{...})^2$
	α	$l_1 - 1$	$\sum_i r l_2 (\bar{y}_{.i.} - \bar{y}_{...})^2$
	resid	$(r-1)(l_1-1)$	difference
	corrected total	$r l_1 - 1$	$\sum_{it} l_2 (\bar{y}_{ti.} - \bar{y}_{...})^2$
split-plot	blocks	$r l_1 - 1$	$\sum_{it} l_2 (\bar{y}_{ti.} - \bar{y}_{...})^2$
	β	$l_2 - 1$	$\sum_j r l_1 (\bar{y}_{..j} - \bar{y}_{...})^2$
	$(\alpha\beta)$	$(l_1-1)(l_2-1)$	$\sum_{ij} r (\bar{y}_{.ij} - \bar{y}_{.i.} - \bar{y}_{..j} + \bar{y}_{...})^2$
	resid	$(r-1)l_1(l_2-1)$	difference
	corrected total	$r l_1 l_2 - 1$	$\sum_{tij} (y_{tij} - \bar{y}_{...})^2$

10.4 More than two experimental factors

To this point, our focus has been on explaining how split-plot designs are organized when one factor is applied at each of two strata of experimental material, referred to as "plots" and "units." However, more than one factor may be applied at either or both strata in experiments designed to investigate more than two factors. For example, in the fabric experiment of Rong et al. (2005) (subsection 10.1.1), temperature is the single whole-plot factor, while binder fiber and binder content are both applied in the split-plot stratum. The ANOVA decomposition of this experiment would be organized so as to compare the temperature main effect to whole-plot residual (groups of six runs within temperature levels), and all other factorial effects to split-plot residual (associated with individual test specimens). It should be fairly obvious why binder fiber and binder content main effect are tested in the split-plot portion of the table, since these factors are exercised "within blocks." Writing out the contrasts associated with any interaction will reveal that they are also contrasts *within blocks*, and so should be evaluated against the same measure of noise as the binder fiber and binder content main effects.

Suppose the fabric experiment had actually been organized so that both temperature and binder fiber had been held constant within each random block, and only binder content varied within blocks. Maintaining the number of levels of each factor, this means that only three specimens would appear in each block – one assigned to each binder content level – and eight kinds of blocks would be present in the design, one kind for each combination of temperature and binder fiber levels. In this case, the temperature and binder fiber main effects *and* the temperature-by-binder fiber interaction would be analyzed as whole-plot effects (i.e., compared to whole-plot residual variation) because any data contrast associated with these effects is constant within blocks, and its uncertainty is reflected by the block-within-temperature-and-binder fiber variation. The factorial effects estimated with split-plot precision (or tested against split-plot residual variation) would be the binder content main effect, and all interactions involving this factor.

The general rule for determining the organization of a split-plot ANOVA is that a factorial effect appears in the whole-plot section if *all* involved factors are applied at the whole-plot stratum (i.e., change levels only between random blocks), and appears in the split-plot section if *any* involved factor is applied at the split-plot stratum. Table 10.5 offers a final demonstration of this decomposition of variance for a four-factor experiment in which the first two factors (A and B) are applied to whole-plots, and the second two (C and D) are applied to split-plots.

10.5 More than two strata of experimental units

Split-plot designs can often be identified easily because the experimental material is organized in a *nested* structure, and different factors or groups of factors are applied at different strata of that structure. The designs we have

Table 10.5 ANOVA Format for a Four-Factor SPD(R,B) with Two Factors Applied in Each Stratum*

Stratum	Source	Degrees of Freedom	Sum of Squares
whole-plot	α	$l_1 - 1$	$\sum_i l_2 r l_3 l_4 (\bar{y}_{i....} - \bar{y}_{.....})^2$
	β	$l_2 - 1$	$\sum_j l_1 r l_3 l_4 (\bar{y}_{.j...} - \bar{y}_{.....})^2$
	$(\alpha\beta)$	$(l_1 - 1)(l_2 - 1)$	difference
	resid	$l_1 l_2 (r - 1)$	$\sum_{ijt} l_3 l_4 (\bar{y}_{ijt..} - \bar{y}_{ij...})^2$
	corrected total	$l_1 l_2 r - 1$	$\sum_{ijt} l_3 l_4 (\bar{y}_{ijt..} - \bar{y}_{.....})^2$
split-plot	blocks	$l_1 l_2 r - 1$	$\sum_{ijt} l_3 l_4 (\bar{y}_{ijt..} - \bar{y}_{.....})^2$
	γ	$l_3 - 1$	$\sum_u l_1 l_2 r l_4 (\bar{y}_{...u.} - \bar{y}_{.....})^2$
	δ	$l_4 - 1$	$\sum_u l_1 l_2 r l_3 (\bar{y}_{....v} - \bar{y}_{.....})^2$
	all interactions except $(\alpha\beta)$	$l_1 l_2 l_3 l_4 - l_1 l_2 - l_3 - l_4 + 3$	ignore blocking structure
	resid	$l_1 l_2 l_3 l_4 (r - 1) - l_1 l_2 (r - 1)$	difference
	corrected total	$l_1 l_2 r l_3 l_4 - 1$	$\sum_{ijtuv} (y_{ijtuv} - \bar{y}_{.....})^2$

* Data notation is y_{ijtuv} where $i = 1 \ldots l_1$ denotes the level of factor A, $j = 1 \ldots l_2$ denotes the level of factor B, $t = 1 \ldots r$ denotes the plot assigned to a specified level of factors A and B, $u = 1 \ldots l_3$ denotes the level of factor C, and $v = 1 \ldots l_4$ denotes the level of factor D.

considered so far have employed two strata of experimental material, sometimes referenced as plots and units. In the fabric example of Rong et al. (2005), the hierarchy of experimental material was individual samples (or units) to which binder fiber and binder content were applied, and temporally sequential groups of six samples (or blocks) to which temperature was applied.

In some experiments, the available experimental material has a more extensive hierarchical structure involving more than two strata. As an example, consider a hypothetical alternative form of the concrete experiment of Soudki et al. (2001) presented in subsection 9.1.1. Suppose that it is most convenient to mix coarse and fine aggregate together in quantities of 2 cubic yards. The bulk aggregate can then be divided into four subquantities, each of 0.5-cubic yards, and these randomly assigned to the four levels of total aggregate/cement ratio (i.e., a different quantity of cement added to each). Finally, the 0.5 cubic-yard units of aggregate with added concrete are subdivided into three equal quantities, and each randomly assigned a level of the water/cement ratio, and samples produced and tested as described in subsection 9.1.1.

Note that the factorial treatment structure in this hypothetical experiment is exactly as described in subsection 9.1.1, but the process of producing experimental units now results in a three-stratum nested hierarchy – preparations of coarse/fine aggregate, subdivided into preparations of aggregate/concrete, subdivided into the final aggregate/concrete/water samples. Hence the significance of the aggregate ratio factor should be assessed relative to variation among the largest batches, while aggregate/cement differences should be compared to variation among the intermediate batches, and variation among final samples is the noise against which the differences among levels of the water/cement ratio should be compared.

The "general rule" cited in the last section for determining the organization of a two-stratum split-plot ANOVA can be easily generalized to an arbitrary number of hierarchical strata of blocking. A factorial effect appears in the section of the table associated with the lowest stratum (often the the smallest experimental unit) to which any involved factor is applied. Hence in a three-stratum system of units, if factor A is applied at the top (largest) stratum, and factors B and C are applied at the intermediate stratum, then the associated three-factor interaction is tested in the intermediate section of the table. If factor D is applied at the bottom (smallest) stratum of experimental unit, the associated main effect and any interaction involving factor D is tested at the bottom section of the table. Table 10.6 offers a demonstration of this decomposition of variance for a four-factor SP(R,B,B) experiment arranged in this way.

10.6 Conclusion

Split-plot designs are factorial plans in which a hierarchy of nested experimental units is used, and the factors are not all applied at the same stratum of units. This occurs as the natural consequence of operational constraints in a number of contexts, e.g.,

- where the levels of one factor are easy to change quickly, but changes in the levels of another are much more difficult or time-consuming,

- where treatments are applied to individuals, and this is most easily done by assigning any given individual to only one level of one factor, but measurements representing all levels of another factor are collected on each individual, or

- where the physical material constituting an experimental unit is sequentially divided or subsampled as time progresses, and the levels of different factors are most easily applied to the experimental material at different stages of this process.

Because the random noise in an experiment is largely associated with the differences among experimental units, the "signal-to-noise" comparisons made in split-plot studies must be made with multiple components of residual variation. The inferences that can be drawn concerning main effects associated with among-plots factors are of generally lower quality (power or precision) than those associated with within-plots factors, or the interactions between among- and within-plots factors, because the benchmark for the former is variation associated with plots while that for the latter is associated with units-within-plots.

Split-plot experiments can be organized using more than two strata of experimental units and/or more than two treatment factors. In each case, so long as the design remains balanced and the units at each hierarchical stratum receive

Table 10.6 ANOVA Format for a Four-Factor SPD(R,B,B) with One Factor Applied in Each of the Top and Bottom Strata, and Two Factors Applied in the Intermediate Stratum*

Stratum	Source	Degrees of Freedom	Sum of Squares
largest plots	α	l_1-1	$\sum_i r_1 l_2 l_3 r_2 l_4 (\bar{y}_{i.....} - \bar{y}_{......})^2$
	resid	$l_1(r_1-1)$	$\sum_{is} l_2 l_3 r_2 l_4 (\bar{y}_{is....} - \bar{y}_{i.....})^2$
	corrected total	$l_1 r_1 - 1$	$\sum_{is} l_2 l_3 r_2 l_4 (\bar{y}_{is....} - \bar{y}_{......})^2$
intermediate plots	blocks	$l_1 r_1 - 1$	$\sum_{is} l_2 l_3 r_2 l_4 (\bar{y}_{is....} - \bar{y}_{......})^2$
	β	l_2-1	$\sum_j l_1 r_1 l_3 r_2 l_4 (\bar{y}_{.j...} - \bar{y}_{......})^2$
	γ	l_3-1	$\sum_u l_1 r_1 l_2 r_2 l_4 (\bar{y}_{..u..} - \bar{y}_{......})^2$
	$(\beta\gamma)$	$(l_2-1)(l_3-1)$	$\sum_{ju} l_1 r_1 r_2 l_4 (\bar{y}_{.ju..} - \bar{y}_{.j...} - \bar{y}_{..u..} + \bar{y}_{......})^2$
	$(\alpha\beta)$	$(l_1-1)(l_2-1)$	$\sum_{ij} l_3 r_2 l_4 (\bar{y}_{ij.j.} - \bar{y}_{i.....} - \bar{y}_{.j...} + \bar{y}_{......})^2$
	$(\alpha\gamma)$	$(l_1-1)(l_3-1)$	$\sum_{iu} r_1 l_2 r_2 l_4 (\bar{y}_{i.u..} - \bar{y}_{i.....} - \bar{y}_{..u..} + \bar{y}_{......})^2$
	$(\alpha\beta\gamma)$	$(l_1-1)(l_2-1)(l_3-1)$	
	resid	$l_1 l_2 l_3 (r_1 r_2 - 1) - l_1(r_1-1)$	difference
	corrected total	$l_1 r_1 l_2 l_3 r_2 - 1$	$\sum_{isjut} l_4 (\bar{y}_{isjut.} - \bar{y}_{isju.})^2$
smallest plots	blocks	$l_1 r_1 l_2 l_3 r_2 - 1$	$\sum_{isjut} l_4 (\bar{y}_{isjut.} - \bar{y}_{isju.})^2$
	δ	$l_4 - 1$	$\sum_v l_1 r_1 l_2 l_3 r_2 (\bar{y}_{.....v} - \bar{y}_{......})^2$
	interactions involving δ	$(l_4-1)(l_1 l_2 l_3 - 1)$	ignore blocking structure
	resid	$l_1 l_2 l_3 l_4 (r_1 r_2 - 1) - l_1 l_2 l_3 (r_1 r_2 - 1)$	difference
	corrected total	$l_1 r_1 l_2 l_3 r_2 l_4 - 1$	$\sum_{isjutv} (y_{isjutv} - \bar{y}_{......})^2$

* Data notation is y_{isjutv} where $i = 1 \ldots l_1$ denotes the level of factor A, $s = 1 \ldots r_1$ denotes the (highest-stratum) plot assigned to a specified level of factor A, $j = 1 \ldots l_2$ denotes the level of factor B, $u = 1 \ldots l_3$ denotes the level of factor C, $t = 1 \ldots r_2$ denotes the (intermediate-stratum) plot assigned to a specified level of factors B and C, and $v = 1 \ldots l_4$ denotes the level of factor D.

only one level of some factors and all levels of the others, estimable functions and ANOVA components associated with factorial effects are computed as in a CRD. The critical difference is that hypothesis tests and confidence intervals for each factorial effect must be constructed using the appropriate residual mean square. Generally, this is identified as:

- for main effects, the residual mean square corresponding to the units to which the associated factor levels were applied, and

- for interactions, the residual mean square corresponding to the smallest units (those resulting in the smallest number of data values) to which the levels of any associated factor were applied.

As with all controlled experiments, randomization is an important part of the execution of a split-plot design. And, as with all other blocked designs, an appropriate randomization procedure gives equal probability to all possible unit-to-treatment assignments that are consistent with the constraints required by the blocking structure.

10.7 Exercises

1. The data reported in subsection 10.1.1 for the fiber experiment of Rong et al. (2005) are summary averages for responses for each of the $4 \times 2 \times 3 = 24$ treatments. Suppose that:

   - For each temperature, three "batches" of runs were made, where each batch contains one sample of fabric made for each of the six possible combinations of binder fiber and binder content.

   - The between-batch (within-temperature) sum of squares is 0.300.

   - The residual sum of squares for a model accounting for all experimental factors (main effects and all interactions) and batch-within-temperature is 3.000.

   Based on this information, and assuming that effects associated with batches can be regarded as random, construct a complete ANOVA table for the experiment, including F-statistics for testing each group of main effects and interactions in the model.

2. The following table contains data collected from a factorial experiment involving a three-level factor A, and a four-level factor B. Fifteen batches of experimental material were drawn randomly from those available, and were split into three groups of equal size. The first level of factor A was applied to each batch in the first group, the second level to the second group, and the third level to the third group. Then, each batch was divided into four separate sub-batches; the sub-batches were numbered 1-4, and each was treated with the corresponding level of factor B.

   (a) Using a statistical analysis program, fit a fixed-effects model with an intercept and terms for factor A, batch-within-A, factor B, and A-by-B interaction ONLY.

(b) From the five sums-of-squares (including the residual) reported in part (a), construct a split-plot ANOVA decomposition, with F-statistics for A and B main effects, and the A-by-B interaction.

(c) What is unusual about the numbers in your ANOVA decomposition, and what assumption should this lead you to question?

(d) Refit a fixed-effects model with terms for only A and B main effects, and the A-by-B interaction (e.g., no batch this time), and calculate the 60 residual values. How can you plot these residuals to further check the assumption in question? (Hint: It may be easier/clearer to do this using three plots.)

Batch	A	B	Response	Batch	A	B	Response	Batch	A	B	Response
1	1	1	40.98	6	2	1	36.50	11	3	1	50.89
1	1	2	38.37	6	2	2	47.10	11	3	2	50.07
1	1	3	47.48	6	2	3	51.21	11	3	3	58.49
1	1	4	48.36	6	2	4	62.11	11	3	4	58.56
2	1	1	42.00	7	2	1	36.92	12	3	1	53.70
2	1	2	39.16	7	2	2	46.87	12	3	2	49.66
2	1	3	52.40	7	2	3	52.09	12	3	3	61.02
2	1	4	46.93	7	2	4	60.75	12	3	4	60.37
3	1	1	49.11	8	2	1	44.43	13	3	1	53.49
3	1	2	47.31	8	2	2	56.74	13	3	2	50.08
3	1	3	45.89	8	2	3	46.38	13	3	3	60.93
3	1	4	41.92	8	2	4	58.70	13	3	4	59.61
4	1	1	43.04	9	2	1	42.97	14	3	1	51.37
4	1	2	41.26	9	2	2	54.95	14	3	2	49.37
4	1	3	50.34	9	2	3	45.93	14	3	3	59.51
4	1	4	46.50	9	2	4	57.17	14	3	4	57.57
5	1	1	36.59	10	2	1	45.51	15	3	1	51.12
5	1	2	32.63	10	2	2	54.93	15	3	2	51.38
5	1	3	57.48	10	2	3	48.88	15	3	3	58.95
5	1	4	52.25	10	2	4	57.83	15	3	4	57.58

3. A consumer products research firm carries out tests of roofing material by installing it on rural buildings throughout the country and measuring wear and other qualities after five years of exposure. In one test, they compare five kinds of roofing shingles by installing sections of all five kinds on each building used. Because climate has an important effect on roofing wear, they test product in southern Arizona, western Massachusetts, and northern Minnesota. Four buildings are used in each region (for a total of 12), and some differences in roofing durability can be expected because of building differences (e.g., ventilation, roof pitch, et cetera). The investigators regard *climate* and *shingle type* as the two factors of interest.

(a) Carefully define blocks, units, and treatments for this study. Although studies like this are often analyzed as though they are experiments, why is this study *not* an experiment in the strict sense?

(b) Explain in physical (not mathematical) terms why the random "noise" associated with climate differences might be different from that associated with single-type differences.

(c) Calculate degrees of freedom for an ANOVA decomposition for this study.

(d) Suppose this experiment were analyzed as CRD. What hypothesis tests would be affected by this decision, and what are the likely consequences?

4. In a clinical trial organized to compare three pain medications, 48 patients who suffer from chronic pain were recruited. The patients were randomly divided into two groups (A and B) of equal size. Over a 4-week period, each patient received each of the drugs for one week, and was treated with no drug (a control condition) for one week. The order of the four medical treatments was independently randomized for each patient. The patients in group A were given the drugs on a rigid schedule, while those in group B took the drugs "as needed." Experimental interest centered on understanding the effects of the drug regimens defined by the four levels of *pain medication* and the two levels of *dosing*.

(a) Carefully define blocks, units, and treatments for this experiment.

(b) Explain in physical (not mathematical) terms why the random "noise" associated with dosing differences might be different from that associated with medication differences.

(c) Calculate degrees of freedom for an ANOVA decomposition for this experiment.

(d) Suppose this experiment were analyzed as CRD. What hypothesis tests would be affected by this decision, and what are the likely consequences?

(e) Ignoring blocking and units (i.e., thinking only about treatments), what is unusual about this 2×4 factorial treatment structure; are there simplifications in the model that might be reasonable because of this?

5. The production of semiconductors involves many steps (acid washes, etchings, coatings, et cetera) that are performed in a particular sequence. At the beginning of the sequence, the raw material is a relatively large "wafer" of silicon, which is subdivided into smaller and smaller pieces at specified points in the process, with the final (smallest) pieces being the individual semiconductors. Suppose the research division of a semiconductor manufacturing company planned an experiment to compare different "versions" of a process for making a particular kind of semiconductor, where a version is defined by the selection of one of three acids applied to entire wafers (at the beginning of the process), one of three etching techniques applied at an interim point, and one of three coatings applied to individual semiconductors (at the end of the process). Hence the treatment structure of

interest is a 3^3 factorial arrangement. Suppose the study is actually carried out by:

- selecting 15 wafers from production, randomly dividing them into three groups of five, and treating all wafers in one group with one of the acids,

- from each acid-treated wafer, selecting 15 interim-size pieces, randomly dividing them into three groups of five, and treating all pieces in one group with one of the etching techniques, and

- from each acid-and-etching-treated interim pieces, selecting 15 individual semiconductors, randomly dividing them into three groups of five, and treating all semiconductors in one group with one of the coatings.

Hence the final data set will have measurements of 3375 semiconductors!

(a) Carefully define blocks, units, and treatments for this experiment.

(b) Explain in physical (not mathematical) terms why the random "noise" associated with acid differences might be different from that associated with etching technique differences, and why both might be different from that associated with coating differences.

(c) Calculate degrees of freedom for an ANOVA decomposition for this experiment.

(d) Suppose this experiment were analyzed as CRD. What hypothesis tests would be affected by this decision, and what are the likely consequences?

6. In a popular classroom demonstration of factorial experiments, students make paper helicopters that are variants of a standard design, differing in the length of "tail" and the number of paperclips attached at the bottom for ballast. A simple version of this experiment is conducted as a CRD. Some number of helicopters are made according to each of all possible "recipes" associated with the two factors, and data are collected on "flight times" for each helicopter under standard conditions. Consider an extension of this experiment in which four helicopters are made according to each of the four "recipes," defined by "short" or "long" tails, and one or two paperclips. Now suppose each helicopter is "flown" four times under standard conditions *except that* the four flights are made under four sets of "environmental conditions" defined by two additional two-level factors; a fan is set up 10 feet away from the "drop zone" and is either running or not, and flights are launched with the room lights on or off. Using the 64 data values collected, interest is in comparing the treatments defined by the four two-level factors.

(a) Carefully define blocks, units, and treatments for this experiment.

(b) Calculate degrees of freedom for an ANOVA decomposition for this experiment.

   (c) Suppose this experiment were analyzed as CRD. What hypothesis tests would be affected by this decision, and what are the likely consequences?

7. "Split-unit" designs can be constructed with more complicated block structures than those described in this chapter. Consider a situation based on a replicated Latin Square design. Suppose the levels of factor A (with l_1 levels) are applied to *entire Latin Squares*; we'll say that r/l_1 must be an integer so that each level of A is applied to the same number of squares. Individual $l_2 \times l_2$ Latin Squares are set up to compare the levels of a second factor B (with l_2 levels), while simultaneously controlling for two kinds of blocks. Suppose the row-blocks and column-blocks are different physical entities in each Latin Square.

   (a) Calculate degrees of freedom for an ANOVA decomposition for this experiment.

   (b) Suppose this experiment were analyzed as CRD. What hypothesis tests would be affected by this decision, and what are the likely consequences?

8. In repeated measures studies, the "within subjects" (or split-plot) factor is often time, while the "among subjects" (or whole-plot) factor(s) represent treatments applied at an initial, or reference, timepoint. Suppose the among subjects factor of such a study has two levels, say two competing drugs, and that r subjects receive each drug. Responses, say the blood concentration of the drug, are then measured in each subject at $1, 2, 3, \ldots, 8$ hours after administration. So the treatment structure is a 2 (drug) by 8 (timepoint) factorial. Letting y_{ijk} represent the drug concentration in the jth subject given drug i at the kth timepoint, interest might center on estimates of the eight quantities:

$$\theta_k = E(\bar{y}_{1.k} - \bar{y}_{2.k}), \quad k = 1, 2, 3, \ldots, 8$$

and tests of the eight hypotheses:

$$\text{Hyp}_{0k} : \theta_k = 0, \quad k = 1, 2, 3, \ldots, 8.$$

If the variance of measurements for a single subject is σ^2, and the variance of subject effects is σ_δ^2, write:

   (a) The formula for the expected squared length of a 95% two-sided confidence interval for θ_k.

   (b) The parameters of the distribution of an F-statistic used to test Hyp_{0k}.

Two-level factorial experiments: basics

11.1 Introduction

In many applications, experiments designed to examine the effects of two-level factors are especially common; Mee (2009) is an excellent reference focused entirely on designs for these situations. Because the minimum meaningful number of levels a factor can take is two, restricting factors to two levels minimizes the number of treatments that must be considered for a given number of factors. Conversely, in preliminary or screening contexts, representing each factor at only two levels maximizes the number of factors that can be examined in a factorial experiment of a given number of included treatments. Of course, the nature of the application-specific questions is the most important issue to be considered in selecting the number of levels for each factor. For example, if one experimental goal is to learn about the relative merits of three different catalysts in a chemical process, the catalyst factor cannot be restricted to two levels without compromising some of the experimental goals. But two-level factorial experiments do provide an efficient means of collecting useful information in a wide variety of applied contexts, and this justifies our in-depth treatment of this special case.

In this context, each of the 2^f treatments can be completely identified by an ordered string or vector of f binary "bits," each symbolized by (0,1), (1,2), $(-,+)$ or ("low","high"), each designating the level to be assigned to the corresponding factor. As in previous chapters, our discussion here will allow the two levels of each factor to be either qualitative (e.g., choice of catalyst 1 or catalyst 2), or quantitative (e.g., 20 mg of additive or 50 mg of additive in a standard solution). Following a standard convention, we will often use "low" and "high" as arbitrary labels here; where the levels of a factor are values on an ordinal, interval, or ratio scale, it will be natural to let "high" refer to the larger one, but no order is necessarily implied so long as the level labels are used consistently throughout the design and analysis of an experiment.

11.2 Example: bacteria and nuclease

Jepsen, Riise, Biedermann, Kristensen, and Emborg (1987) performed a laboratory experiment to investigate the influence of three factors on the amount of nuclease produced in cultures of bacteria. For each strain investigated, two replicate flasks of culture were prepared at each of eight experimental

Table 11.1 Nuclease Concentration for *Serratia Marcescens* W280, Jepsen et al. (1987)

Initial pH	Temp. (°C)	Flask Baffles	Nuclease (U/ml)	
7.4	30	0	340	310
7.4	37	0	70	80
8.2	30	0	700	440
8.2	37	0	260	200
7.4	30	2	770	-
7.4	37	2	130	70
8.2	30	2	1270	1380
8.2	37	2	470	570

conditions determined by initial pH (7.4 or 8.2), temperature (30 °C or 37 °C), and the number of baffles in the flask (0 or 2). The culture containers used were "shake flasks," and the baffles controlled the amount of aeration that took place in each culture. Hence sixteen flasks (two replicates for each of eight experimental conditions) were prepared for each strain examined. The response variables measured on each flask at the end of a standard protocol were units of nuclease produced, and number of viable cells produced, each per ml of culture. Table 11.1 contains the values of nuclease produced in the experiment performed using the bacterium strain *Serratia marcescens* W280. Note that for this strain, only one value was reported for the treatment defined by pH 7.4, 30 °C, and two baffles, leading to a mild imbalance in the design and $N = 15$.

11.3 Two-level factorial structure

Two-level experimental designs are often depicted "spatially" by representing treatments as the corners of a square ($f = 2$), cube ($f = 3$), or hyper-cube ($f > 3$), where each spatial dimension is associated with a factor. Figure 11.1 presents such a representation for the cell biology experiment just described.

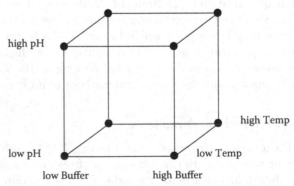

Figure 11.1 Geometric representation of a 2^3 factorial experiment.

Following the notation of Chapter 9, we may define a cell means model for an equally replicated, two-level, three-factor experiment as:

$$y_{ijkt} = \mu_{ijk} + \epsilon_{ijkt} \qquad t = 1, \ldots, r,$$

where each of i, j, and k take values of 1 or 2 to indicate the level of the corresponding factor, and t indexes the within-treatment observation. An equivalent factorial effects model of overparameterized form (Section 9.2) can be written as:

$$y_{ijkt} = \mu + \dot{\alpha}_i + \dot{\beta}_j + \dot{\gamma}_k + (\dot{\alpha\beta})_{ij} + (\dot{\alpha\gamma})_{ik} + (\dot{\beta\gamma})_{jk} + (\dot{\alpha\beta\gamma})_{ijk} + \epsilon_{ijkt}$$

where the four indices take the same values and have the same meanings as in the cell means model. Writing this model in matrix notation, each factorial effect would be associated with a model matrix column containing values of an indicator variable (0's and 1's). The corresponding parameter is interpreted as the additive increment in expected response for conditions in which that indicator is "1", compared to conditions in which it is "0", after considering all effects of lower order. For example, Figure 11.2 graphically identifies the treatments (model matrix rows) that are coded with 0's and 1's for the main effect associated with the "low" level of pH, $\dot{\alpha}_1$, the two-factor interaction associated with the "low" levels of pH and temperature, $(\dot{\alpha\beta})_{11}$, and the three-level interaction associated with the "low" levels of pH, temperature, $(\dot{\alpha\beta\gamma})_{111}$, and number of flask baffles in the cell biology experiment.

For analysis purposes, a more convenient parameterization for two-level factorials is a special case of the full-rank form introduced in Section 9.3. Recall that in the general case, the number of parameters required to express a main effect is one less than the number of levels associated with that factor, and that an interaction for any collection of factors is the product of numbers, each one less than the number of levels for one of the involved factors, e.g., for the interaction associated with the first three factors, $(l_1 - 1)(l_2 - 1)(l_3 - 1)$. Where all factors are represented at two levels, this means that each main effect and interaction (of any order) can be represented by a single parameter. The only contrast needed in defining the model for this parameterization is $\mathbf{F'} = (-1, +1)$, leading to an effects model of full rank that can be easily defined in the following way.

Let $(-)_{121\ldots2}$ represent any qth order factorial effect from the overparameterized factorial effects model, where the q subscripts identify the levels of the q associated factors. Denote by $(-)$ a new qth order effect that is identified only by q factors, not their levels. Then a full-rank effects model can be written by replacing each $(-)_{121\ldots2}$ in the overparameterized model by $(-) \times (-1)^t$, where t is the number of 1's in subscript string. That is, the sign of $(-)$ is reversed for treatments in which an odd number of associated factors are set at their respective "low" levels, relative to the value it takes when an even number (including 0) of factors are set to "low" levels. For example, in

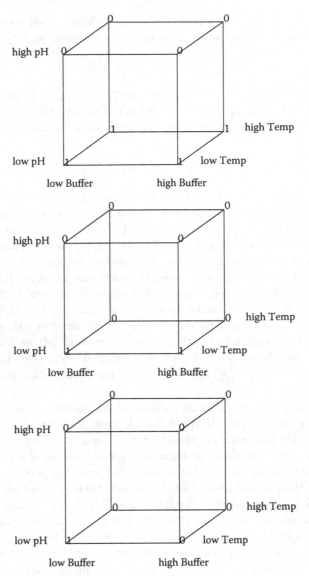

Figure 11.2 Geometric representation of model matrix column entries corresponding to $\dot\alpha_1$, $(\alpha\beta)_{11}$, and $(\alpha\beta\gamma)_{111}$.

a 2^3 factorial experiment, the cell mean μ_{121} can be written as either:

$$\mu_{121} = \mu + \dot\alpha_1 + \dot\beta_2 + \dot\gamma_1 + (\dot{\alpha\beta})_{12} + (\dot{\alpha\gamma})_{11} + (\dot{\beta\gamma})_{21} + (\alpha\beta\gamma)_{121}$$

using the overparameterized effects model, or

$$\mu_{121} = \mu - \alpha + \beta - \gamma - (\alpha\beta) + (\alpha\gamma) - (\beta\gamma) + (\alpha\beta\gamma)$$

using the full-rank parameterization just described. The full-rank parameterization requires a total of 2^f parameters, the same as the number of cell means. Furthermore, all parameters are used in the representation of any cell mean; only the plus and minus signs differ to accommodate the specific treatment. Finally, each parameter (except for μ) in this representation corresponds to a *contrast* in all 2^f cell means, where the coefficients for half (2^{f-1}) of the cell means are -2^{-f} and the other half are $+2^{-f}$. These contrasts are displayed graphically in Figure 11.3 for α, $(\alpha\beta)$, and $(\alpha\beta\gamma)$ from a three-factor model.

The necessary pluses and minuses needed to complete the model in this parameterization can be represented by defining an "independent variable" for each factor – essentially a recoding of the subscript identifying the level of that factor. For $f = 3$ factors, define:

$$x_1 = \begin{cases} -1 & i = 1 \\ +1 & i = 2 \end{cases} \quad x_2 = \begin{cases} -1 & j = 1 \\ +1 & j = 2 \end{cases} \quad x_3 = \begin{cases} -1 & k = 1 \\ +1 & k = 2 \end{cases}.$$

Then any cell mean can be expressed as:

$$\mu_{ijk} = \mu + x_1\alpha + x_2\beta + x_3\gamma + x_1x_2(\alpha\beta) + x_1x_3(\alpha\gamma) + x_2x_3(\beta\gamma) + x_1x_2x_3(\alpha\beta\gamma).$$

Hence we have an expression in the form of a regression model with x's that are always either -1 or $+1$. In matrix form, the data model for the complete, unreplicated 2^3 factorial design (with obvious extension to the general 2^f case) can be written as:

$$\mathbf{y} = \begin{pmatrix} y_{111} \\ y_{112} \\ y_{121} \\ y_{122} \\ y_{211} \\ y_{212} \\ y_{221} \\ y_{222} \end{pmatrix} = \begin{pmatrix} + & - & - & - & + & + & + & - \\ + & - & - & + & + & - & - & + \\ + & - & + & - & - & + & - & + \\ + & - & + & + & - & - & + & - \\ + & + & - & - & - & - & + & + \\ + & + & - & + & - & + & - & - \\ + & + & + & - & + & - & - & - \\ + & + & + & + & + & + & + & + \end{pmatrix} \begin{pmatrix} \mu \\ \alpha \\ \beta \\ \gamma \\ (\alpha\beta) \\ (\alpha\gamma) \\ (\beta\gamma) \\ (\alpha\beta\gamma) \end{pmatrix} + \epsilon. \quad (11.1)$$

When $r > 1$, the unique rows of $\mathbf{X}$ are as given above, but each appears r times in the matrix. The number of parameters is 2^f because each is specified by the presence or absence of each of f symbols $(\alpha, \beta, \dots)$.

Note that, apart from a factor of $\sqrt{N}$, $\mathbf{X}$ is an orthogonal matrix:

$$\mathbf{X}'\mathbf{X} = r2^f\mathbf{I}_{2^f \times 2^f} = N\mathbf{I}_{2^f \times 2^f}$$

so

$$(\mathbf{X}'\mathbf{X})^{-1} = \frac{1}{N}\mathbf{I}, \quad \hat{\theta} = \frac{1}{N}\mathbf{X}'\mathbf{y}.$$

If $r \geq 1$

$$\hat{\theta} = 2^{-f}\mathbf{M}'\bar{\mathbf{y}}$$

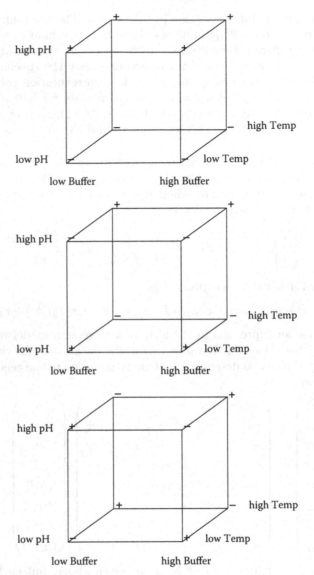

Figure 11.3 Geometric representation of model matrix column entries corresponding to α, $(\alpha\beta)$, and $(\alpha\beta\gamma)$.

where $\mathbf{M}$ is the $2^f \times 2^f$ model matrix for an unreplicated design, $\bar{\mathbf{y}}$ is the vector of treatment-specific averages, and $\boldsymbol{\theta}$ is the parameter vector in model (11.1). *Any* linear combination of elements of $\boldsymbol{\theta}$ is estimable because $\mathbf{M}$ is square and of full rank. (That is, any $\mathbf{c}$ of dimension 2^f can be expressed as a linear combination of the rows of $\mathbf{M}$.) But we generally don't regard anything involving the first element as meaningful since it includes experiment-specific effects.

Note that model (11.1) is written in unpartitioned notation; $\boldsymbol{\theta}$ actually contains one nuisance parameter, μ. We could partition the model as in previous chapters, to emphasize which parameters are experimentally meaningful and which are not:

$$\mathbf{y} = \mathbf{1}\mu + \mathbf{X}_2\boldsymbol{\phi} + \boldsymbol{\epsilon},$$

but since all columns of $\mathbf{X}_2$ have zero sums,

$$\mathbf{H}_1\mathbf{X}_2 = \mathbf{0}, \quad \mathbf{X}_{2|1} = \mathbf{X}_2$$

and this means that "correction for the mean" or (thinking of this form as a regression model) intercept is automatic in this parameterization. As a result,

$$\mathbf{X}_{2|1}'\mathbf{X}_{2|1} = r2^f \mathbf{I}_{(2^f-1)\times(2^f-1)} = N\mathbf{I}_{(2^f-1)\times(2^f-1)}$$

so

$$\mathcal{I}^{-1} = \frac{1}{N}\mathbf{I}, \quad \hat{\boldsymbol{\phi}} = \frac{1}{N}\mathbf{X}_2'\mathbf{y}.$$

11.4 Estimation of treatment contrasts

11.4.1 Full model

Suppose we order both the treatment mean vector and the corresponding data average vector lexicographically by their indexes so that:

$$\boldsymbol{\mu}' = (\mu_{1\ldots11}, \mu_{1\ldots12}, \mu_{1\ldots21}, \mu_{1\ldots22}, \ldots, \mu_{2\ldots22})$$

and $E(\bar{\mathbf{y}}) = \boldsymbol{\mu}$. By our definition of factorial effects, $\boldsymbol{\mu} = \mathbf{M}\boldsymbol{\theta}$. Because $\mathbf{M}$ is square and of full rank, there is a one-to-one relationship between $\boldsymbol{\mu}$ and $\boldsymbol{\theta}$; we can discuss estimation of these vectors interchangeably depending on whether we want to focus on the factorial effects (elements of $\boldsymbol{\theta}$) or more specific functions of cell means (elements of $\boldsymbol{\mu}$). So:

$$\widehat{\mathbf{c}'\boldsymbol{\mu}} = \mathbf{c}'\widehat{\mathbf{M}\boldsymbol{\theta}} = \mathbf{c}'\mathbf{M}\hat{\boldsymbol{\theta}} = \mathbf{c}'\mathbf{M}2^{-f}\mathbf{M}'\bar{\mathbf{y}} = \mathbf{c}'\bar{\mathbf{y}}$$

with the last equation holding because $\mathbf{M}\mathbf{M}' = 2^f\mathbf{I}$. As an immediate result, $E[\widehat{\mathbf{c}'\boldsymbol{\mu}}] = \mathbf{c}'\boldsymbol{\mu}$, and $Var[\widehat{\mathbf{c}'\boldsymbol{\mu}}] = (\sigma^2/r)\mathbf{c}'\mathbf{c}$. That is, we can unbiasedly estimate linear combinations of the elements of $\boldsymbol{\mu}$, with precision that is worse with more noise (σ^2) or larger coefficients ($\mathbf{c}'\mathbf{c}$), and better with more replication (r). Note that while this shows we *can* estimate any linear combination of $\boldsymbol{\mu}$ under this model, those that are not contrasts would not be considered interesting from a strict experimental perspective.

11.4.2 Reduced model

Now return to the $f = 3$ example, and suppose we know or are willing to assume that interactions involving γ don't exist, that is, that:

$$(\alpha\gamma) = (\beta\gamma) = (\alpha\beta\gamma) = 0.$$

We can partition the model as:

$$\boldsymbol{\mu} = \mathbf{X}_1\boldsymbol{\theta}_1 + \mathbf{X}_2\boldsymbol{\theta}_2$$

where $\boldsymbol{\theta}_2$ contains the p_2 terms we have decided to omit, and $\boldsymbol{\theta}_1$ contains the p_1 terms that remain in the model $(p_1 + p_2 = 2^f)$. Under our assumption, the second term in the partition $(\mathbf{X}_2\boldsymbol{\theta}_2)$ is zero, so we fit the model of form

$$\mathbf{y} = \mathbf{X}_1\boldsymbol{\theta}_1 + \boldsymbol{\epsilon}$$

leading to $\hat{\boldsymbol{\theta}}_1 = \frac{1}{N}\mathbf{X}_1'\mathbf{y} = \frac{1}{2^f}\mathbf{M}_1'\bar{\mathbf{y}}$, where $\mathbf{M} = (\mathbf{M}_1|\mathbf{M}_2)$ is the full model matrix for an unreplicated design, and $\bar{\mathbf{y}}$ is the vector of treatment-specific averages, with information matrix $\mathcal{I} = N\mathbf{I}$. Then, according to our assumptions:

$$\boldsymbol{\mu} = \begin{pmatrix} + & - & - & - & + \\ + & - & - & + & + \\ + & - & + & - & - \\ + & - & + & + & - \\ + & + & - & - & - \\ + & + & - & + & - \\ + & + & + & - & + \\ + & + & + & + & + \end{pmatrix} \begin{pmatrix} \mu \\ \alpha \\ \beta \\ \gamma \\ (\alpha\beta) \end{pmatrix} = \mathbf{M}_1\boldsymbol{\theta}_1.$$

In this case, $\widehat{\mathbf{c}'\boldsymbol{\mu}} = \mathbf{c}'\widehat{\mathbf{M}_1\boldsymbol{\theta}_1} = \mathbf{c}'\mathbf{M}_1\hat{\boldsymbol{\theta}}_1$, but this is not necessarily the same thing as $\mathbf{c}'\bar{\mathbf{y}}$, as follows when the full model is used. Instead,

$$\widehat{\mathbf{c}'\boldsymbol{\mu}} = \mathbf{c}'\mathbf{M}_1(\mathbf{M}_1'\mathbf{M}_1)^{-1}\mathbf{M}_1'\bar{\mathbf{y}} = 2^{-f}\mathbf{c}'\mathbf{M}_1\mathbf{M}_1'\bar{\mathbf{y}}.$$

The difference here is that $\mathbf{M}_1\mathbf{M}_1'$ is not equal to $\mathbf{MM}' = 2^f\mathbf{I}$. Hence

$$E(\widehat{\mathbf{c}'\boldsymbol{\mu}}) = 2^{-f}\mathbf{c}'\mathbf{M}_1\mathbf{M}_1'(\mathbf{M}_1\boldsymbol{\theta}_1) = 2^{-f}2^f\mathbf{c}'\mathbf{M}_1\boldsymbol{\theta}_1 = \mathbf{c}'\boldsymbol{\mu}$$

if the reduced model is correct. Whether the reduced model is correct or not,

$$\begin{aligned} Var(\widehat{\mathbf{c}'\boldsymbol{\mu}}) &= \mathbf{c}'\mathbf{M}_1 \, Var(\hat{\boldsymbol{\theta}})\mathbf{M}_1'\mathbf{c} \\ &= \mathbf{c}'\mathbf{M}_1(\mathbf{M}_1'\mathbf{M}_1)^{-1}\mathbf{M}_1' \, Var(\bar{\mathbf{y}}) \, \mathbf{M}_1(\mathbf{M}_1'\mathbf{M}_1)^{-1}\mathbf{M}_1'\mathbf{c} \\ &= (\sigma^2/r)2^{-2f}\mathbf{c}'\mathbf{M}_1\mathbf{M}_1'\mathbf{M}_1\mathbf{M}_1'\mathbf{c} \\ &= (\sigma^2/r)2^{-f}\mathbf{c}'\mathbf{M}_1\mathbf{M}_1'\mathbf{c}. \end{aligned}$$

Recall that for the full model, $\mathbf{MM}'$ is $2^f\mathbf{I}$, so $Var(\widehat{\mathbf{c}'\boldsymbol{\mu}}) = (\sigma^2/r)\mathbf{c}'\mathbf{c}$, which we can write as:

$$\begin{aligned} Var(\widehat{\mathbf{c}'\boldsymbol{\mu}})\text{(full model)} &= (\sigma^2/r)2^{-f}\mathbf{c}'\mathbf{MM}'\mathbf{c} \\ &= (\sigma^2/r)2^{-f}(\mathbf{c}'\mathbf{M}_1|\mathbf{c}'\mathbf{M}_2)\begin{pmatrix} \mathbf{M}_1'\mathbf{c} \\ \mathbf{M}_2'\mathbf{c} \end{pmatrix} \\ &= (\sigma^2/r)2^{-f}\mathbf{c}'\mathbf{M}_1\mathbf{M}_1'\mathbf{c} + (\sigma^2/r)2^{-f}\mathbf{c}'\mathbf{M}_2\mathbf{M}_2'\mathbf{c} \\ &= Var(\widehat{\mathbf{c}'\boldsymbol{\mu}})\text{(reduced model)} + \text{a non-negative quantity.} \end{aligned}$$

So estimates based on a reduced model have variance no greater than, and sometimes less than, estimates of the same quantity derived under the full model.

11.4.3 Examples

The following examples demonstrate how the particular treatment combination being estimated influences the amount of variance reduction that can be achieved by using a reduced model.

Single cell mean: In this case, $\mathbf{c} = (0, 0, 0 \ldots 1 \ldots 0)'$, i.e., $\mathbf{c}'\boldsymbol{\mu}$ is a single element of $\boldsymbol{\mu}$. It follows that $\mathbf{c}'\mathbf{M}_1$ is a single row from $\mathbf{M}_1$, and therefore $\mathbf{c}'\mathbf{M}_1\mathbf{M}_1'\mathbf{c}$ is the number of parameters included in the reduced model, p_1. But $\mathbf{c}'\mathbf{M}\mathbf{M}'\mathbf{c} = 2^f$ for the full model, so the reduced model yields a sampling variance that is smaller by a factor of $p_1/2^f$ than the full model. (Note that this example is of mostly academic interest, since the estimate being considered is not in the form of a treatment contrast.)

Treatment contrast corresponding to a reduced model effect: Here, let $\mathbf{c}$ be a column from $\mathbf{M}_1$, i.e., a treatment contrast associated with a factorial effect retained in the reduced model. In this case, $\mathbf{c}'\mathbf{M}_1 = (0, 0, 0 \ldots 2^f \ldots 0)$ because the collection of columns in $\mathbf{M}_1$ are orthogonal. Therefore, $\mathbf{c}'\mathbf{M}_1\mathbf{M}_1'\mathbf{c} = \mathbf{c}'\mathbf{M}\mathbf{M}'\mathbf{c} = (2^f)^2$, so the sampling variance *is the same* for the contrast estimate constructed under either the full or reduced model. In fact, the estimate itself is the same function of the data in either case.

Treatment contrast corresponding to an effect removed from the reduced model: In this case, $\mathbf{c}$ is a column from $\mathbf{M}_2$, i.e., a treatment contrast associated with a factorial effect excluded from the reduced model. Because all effect contrasts are orthogonal, this means that $\mathbf{c}'\mathbf{M}_1 = (0, 0, 0, \ldots, 0)$, and that the sample variance under the reduced model is zero! But this is exactly what we should expect because "excluding an effect from the model" is the same as adding a modeling assumption that the effect is (exactly) zero – there is no uncertainty in its value.

A more general case: Consider the treatment contrast $\eta = 2\mu_{22} - \mu_{12} - \mu_{21}$ in cell-means notation for a 2^2 factorial experiment. This does not correspond to a single factorial effect, but can be written as $\eta = 2\alpha + 2\beta + 4(\alpha\beta)$. Under the full model:

$$Var(\hat{\eta}) = \frac{\sigma^2}{4r}\mathbf{c}'\mathbf{M}\mathbf{M}'\mathbf{c} = \frac{\sigma^2}{4r}(0, 2, 2, 4)'\begin{pmatrix} 0 \\ 2 \\ 2 \\ 4 \end{pmatrix} = 6\frac{\sigma^2}{r}.$$

But if we can justify a reduced model that does not include $(\alpha\beta)$,

$$Var(\hat{\eta}) = \frac{\sigma^2}{4r}\mathbf{c}'\mathbf{M}_1\mathbf{M}_1'\mathbf{c} = \frac{\sigma^2}{4r}(0, 2, 2)'\begin{pmatrix} 0 \\ 2 \\ 2 \end{pmatrix} = 2\frac{\sigma^2}{r}.$$

In summary, the potential benefits of using a reduced model include improved precision associated with some treatment contrasts. The degree of improvement depends on the contrast of interest and the form of the reduced model, i.e., the selection of retained effects. *But* there is also risk in adopting a reduced model, specifically that the estimates will be biased and tests potentially invalid if the omitted effects are actually present (i.e., nonzero).

11.5 Testing factorial effects

In some cases, it may actually be known, either from theory of the system under study or long experience with many similar experiments, that certain model terms can safely be assumed to be zero and thus removed from the model. However in *most* cases, such decisions come about through analysis of current experimental data, perhaps guided by system knowledge or experience. Formal tests of the hypothesis that an effect, or a group of related effects, can be removed from the model are valuable procedures in this process.

11.5.1 Individual model terms, experiments with replication

Suppose we wish to test $\text{Hyp}_0 : (-) = 0$, where "$(-)$" represents any factorial effect, e.g., α or $(\alpha\beta\gamma)$. The least-squares estimate of $(-)$ is $\widehat{(-)} = (\mathbf{M'M})^{-1}$ $\mathbf{m}'_{(-)}\bar{\mathbf{y}} = 2^{-f}\mathbf{m}'_{(-)}\bar{\mathbf{y}}$, where $\mathbf{m}_{(-)}$ is the column from $\mathbf{M}$ that corresponds to the parameter $(-)$. For any $(-)$,

$$Var[\widehat{(-)}] = 2^{-2f}\mathbf{m}'_{(-)} \; Var(\bar{\mathbf{y}}) \; \mathbf{m}_{(-)} = 2^{-2f}(\sigma^2/r)\mathbf{m}'_{(-)}\mathbf{m}_{(-)} = \sigma^2/N$$

where the last inner product is 2^f since this is true for every column in $\mathbf{M}$. The unbiased estimate of σ^2 is the mean squared error (MSE) – in this case a pooled sample variance computed from within-treatment variability of measurements:

$$\hat{\sigma}^2 = s_{pooled}^2 = \sum_{ij\ldots} s_{ij\ldots}^2/2^f,$$

where $s_{ij\ldots}^2$ is the sample variance of the r data values collected under the treatment identified by its subscripts, a statistic that is independent of $\widehat{(-)}$. A standard test statistic that follows a t-distribution under Hyp_0 can then be computed by dividing the estimate by its standard error:

$$t = \widehat{(-)}/\sqrt{s_{pooled}^2/N}$$

again, noting that this form holds for any factorial effect $(-)$, and the two-sided test is completed by comparing this test statistic to its critical value, $t_{1-\alpha/2}(2^f(r-1))$.

11.5.2 Multiple model terms, experiments with replication

Now consider the more general challenge of testing:

$$\text{Hyp}_0: \quad \boldsymbol{\mu} = \mathbf{M}_1\boldsymbol{\theta}_1$$

$$\text{Hyp}_A: \quad \boldsymbol{\mu} = \mathbf{M}\boldsymbol{\theta} = \mathbf{M}_1\boldsymbol{\theta}_1 + \mathbf{M}_2\boldsymbol{\theta}_2.$$

Especially with larger experiments and models, group tests of this form are useful because they reduce inflated type-I error rates that would result from multiple testing of individual parameters, and allow summary tests that may be physically meaningful or interesting, e.g., for "all interactions involving factor 3."

For the full-rank factorial model with $\pm$ "coding," for any included collection of factorial effects,

$$
\begin{aligned}
SSE &= (\mathbf{y} - \mathbf{X}\hat{\boldsymbol{\theta}})'(\mathbf{y} - \mathbf{X}\hat{\boldsymbol{\theta}}) \\
&= \mathbf{y}(\mathbf{I} - \mathbf{X}(\mathbf{X}'\mathbf{X})^-\mathbf{X}')\mathbf{y} \\
&= \mathbf{y}'\mathbf{y} - (\mathbf{y}'\mathbf{X}(\mathbf{X}'\mathbf{X})^-\mathbf{X}')(\mathbf{X}(\mathbf{X}'\mathbf{X})^-\mathbf{X}'\mathbf{y}) \\
&= \mathbf{y}'\mathbf{y} - (\hat{\boldsymbol{\theta}}'\mathbf{X}')(\mathbf{X}\hat{\boldsymbol{\theta}}) \\
&= \mathbf{y}'\mathbf{y} - r2^f\hat{\boldsymbol{\theta}}'\hat{\boldsymbol{\theta}}.
\end{aligned}
$$

In this case, it follows that the component of the treatment sum of squares associated with θ_2 is

$$
\begin{aligned}
SST_2 &= SSE(\text{Hyp}_0) - SSE(\text{Hyp}_A) = \mathbf{y}'\mathbf{y} - N\hat{\boldsymbol{\theta}}_1'\hat{\boldsymbol{\theta}}_1 - \mathbf{y}'\mathbf{y} + N\hat{\boldsymbol{\theta}}'\hat{\boldsymbol{\theta}} \\
&= N(\hat{\boldsymbol{\theta}}'\hat{\boldsymbol{\theta}} - \hat{\boldsymbol{\theta}}_1'\hat{\boldsymbol{\theta}}_1) = N\hat{\boldsymbol{\theta}}_2'\hat{\boldsymbol{\theta}}_2
\end{aligned}
$$

because omitting some parameters from the model does not change the estimates of the remaining parameters *in this case* since the columns of $\mathbf{M}$ are orthogonal. The relevant F-statistic is then the ratio of the mean square corresponding to the omitted model terms, and the residual mean square based on the full model:

$$F = (N\hat{\boldsymbol{\theta}}_2'\hat{\boldsymbol{\theta}}_2/p_2)/s_{pooled}^2$$

and the test is performed by comparing the computed statistic to the critical value, $F_{1-\alpha}(p_2, 2^f(r-1))$. For example, in testing $\text{Hyp}_0 : (\alpha\gamma) = (\beta\gamma) = 0$ the numerator mean square is $N[\widehat{(\alpha\gamma)}^2 + \widehat{(\beta\gamma)}^2]/2$. Because the information matrix for *any* subset of model parameters is $\mathcal{I} = N\mathbf{I}$, the power of the test is the probability that a random variable distributed as $F'(p_2, 2^f(r-1), N\boldsymbol{\theta}_2'\boldsymbol{\theta}_2/\sigma^2)$ is larger than the critical value of the test.

11.5.3 Experiments without replication

Unreplicated studies ($r = 1$) are not uncommon with two-level factorials, especially when the number of factors is large, and/or the experiment is viewed as "preliminary" with the goal of identifying the potentially most important factors for more detailed study later. If the full model is correct and ϵ is normally distributed, each effect estimate is distributed as:

$$\widehat{(-)} \sim N((-),\ \sigma^2/N)$$

and the estimates are independent because $Var(\hat{\boldsymbol{\theta}}) = \frac{\sigma^2}{N}\mathbf{I}$. Coefficient estimates are linear statistics giving equal weight to each data value, and so are approximately normally distributed (by the Central Limit Theorem) even if ϵ is not normal. For effects that are actually zero:

$$\widehat{(-)} \sim N(0,\ \sigma^2/N).$$

If *most* effects are actually zero – a condition Box and Meyer (1986) called *effect sparsity* – procedures developed to detect *outliers* can be used to identify the relatively few effects that appear to be "real."

Graphical procedures

Normal plots and *half-normal plots* of the effect estimates are helpful in this regard, and can be constructed as follows. For any group of effects, $(-)_i, i = 1\ldots p$, usually excluding μ, order the corresponding estimates from least to greatest, and refer to them as $\widehat{(-)}_{[1]} \leq \widehat{(-)}_{[2]} \leq \cdots \leq \widehat{(-)}_{[p]}$. Then a normal plot is constructed by plotting the sorted $\widehat{(-)}_{[i]}$ versus quantile values from a standard normal distribution, $\Phi^{-1}((i - \frac{1}{2})/p)$, $i = 1, 2, 3, \ldots, p$, where $\Phi^{-1}(-)$ is the inverse of the standard normal cumulative distribution function. Note that because the effect estimates are ordered, and that the same order relationship holds for the corresponding plotted Φ^{-1} values, the series of points plotted must increase together along the two axes. Half-normal plots, introduced by Daniel (1959), are constructed by plotting the sorted absolute values of effect estimates, $|\widehat{(-)}|_{[i]}$, versus quantiles from the "positive half" of a standard normal distribution, $\Phi^{-1}(\frac{1}{2} + (i - \frac{1}{2})/(2p))$, $i = 1, 2, 3, \ldots, p$. With either plot, the idea is that if all parameter estimates actually have expectation zero, the plotted statistics are actually an i.i.d. sample from a population with zero mean and standard deviation $\sigma/\sqrt{N}$, and the plotted points should lie approximately along a straight line. Any "real" effects (those that are not zero) tend to appear as "outliers," typically below the line at the left side of the normal plot, or above the line at the right side of the normal or half-normal plot. The normal plot is preferred by some investigators because there is some loss of information involved in constructing the half-normal version, i.e., the latter can be constructed from the former, but not vice versa. However, half-normal plots have the advantage of being invariant to changes in the assignment of "high" and "low" labels to factor levels.

To demonstrate the use of a half-normal plot, data were simulated for an unreplicated 2^4 factorial experiment, where all factorial effects are actually zero except:

$$\alpha = 6, \quad \beta = -4, \quad (\alpha\beta) = 2, \quad \text{and} \quad \sigma = 3$$

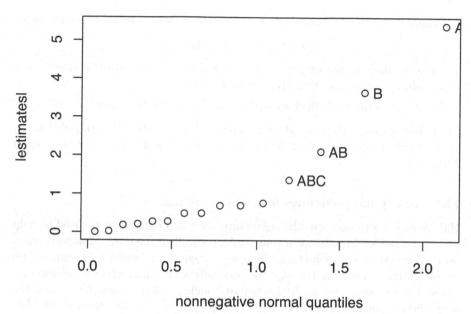

Figure 11.4 Half-normal plot of estimated effects from an unreplicated 2^4 experiment. (R11.1).

and the resulting plot computed (R11.1) and displayed in Figure 11.4. In this case, we do visually detect the three "real" effects. But if we also classify the 4th point from the right, representing $(\widehat{\alpha\beta\gamma})$, as begin "above the line," we would also have one "false alarm."

Finally, normal plots and half-normal plots can be useful techniques *only* under effect sparsity, because they rely on the few "real" effects (outliers) to be different from the many "absent" effects. When too many factorial effects are nonzero, the "reference line" is not visually clear, and classification of effects as "real" or "absent" becomes ambiguous.

Lenth's method

Lenth's Method (1989) is an algorithmic process for performing the visual half-normal plot analysis in a more automatic and objective manner:

1. Denote by $\mathcal{B}$ the set of absolute values of estimated coefficients of interest.

2. Compute an initial robust estimate of $\sigma/\sqrt{N}$:

$$s_0 = 1.5 \times \text{median}(\mathcal{B}).$$

3. Let $\mathcal{B}^*$ be the subset of $\mathcal{B}$ containing elements that are less than $2.5 \times s_0$ (that is, remove those estimates for which the effects seem clearly nonzero by this rule).

4. Compute a refined estimate of $\sigma/\sqrt{N}$, the so-called "pseudo standard error"

$$PSE = 1.5 \times \text{median}(\mathcal{B}^*).$$

Here, we depend on robustness of the median to minimize bias due to any remaining estimates of "active" effects in $\mathcal{B}^*$.

5. Treat any estimated effect as significant if it is greater than $t \times PSE$.

In the last step, t is the critical value used in the procedure. Lenth published a table of values of t; for $\alpha = 0.05$, and $\mathcal{B}$ of at least moderate size, t is between 2 and 2.5.

11.6 Additional guidelines for model editing

Half-normal plots and test-like algorithms such as Lenth's can be used to help you find a reduced model form that conforms well to the data. However, when used alone, they can sometimes lead to a suggested model that "makes no sense" in the context of the way we generally view main effects and interactions. For example, we might be led to conclude that α and $(\beta\gamma\delta)$ are the only effects that are nonzero in a four-factor experiment, but then worry that such a model has little physical credibility because the interaction does not involve the only factor with an apparently active main effect. Two model-editing "principles" have been suggested as guidelines for selecting additional model terms to remedy this situation.

Effect Hierarchy Principle: If an interaction involving a given set of factors is included in the model, *all* main effects and interactions involving subsets of these factors should also be included.

Effect Heredity Principle: (Hamada and Wu, 1992) If an interaction involving a given set of factors is included in the model, *at least one effect of the next smallest order* involving a subset of these factors should also be included.

The Effect Heredity Principle is actually used recursively. For example, including $(\alpha\beta\gamma\delta)$ implies that at least one of the four three-factor interactions involving these factors must also be included. But if we choose to include $(\alpha\beta\gamma)$, *this* implies that at least one of $(\alpha\beta)$, $(\alpha\gamma)$, or $(\beta\gamma)$ must be included. And given our selection of a two-factor interaction, at least one of the two related main effects must be included.

For example, if a graphical analysis suggests that main effects associated with factors 1 and 2 are the only effects necessary in a reduced model, neither principle would require the addition of any further terms. If the analysis suggests that α and the $(\alpha\beta)$ interaction are needed, the (weaker) Effect Heredity Principle requires no additional terms, but the Hierarchy Principle requires the addition of the main effect for factor 2. If the analysis suggests that only the $(\alpha\beta)$ interaction is needed, the Effect Heredity Principle requires that one of the main effects associated with factors 1 and 2 be added, while the Effect

Hierarchy Principle requires that they both be included. Finally, suppose an unreplicated experiment in four two-level factors leads to a half-normal plot suggesting that only the $(\alpha\beta\gamma)$ and $(\alpha\delta)$ interactions as active. Incorporating these effects, and following the Hierarchy Principle, the α, β, γ, δ, $(\alpha\beta)$, $(\alpha\gamma)$, and $(\beta\gamma)$ effects would also be required in the model. Following the Heredity Principle, α and $(\alpha\beta)$ (minimally) could be added.

A heuristic motivation for these ideas might be suggested by the following. If the $(\alpha\beta)$ interaction is present, factors 1 and 2 are clearly important, and $E(y)$ changes when the levels of either factor are individually changed, at least in some cases. Given the right collection of factor levels (even if this is not the collection of levels used in the experiment at hand), or the right set of factors (even if some of these are not varied in the experiment at hand), at least one of the two main effects could well be necessary. The Hierarchy Principle (relatively more conservative) is generally more popular than the Heredity Principle, except when considerable knowledge about the action of factors is available so that informed choices can be made about which terms to include and exclude.

11.7 Conclusion

Two-level experiments are popular in many applications, especially when a large number of factors are involved. They are also frequently used in preliminary "screening" studies where the primary goal is to determine which factors may have the greatest influence on the response, but not necessarily what level or combination of levels might be "optimal." The algebra associated with modeling, testing, and estimation can be simplified by formulating a full-rank effects model based on "regressors" that take values of ± 1.

In many applications, the development of a reduced model that contains far fewer than all possible factorial effects, but still accurately represents the structure of the data, is an important experimental goal. This chapter covers the basic elements of inference for estimates and tests for replicated ($r > 1$) two-level experiments, and also discusses how normal and half-normal plots, and Lenth's Method, can be used in unreplicated experiments to identify "active" effects when most factorial effects are zero or negligible. In conjunction with the Effect Hierarchy and Heredity Principles, these analysis techniques can be used to develop scientifically sensible, statistically efficient models of the factors-response relationship.

11.8 Exercises

1. The following table contains data from a 2^4 experiment, with $r = 2$ observations collected at each set of treatment conditions.

Factor Levels				Response	
A	B	C	D	1	2
1	1	1	1	10.7	11.0
1	1	1	2	8.7	8.0
1	1	2	1	10.1	11.7
1	1	2	2	7.9	9.5
1	2	1	1	9.8	9.5
1	2	1	2	9.2	8.2
1	2	2	1	11.2	11.0
1	2	2	2	10.3	9.6
2	1	1	1	10.8	11.6
2	1	1	2	10.2	8.6
2	1	2	1	11.9	11.4
2	1	2	2	10.8	9.4
2	2	1	1	11.6	11.5
2	2	1	2	10.4	11.5
2	2	2	1	11.2	11.4
2	2	2	2	11.8	8.9

(a) Compute the 15 factorial effect estimates and MSE (the estimate of σ^2) under the full model.

(b) Using F-statistics and the appropriate $\alpha = 0.05$ critical values, test:

- $\text{Hyp}_0 : (\alpha\beta\gamma\delta) = 0$
- $\text{Hyp}_0 :$ all three-factor interactions $= 0$
- $\text{Hyp}_0 :$ all two-factor interactions $= 0$

(c) Suppose that, in fact, two of the three-factor interactions have (true) values of $\frac{1}{3}\sigma$, while the other two are zero. At level 0.05, what is the power of the second test specified in part (b)?

2. Now, suppose that the experiment described in exercise 1 had actually been unreplicated, and that the single data value generated for each treatment had actually been the average of the two values given in the table.

(a) Compute the 15 factorial effect estimates.

(b) Construct a half-normal plot of these effect estimates and interpret it. Use Lenth's Method to confirm (or otherwise) your visual analysis of the graph. (Use $t = 2.5$.)

(c) Suppose your analysis of the above plot had led you to the conclusion that α, $(\alpha\beta)$, and $(\beta\gamma)$ were nonzero. In addition to these three factorial effects, what (if any) additional effects should be included in a tentative model:

- to satisfy the requirement of effect heredity?
- to satisfy the requirement of effect hierarchy?

3. Consider an unreplicated 2^6 complete factorial experiment. The corrected total sum of squares,

$$SSCT = \sum_{ijklmn} (y_{ijklmn} - \bar{y}_{......})^2$$

has the value of 2856. Using Lenth's method, an informal analysis of the data suggests that there are only three "active" factorial effects, with least-squares estimates: $\hat{\alpha} = 3$, $\widehat{(\alpha\beta)} = 4$, and $\widehat{(\alpha\beta\gamma)} = 2$.

(a) If a factorial model including only an intercept and these three effects is fitted to the data, what is the value of the residual sum of squares?

(b) What minimal collection of additional effects, besides the three listed above, would have to be reintroduced into the model in order to satisfy:

   • the effect heredity principle?
   • the effect hierarchy principle?

(c) Suppose you are given that:

$$\bar{y}_{..21..} = 14, \quad \bar{y}_{..11..} = 15, \quad \bar{y}_{..22..} = 13, \quad \text{and} \quad \bar{y}_{..12..} = 12.$$

Identify and give the value of any additional factorial effect estimates that can be computed based on this information.

4. Consider a 2^6 factorial study with each treatment applied to two units. Suppose the analyst determines that a model containing an intercept, all main effects, and all two-factor interactions involving factor A is adequate. Put another way, the assumption is made that all two-factor interactions *not* involving factor A, and all interactions involving three or more factors, can be eliminated. Based on this model, and apart from a factor of σ^2:

(a) What is the variance of the estimates of main effects and two-factor interactions? (Hint: The variances of all these terms are equal.)

(b) What is the variance of the least-squares estimate of $E(y_{111222})$? (Hint: The answer would be the same for any collection of subscripts on y.) Why would this estimate not be of particular interest in a "true" experimental setting?

(c) What is the variance of the least-squares estimate of $E(y_{111222}) - E(y_{111111})$ under this model? (Hint: The answer would *not* necessarily be the same for another pair of treatments.)

5. Consider a 2^3 experiment, in two replicates ($r = 2$), without any blocking.

(a) The investigator has particular interest in estimating $\mu_{111} - \mu_{122}$. If the full model (7 factorial effects plus the intercept) is assumed, rewrite this quantity in terms of factorial effects (i.e., α, $(\alpha\beta)$, et cetera), and give the variance of its least-squares estimate in terms of the unknown value of σ^2.

(b) Suppose instead that a reduced model containing only the intercept and main effects is adequate. Rewrite the difference of the two cell means from part (a) in terms of factorial effects included in *this* model, and give the variance of its least-squares estimate in terms of the unknown value of σ^2.

(c) Finally, return to an analysis based on the full model, and suppose that analysis of the data yields:

$$\widehat{(\alpha\beta)} = 3 \quad \widehat{(\alpha\beta\gamma)} = 2 \quad SSE = 16$$

In order to perform an F-test for:

$$\text{Hyp}_0 : (\alpha\beta) = (\alpha\beta\gamma) = 0$$
$$\text{Hyp}_A : (\alpha\beta) \neq 0 \text{ or } (\alpha\beta\gamma) \neq 0$$

calculate the value of the:

- numerator degrees of freedom
- numerator mean square
- denominator degrees of freedom
- denominator mean square

6. Four synthetic rubber compounds can be formulated by using one of two curing temperatures (360 or 420 degrees), and one of two different concentrations of oil (3% or 5% by weight). A chemist would like to perform an experiment to compare properties of these formulations. However, there are operational constraints imposed by the equipment available in his laboratory. First of all, the oven he must use has enough room to make only two batches of rubber at a time. Second, there is only enough time in a day to use the oven twice. So, four batches of rubber can be made each day, one for each treatment, and in each operation of the oven (i.e., one of the temperature levels), one batch can be produced at each of the oil concentration levels. It is anticipated that there will be potential differences associated with oven runs and with days, and that both can be considered as random effects.

(a) Compute degrees of freedom for an appropriate split-plot ANOVA decomposition for this problem, for only one day's experimentation. Include lines for block (day), residual, and corrected total in the whole-plot (oven run) stratum; block (oven run), residual, and corrected total in the split-plot (batch) stratum; and include appropriate factorial effects in each. (Remember, there are only four data values collected in a day. None of the factorial effects of interest will be testable!)

(b) Suppose now that this design is run over $r > 1$ days (the same thing is done each day), and that the effects of "day" are assumed to be additive (e.g., do not interact with the treatments of interest). Compute the degrees of freedom for an appropriate split-plot ANOVA decomposition for this experiment.

(c) Suppose the true main effect associated with changing oil content from 3% to 5% is exactly the same as the main effect associated with changing the curing temperature from 360 degrees to 420 degrees. Give *two* reasons why, for any $r > 1$ days, the test for the oil concentration main effect will be more powerful than the test for the temperature main effect.

7. Consider a two-level factorial experiment in $f > 2$ factors, where it is understood that interactions of order 3 or higher are probably negligible. The experimenter is especially interested in learning whether, for factors 1 and 2:
$$\alpha = \beta = (\alpha\beta) \neq 0.$$
In fact, he would like to know whether any pair of factors in the experiment have this property, i.e., the main effects parameters and the two-factor interaction parameter are all equal to some nonzero quantity. What does this imply about two such factors? Can you think of physical situations in which this could arise?

8. Consider a 2^f factorial treatment structure and 2^f treatment contrasts that compare the expected response at a specified treatment with the average of expected responses at the f "neighboring" treatments, each specified by changing the level of one factor. For example:

$$\theta_{121...2} = E(y_{121...2}) - \frac{1}{f}[E(y_{221...2}) + E(y_{111...2}) + E(y_{122...2}) + \cdots + E(y_{121...1})].$$

Such a function can be interpreted as a "local" measure of the joint effect of all factors. For a CRD with r units assigned to each treatment:

(a) What is the variance of the least-squares estimate of any such function?

(b) Specifically for $f = 4$, rewrite θ_{1221} in terms of factorial effects. Use this expression to determine the variance of $\hat{\theta}_{1221}$, and check your answer against the more general result from part (a).

CHAPTER 12

Two-level factorial experiments: blocking

12.1 Introduction

As described in Section 9.6, blocking is often required in factorial experiments because the number of treatments can be quite large for even a moderate number of experimental factors. For two-level factorial studies, f factors lead to a study of size $N = r \times t = r2^f$ if each possible treatment is replicated r times. If this is so large that consistent experimental control cannot be exerted throughout all runs, or if several batches of experimental material must be used to complete the study, blocking the experiment into a few or several subexperiments may account for substantial uncontrolled variation and so improve the quality of inferences that can be drawn from the data.

12.1.1 Models

An overparameterized effects model for a two-level blocked experiment can be written as:

$$y_{mijl...} = \mu + \theta_m + [\dot{\alpha}_i + \dot{\beta}_j + \cdots + (\alpha\dot{\beta}...)_{ij...}] + \epsilon_{mij...}$$

where θ_m denotes the additive effect associated with block m. A corresponding full-rank effects model can be written as:

$$y_{mijl...} = \mu + \theta_m + [x_1\alpha + x_2\beta + \cdots + x_1x_2...(\alpha\beta...)] + \epsilon_{mij...}$$

where, as described in the models discussed in Chapters 9 and 11, $x_i = \pm 1$ indicates that factor i is at its high or low level. Because the blocks form a partition of the experimental runs, an equivalent form for the full-rank parameterization is:

$$y_{mijl...} = \theta_m + [x_1\alpha + x_2\beta + ...x_1x_2...(\alpha\beta...)] + \epsilon_{mij...}.$$

Using this model, a matrix form for the entire blocked experiment is:

$$\mathbf{y} = \mathbf{X}_1\boldsymbol{\theta} + \mathbf{X}_2\boldsymbol{\phi} + \boldsymbol{\epsilon}$$

where, if each block contains k observations and the elements of $\mathbf{y}$ are ordered by block,

$$\mathbf{X}_1 = \begin{pmatrix} \mathbf{1}_k & \mathbf{0}_k & \cdots & \mathbf{0}_k \\ \mathbf{0}_k & \mathbf{1}_k & \cdots & \mathbf{0}_k \\ \cdots & \cdots & \cdots & \cdots \\ \mathbf{0}_k & \mathbf{0}_k & \cdots & \mathbf{1}_k \end{pmatrix} \qquad \mathbf{H}_1 = \frac{1}{k} \begin{pmatrix} \mathbf{J}_{k \times k} & \mathbf{0}_{k \times k} & \cdots & \mathbf{0}_{k \times k} \\ \mathbf{0}_{k \times k} & \mathbf{J}_{k \times k} & \cdots & \mathbf{0}_{k \times k} \\ \cdots & \cdots & \cdots & \cdots \\ \mathbf{0}_{k \times k} & \mathbf{0}_{k \times k} & \cdots & \mathbf{J}_{k \times k} \end{pmatrix}.$$

If a complete block design (CBD) is used, $k = 2^f$ and the rows of $\mathbf{X}_2$ corresponding to each k-row section of $\mathbf{X}_1$ "code for" all 2^f factorial treatments. If incomplete blocks are used, $k < 2^f$ and the submatrices of $\mathbf{X}_2$ corresponding to blocks differ depending on the subset of treatments included.

12.1.2 Notation

While the notation we have introduced is adequate to handle our discussion of designs for two-level experiments, the special structure of these studies has led to some convenient and commonly used simpler forms. *Factors* are often denoted by upper-case letters: "A","B", ... Also, *columns* of the model matrix are sometimes denoted using upper-case letters; e.g., "ABC" can refer to the column vector of $x_1 x_2 x_3$ values, the "regressor" associated with the parameter $(\alpha\beta\gamma)$. The upper-case "I" is used to represent the column of 1's associated with μ. (This reflects its role as the *identity element* in a more formal algebraic treatment.) A treatment is sometimes designated by listing lower-case letters associated with factors set to level 2, e.g.,

- "ac" = the treatment defined by setting factors A and C at level 2, and others at level 1
- "abcd" = the treatment defined by setting factors A–D at level 2, and others at level 1
- "(1)" = the treatment defined by setting all factors at level 1

Following common convention, most of the presentation in Chapters 12 and 13 will follow this pattern. In order to avoid confusion, the reader should carefully note the important difference between how letters of upper and lower case are used; for example "a" refers to a particular *treatment*, while "A" denotes a *factor*. And upper-case letters are associated with *columns*, while lower-case letters are associated with *rows*, of $\mathbf{X}$. There is an ambiguity in this system since "A" can stand either for a factor, or for the "regressor" associated with its main effect (α), but the distinction is always made clear by the context.

12.2 Complete blocks

The most straightforward form of blocking in 2^f experimentation is the CBD, calling for a complete replicate of all 2^f treatments to be applied in each block. For example, a CBD for $f = 2$ factors, arranged in r complete blocks, can be depicted as:

$$
\begin{array}{ccccc}
\text{block 1} & \text{block 2} & \cdots & \text{block } r \\
\boxed{\begin{array}{c}(1)\\ a\\ b\\ ab\end{array}} & \boxed{\begin{array}{c}(1)\\ a\\ b\\ ab\end{array}} & \cdots & \boxed{\begin{array}{c}(1)\\ a\\ b\\ ab\end{array}}
\end{array}
$$

Now, let $\mathbf{M}_2$ be the $2^f \times (2^f - 1)$ model matrix associated with just the factorial effects (i.e., excluding μ) for an unreplicated 2^f experiment. If the elements of $\mathbf{y}$ are ordered by block, and by the same treatment sequence within each block, $\mathbf{X}_2$ can be written as:

$$
\mathbf{X}_2 = \begin{pmatrix} \mathbf{M}_2 \\ \mathbf{M}_2 \\ \cdots \\ \mathbf{M}_2 \end{pmatrix}.
$$

From this,

$$
\mathbf{H}_1 = \frac{1}{2^f} \begin{pmatrix}
\mathbf{J}_{2^f \times 2^f} & \mathbf{0}_{2^f \times 2^f} & \cdots & \mathbf{0}_{2^f \times 2^f} \\
\mathbf{0}_{2^f \times 2^f} & \mathbf{J}_{2^f \times 2^f} & \cdots & \mathbf{0}_{2^f \times 2^f} \\
\cdots & \cdots & \cdots & \cdots \\
\mathbf{0}_{2^f \times 2^f} & \mathbf{0}_{2^f \times 2^f} & \cdots & \mathbf{J}_{2^f \times 2^f}
\end{pmatrix}
$$

and $\mathbf{X}_{2|1} = (\mathbf{I} - \mathbf{H}_1)\mathbf{X}_2 = \mathbf{X}_2 - \mathbf{0} = \mathbf{X}_2$

because the sum of each column of $\mathbf{M}_2$ is zero. Hence the estimates of factorial effects are computed as if the experiment were not blocked.

As with CBDs for unstructured treatments, a standard assumption is that there is no block-by-treatment interaction, leading to an ANOVA decomposition as displayed in Table 12.1. The line for "residual" would be "block-by-treatment interaction" if that term had been included in the model. The sum of squares for "treatments" can be further decomposed into one-degree-of-freedom components for each factorial effect, e.g., $N\hat{a}^2$. Note that this is just as in the unblocked case, where the treatment sum of squares can be

Table 12.1 ANOVA Decomposition for 2^f CBD in r Blocks

Source	Degrees of Freedom	Sum of Squares
blocks	$r - 1$	$\sum_m 2^f (\bar{y}_{m...} - \bar{y}_{...})^2$
treatments	$2^f - 1$	$\sum_{ijl...} r(\bar{y}_{.ijl...} - \bar{y}_{...})^2$
residual	$(2^f - 1)(r - 1)$	difference
corrected total	$r2^f - 1$	$\sum_{mijl...} (y_{mijl...} - \bar{y}_{...})^2$

decomposed into a one-degree-of-freedom sum of squares for each effect. So, for example, for $f = 2$, $SST = N\hat{\alpha}^2 + N\hat{\beta}^2 + N\widehat{(\alpha\beta)}^2$.

12.2.1 Example: gophers and burrow plugs

Werner, Nolte, and Provenza (2005) conducted laboratory experiments to study the effects of various environmental factors on "burrow plugging" behavior of pocket gophers (*Thomomys mazama, Thomomys talpoides*). In one experiment, 24 gophers were individually placed in an artificial burrow system constructed of clear polyvinyl chloride pipe that included a nesting box, a food cache, and four closed 1-m-long "loops" of pipe. Two 2-level factors were employed – intensity of artificial light ("light" at a standard intensity, and "no light") and the presence of small "burrow openings" in the pipe ("openings" of a standard size, and "no openings"). Each of the four factor combinations, ("light," "openings"), ("light," "no openings"), ("no light," "openings"), and ("no light," "no openings") was applied to one of the four loops in the system. A standard quantity of sawdust ("plugging substrate") was placed in the system at the beginning of each trial, and the response measured in each loop was the length of the sawdust "plug" constructed by the gopher in a standard length of time. Hence each gopher produced one measurement for each of the four treatments, and the differences among these 24 blocks were accounted for as resulting from animal-to-animal or other trial-to-trial nuisance variation.

12.3 Balanced incomplete block designs (BIBDs)

The only class of incomplete block designs we have discussed extensively is the BIBD class (Chapter 7). Recall that the structural balance properties of BIBDs result in some loss of efficiency, relative to CRDs of the same size, of treatment contrasts, and that this loss is proportionately the same for all treatment contrasts. As discussed in Section 9.6, this approach can also be taken in constructing blocked factorial designs. For example:

block 1	block 2	block 3	block 4
(1)	(1)	(1)	a
a	a	b	b
b	ab	ab	ab

is a BIBD with $t = 2^2$ treatments, organized in $b = 4$ blocks, each of size $k = 3$. In the notation of Chapter 7, the design is a BIBD because:

- each treatment appears in $r = 3$ blocks, and
- each pair of treatments appears together in $\lambda = 2$ blocks.

Analysis of data from a factorial experiment organized as a BIBD follows easily from the results given in Chapter 7, with the understanding that a factorial representation is really only a reparameterization of the models we would use if we ignored the factorial structure of the experiment. For example, we can arbitrarily assign the 2^2 treatments in the above experiment to:

$$(1) = \text{``treatment 1''}$$
$$a = \text{``treatment 2''}$$
$$b = \text{``treatment 3''}$$
$$ab = \text{``treatment 4''}.$$

Then expressions for treatment effects can be written in either notation:

unstructured		factorial
τ_1	$=$	$-\alpha - \beta + (\alpha\beta)$
τ_2	$=$	$+\alpha - \beta - (\alpha\beta)$
τ_3	$=$	$-\alpha + \beta - (\alpha\beta)$
τ_4	$=$	$+\alpha + \beta + (\alpha\beta)$

This relationship is 4-to-3, and can be inverted to express the factorial effects:

unstructured		factorial
$-\frac{1}{4}\tau_1 + \frac{1}{4}\tau_2 - \frac{1}{4}\tau_3 + \frac{1}{4}\tau_4$	$=$	α
$-\frac{1}{4}\tau_1 - \frac{1}{4}\tau_2 + \frac{1}{4}\tau_3 + \frac{1}{4}\tau_4$	$=$	β
$+\frac{1}{4}\tau_1 - \frac{1}{4}\tau_2 - \frac{1}{4}\tau_3 + \frac{1}{4}\tau_4$	$=$	$(\alpha\beta)$

Hence $\alpha = \mathbf{c}'\boldsymbol{\tau}$ in the notation of Chapter 7, where $\mathbf{c}' = (-\frac{1}{4}, +\frac{1}{4}, -\frac{1}{4}, +\frac{1}{4})$, leading to expressions for $\hat{\alpha}$ and $Var(\hat{\alpha})$.

But while BIBD structure *can* be used in factorial settings, the most commonly used blocking technique in 2^f experiments is fundamentally different. In most factorial experiment settings, some factorial effects (e.g., main effects and low-order interactions) are of substantially more interest than others (most higher-order interactions). As a result, the *regular* blocking schemes most often used in 2^f experimentation sacrifice all information for one or a few selected effects of no or limited interest, while preserving full efficiency for all others. The remainder of this chapter describes regular block designs for complete two-level factorial experiments.

12.4 Regular blocks of size 2^{f-1}

The simplest case of regular blocking in 2^f experimentation specifies that a "half-replicate" – one-half of all possible treatments – be included in each block of size 2^{f-1} so that each treatment appears exactly once in each *pair* of blocks. But this involves some loss of information about treatments; that is, we need to give up one of the "treatment" degrees of freedom to represent possible differences between the two blocks. When the experiment is divided into *regular* blocks, we do this by intentionally *confounding* one of the factorial effects with blocks. This is done by arranging blocks so that, for a selected factorial effect $(-)$,

- $(-)$ is multiplied by $+$ in the effects model representation for all treatments included in one block, and

- $(-)$ is multiplied by $-$ in the effects model representation for all treatments included in the other block.

This implies that $(-)$ is no longer estimable because the contrast associated with it is exactly the same as that associated with the block difference, so we generally want to select an effect that is:

- unlikely to be important or interesting, and/or

- likely to be zero or negligible.

The highest-order interaction is often used. The rationale here is that by intentionally confounding one factorial effect with blocks, full information about all other factorial effects is preserved. This is because every pair of factorial effects is orthogonal (e.g., is associated with two model matrix columns that are orthogonal). If ABC is confounded with the block contrast, i.e., those treatments for which the entry in the ABC column is $+1$ are assigned in one block, while those for which it is -1 are placed in the other, then all other factorial effects are orthogonal to (and therefore unaffected by) the effect of blocking. This situation is sometimes summarized by the *generating relation* "I $= \pm$ ABC," indicating that ABC is either always $+1$ or always -1 within each block. We sometimes say that the parameter $(\alpha\beta\gamma)$ is "confounded with" the block contrast $\theta_2 - \theta_1$ in this case, because either has the same estimate if the other is omitted from the model.

Listing treatments explicitly in each block, and remembering that we *could* have used any factorial effect to "split" the treatments into blocks (e.g., AB rather than ABC) if this had made more sense, we have:

Treatment	I	A	B	C	AB	AC	BC	ABC	Block
(1)	+	−	−	−	+	+	+	−	1
a	+	+	−	−	−	−	+	+	2
b	+	−	+	−	−	+	−	+	2
c	+	−	−	+	+	−	−	+	2
ab	+	+	+	−	+	−	−	−	1
ac	+	+	−	+	−	+	−	−	1
bc	+	−	+	+	−	−	+	−	1
abc	+	+	+	+	+	+	+	+	2

or

	block 1	block 2
	(1)	a
	ab	b
	ac	c
	bc	abc

All pairs of factorial effects are orthogonal, so all except ABC are orthogonal to "ABC+block." Other treatment estimates and sums of squares are unchanged by the introduction of blocks. If the design is executed without replication, there is no component in the analysis of variance decomposition that can be used to estimate σ^2. As a result, we might need to use normal plots omitting $\hat{\mu}$ and $(\widehat{\alpha\beta\gamma})$, the latter of which actually estimates $(\alpha\beta\gamma)+\theta_2-\theta_1$, to determine which effects appear to be nonzero.

Now, suppose we can afford to apply each treatment $r > 1$ times, but must still use blocks of size 2^{f-1}:

block 1	block 2	block 3	block 4	...	block $2r-1$	block $2r$
(1)	a	(1)	a		(1)	a
ab	b	ab	b		ab	b
ac	c	ac	c	...	ac	c
bc	abc	bc	abc		bc	abc

Ordering $\mathbf{y}$ by replicate, and by block within replicate, we can write:

$$\mathbf{H}_1 = \frac{1}{2^{f-1}}$$

$$\times \begin{pmatrix}
\mathbf{J}_{2^{f-1}\times 2^{f-1}} & \mathbf{0}_{2^{f-1}\times 2^{f-1}} & \mathbf{0}_{2^{f-1}\times 2^{f-1}} & \mathbf{0}_{2^{f-1}\times 2^{f-1}} & \cdots & \mathbf{0}_{2^{f-1}\times 2^{f-1}} & \mathbf{0}_{2^{f-1}\times 2^{f-1}} \\
\mathbf{0}_{2^{f-1}\times 2^{f-1}} & \mathbf{J}_{2^{f-1}\times 2^{f-1}} & \mathbf{0}_{2^{f-1}\times 2^{f-1}} & \mathbf{0}_{2^{f-1}\times 2^{f-1}} & \cdots & \mathbf{0}_{2^{f-1}\times 2^{f-1}} & \mathbf{0}_{2^{f-1}\times 2^{f-1}} \\
\mathbf{0}_{2^{f-1}\times 2^{f-1}} & \mathbf{0}_{2^{f-1}\times 2^{f-1}} & \mathbf{J}_{2^{f-1}\times 2^{f-1}} & \mathbf{0}_{2^{f-1}\times 2^{f-1}} & \cdots & \mathbf{0}_{2^{f-1}\times 2^{f-1}} & \mathbf{0}_{2^{f-1}\times 2^{f-1}} \\
\mathbf{0}_{2^{f-1}\times 2^{f-1}} & \mathbf{0}_{2^{f-1}\times 2^{f-1}} & \mathbf{0}_{2^{f-1}\times 2^{f-1}} & \mathbf{J}_{2^{f-1}\times 2^{f-1}} & \cdots & \mathbf{0}_{2^{f-1}\times 2^{f-1}} & \mathbf{0}_{2^{f-1}\times 2^{f-1}} \\
\cdots & \cdots & \cdots & \cdots & \cdots & \cdots & \cdots \\
\mathbf{0}_{2^{f-1}\times 2^{f-1}} & \mathbf{0}_{2^{f-1}\times 2^{f-1}} & \mathbf{0}_{2^{f-1}\times 2^{f-1}} & \mathbf{0}_{2^{f-1}\times 2^{f-1}} & \cdots & \mathbf{J}_{2^{f-1}\times 2^{f-1}} & \mathbf{0}_{2^{f-1}\times 2^{f-1}} \\
\mathbf{0}_{2^{f-1}\times 2^{f-1}} & \mathbf{0}_{2^{f-1}\times 2^{f-1}} & \mathbf{0}_{2^{f-1}\times 2^{f-1}} & \mathbf{0}_{2^{f-1}\times 2^{f-1}} & \cdots & \mathbf{0}_{2^{f-1}\times 2^{f-1}} & \mathbf{J}_{2^{f-1}\times 2^{f-1}}
\end{pmatrix}$$

$$\mathbf{X}_2 = \begin{pmatrix}
\mathbf{M}_{2,1} \\
\mathbf{M}_{2,2} \\
\mathbf{M}_{2,1} \\
\mathbf{M}_{2,2} \\
\cdots \\
\mathbf{M}_{2,1} \\
\mathbf{M}_{2,2}
\end{pmatrix} = (\mathbf{x}_1\ \mathbf{x}_2\ \cdots\ \mathbf{x}_{2^f-1})$$

where $\begin{pmatrix} \mathbf{M}_{2,1} \\ \mathbf{M}_{2,2} \end{pmatrix}$ is a partitioning of $\mathbf{M}_2$ into runs appearing in the two kinds of blocks. $\mathbf{M}_{2,1}$ and $\mathbf{M}_{2,2}$ each have columns with zero sums *except* for the column corresponding to the confounded effect. So, letting $\mathbf{X}_2 = (\mathbf{x}_1, \mathbf{x}_2, \mathbf{x}_3, \ldots, \mathbf{x}_{2^f-2}, \mathbf{x}_{2^f-1})$, if $\mathbf{x}_{2^f-1}$ is the column corresponding to the factorial effect

Table 12.2 ANOVA Decomposition for 2^f in Regular Blocks of Size $2^f - 1$, Blocked by Confounding ABC

Source	Degrees of Freedom	Sum of Squares
blocks	$2r - 1$	$\sum_m 2^{f-1}(\bar{y}_{m...} - \bar{y}_{...})^2$
treatments	$2^f - 2$	$N \sum \widehat{(-)}^2$ excluding $\widehat{(\alpha\beta\gamma)}^2$ from the sum
residual	$(2^f - 2)(r - 1)$	difference
corrected total	$r2^f - 1$	$\sum_{mijl...}(y_{mijl...} - \bar{y}_{...})^2$

confounded with blocks,

$$\mathbf{H}_1\mathbf{X}_2 = (\mathbf{0}\ \mathbf{0}\ \ldots\ \mathbf{0}\ \mathbf{x}_{2^f-1}) \quad \mathbf{X}_{2|1} = (\mathbf{I} - \mathbf{H}_1)\mathbf{X}_1 = (\mathbf{x}_1\ \mathbf{x}_2\ \ldots\ \mathbf{x}_{2^f-2}\ \mathbf{0}),$$

formally showing that treatment comparisons involving this effect are not estimable.

Letting $m = 1, 2, 3, \ldots, 2r$ index blocks in the design, an ANOVA decomposition is shown in Table 12.2. Note that in the corrected total sum of squares, not all possible combinations of index values appear because not all treatments $(ij\ldots)$ appear in each block (m). For $r = 1$, there are no degrees of freedom for residual variation under the full model, but an error estimate *could* come from degrees of freedom in "treatments" corresponding to terms omitted from the model (and not confounded with blocks). For example, with $f = 3$ and $r = 1$, and continuing to confound ABC with blocks, if we have reason to assume $(\alpha\gamma) = (\beta\gamma) = 0$, an alternate ANOVA decomposition is displayed in Table 12.3.

12.4.1 Random blocks

If blocks can be regarded as random entities, an experiment divided into regular blocks *can* be analyzed as a split-plot experiment, with levels of ABC compared *between* blocks, and other factorial effects compared *within* blocks. If ABC is confounded with blocks in each replicate, an ANOVA decomposition can be written as shown in Table 12.4. The last three lines of the ANOVA decomposition – those representing the split-plot portion of the experiment – are

Table 12.3 ANOVA Decomposition for Single-Replicate 2^3 in Regular Blocks of Size 2^2, Blocked by Confounding ABC and Assuming $(\alpha\gamma) = (\beta\gamma) = 0$

Source	Degrees of Freedom	Sum of Squares
blocks	1	$4[(\bar{y}_{1...} - \bar{y}_{....})^2 + (\bar{y}_{2...} - \bar{y}_{....})^2]$
treatments	4	$8[\hat{\alpha}^2 + \hat{\beta}^2 + \hat{\gamma}^2 + \widehat{(\alpha\beta)}^2]$
residual	2	$8[\widehat{(\alpha\gamma)}^2 + \widehat{(\beta\gamma)}^2]$
corrected total	7	$\sum_{mijl}(y_{mijl} - \bar{y}_{....})^2$

Table 12.4 ANOVA Decomposition for 2^f in Regular Blocks of Size 2^{f-1}, ABC Confounded with Random Blocks

Stratum	Source	Degrees of Freedom	Sum of Squares
whole-plot	ABC	1	$N(\widehat{\alpha\beta\gamma})^2$
	blocks|ABC	$2(r-1)$	$\sum_m 2^{f-1}(\bar{y}_{m...} - \bar{y}_{....})^2$ $-N(\widehat{\alpha\beta\gamma})^2$
	corrected total	$2r-1$	$\sum_m 2^{f-1}(\bar{y}_{m...} - \bar{y}_{....})^2$
split-plot	other treatments	$2^f - 2$	$N\sum(\widehat{-})^2$ excluding $(\widehat{\alpha\beta\gamma})^2$
	residual	$(2^f - 2)(r-1)$	difference
	corrected total	$r2^f - 1$	$\sum_{mijl}(y_{mijl} - \bar{y}_{....})^2$

identical to their counterparts in the fixed block ANOVA. But the whole-plot contrast, $(\widehat{\alpha\beta\gamma})$, is compared to the $2(r-1)$ degree-of-freedom "blocks\|ABC" line representing block-to-block variation other than that which can be attributed to the systematic effect of the three-factor interaction.

12.4.2 Partial confounding

The designs described above allow fully efficient estimation of all factorial effects *except* for the effect selected for confounding, for which formal inferences cannot be made if block effects are fixed, and for which generally less informative whole-plot inferences can be made if block effects are random. But what if the situation is such that:

- block effects cannot reasonably be treated as random,
- there is a need to make inferences about all effects, and
- blocks of size 2^{f-1} are required?

One option is *partial confounding*, in which different factorial effects are confounded with blocks in each complete replicate. For example, with 2^3 treatments and blocks of size 4, a design that confounds a different effect with blocks in each of r full replicates might be depicted as:

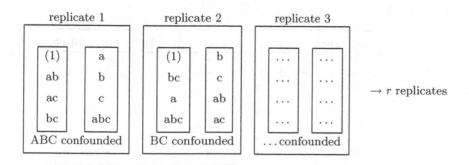

Table 12.5 ANOVA Decomposition for 2^f in Regular Blocks of Size 2^{f-1}, Demonstrating Partial Confounding

Source	Degrees of Freedom	Sum of Squares
blocks	$2r - 1$	$\sum_m 2^{f-1}(\bar{y}_{m...} - \bar{y}_{...})^2$
unconfounded effects	$2^f - 1 - r$	$2^f_r \sum (\hat{-})^2$ excluding $(\widehat{\alpha\beta\gamma})^2$, $(\widehat{\beta\gamma})^2$, ...
confounded effects	r	$2^f(r-1)[(\widehat{\alpha\beta\gamma})^2 + (\widehat{\beta\gamma})^2 + ...]$
residual	$(2^f - 2)(r - 1) - 1$	difference
corrected total	$r2^f - 1$	$\sum_{mijl...}(y_{mijl...} - \bar{y}_{...})^2$

Estimates and sums of squares for nonconfounded effects are computed as usual, e.g., $N\hat{\alpha}^2$ if A is not selected as the confounding effect in any replicate. Estimates and sums of squares for confounded effects are computed using only data from replicates in which they *are not* confounded. For example, in the design depicted above, ABC is confounded with blocks in the first replicate, so $(\widehat{\alpha\beta\gamma})$ is computed from the data in replicates 2 through r only (assuming that ABC is not confounded with blocks again in any of the remaining replicates), as if the design were an unblocked experiment of $(r - 1)2^3$ observations, and the corresponding component of SST would be $(r - 1)2^3(\widehat{\alpha\beta\gamma})^2$. A "sketch" of the ANOVA decomposition, assuming that a different factorial effect is confounded with blocks in each replicate, is shown in Table 12.5.

12.5 Regular blocks of size 2^{f-2}

When blocks of size 2^{f-1} are too large, smaller "quarter replicate" blocks, each containing 1/4 of all possible treatments, can be used. We accomplish this by further splitting of 2^{f-1}-size blocks, by selecting a second factorial effect to confound with blocks. Continuing the previous $f = 3$ example where ABC was chosen to generate the first treatment split, now add BC for the second. This means that in addition to losing the ability to estimate $(\alpha\beta\gamma)$, we now are also sacrificing information about $(\beta\gamma)$. Treatments are now sorted into blocks according to the four possible combinations of values for these two factorial effects:

Treatment	I	A	B	C	AB	AC	BC	ABC	Block
(1)	+	−	−	−	+	+	+	−	1
a	+	+	−	−	−	−	+	+	2
b	+	−	+	−	−	+	−	+	3
c	+	−	−	+	+	−	−	+	3
ab	+	+	+	−	+	−	−	−	4
ac	+	+	−	+	−	+	−	−	4
bc	+	−	+	+	−	−	+	−	1
abc	+	+	+	+	+	+	+	+	2

or,

block 1	block 2	block 3	block 4
(-)	a	b	ab
bc	abc	c	ac

So for this single-replicate design, eight observations are collected, two in each of four blocks. ABC and BC are confounded with blocks, so $(\alpha\beta\gamma)$ and $(\beta\gamma)$ are not estimable. But 3 degrees of freedom are required to represent the systematic differences among blocks. This implies that another contrast among treatments – or a third factorial effect – must also be confounded with blocks.

To identify the additional confounded contrast, note that within each block, some factorial effects are constant. Specifically, ABC $= x_1 x_2 x_3$ is "+" in blocks 2 and 3, and "−" in blocks 1 and 4, and BC $= x_2 x_3$ is "+" in blocks 1 and 2, and "−" in blocks 3 and 4. These both follow immediately from the selections we made in constructing blocks. But this implies that $x_1 x_2 x_3 \times x_2 x_3 = x_1 = A$ is *also* constant within blocks; "−" in blocks 1 and 3, and "+" in blocks 2 and 4, implying that α can't be estimated from within-block data comparisons. Symbolically, within:

$$\text{block } 1, I = -\text{ABC} = +\text{BC} = -\text{A}$$
$$\text{block } 2, I = +\text{ABC} = +\text{BC} = +\text{A}$$
$$\text{block } 3, I = +\text{ABC} = -\text{BC} = -\text{A}$$
$$\text{block } 4, I = -\text{ABC} = -\text{BC} = +\text{A}$$

where $A = x_1$ represents the column from **X** associated with α, and $I = 1$ represents the "regressor" for μ. By choosing to confound ABC and BC with blocks, we also confound their *generalized interaction*, A. As a practical matter, this is a poor choice of design for most purposes since it "sacrifices" information about a main effect. In most cases, a better choice would be specified by the generating relation:

$$I = \pm \text{ AB} = \pm \text{ BC } (= \pm \text{ AC}),$$

block 1	block 2	block 3	block 4
(-)	c	a	b
abc	ab	bc	ac

"$\pm$" notation means that within some blocks AB $= +1$ (or I), and within others AB $= -1$ (or $-$I), but that it is always constant within each block. "$(= \pm$ AC $)$" means that the generalized interaction AC is actually implied by selection of AB and BC as confounded effects. If AB and AC had been selected as the "splitting" effects, BC would have been the generalized interaction; the design would be the same.

Now, suppose we replicate each of these blocks r times for r full replicates of the design in $4r$ blocks, each of size 2. Extending the notation of Section 12.4, and again ordering the elements of $\mathbf{y}$ by replicate, block-within-replicate, and a standard treatment sequence within block, we have:

$$\mathbf{H}_1 = \frac{1}{2^{f-2}} \begin{pmatrix} \mathbf{J}_{2^{f-2} \times 2^{f-2}} & \mathbf{0}_{2^{f-2} \times 2^{f-2}} & \cdots & \mathbf{0}_{2^{f-2} \times 2^{f-2}} \\ \mathbf{0}_{2^{f-2} \times 2^{f-2}} & \mathbf{J}_{2^{f-2} \times 2^{f-2}} & \cdots & \mathbf{0}_{2^{f-2} \times 2^{f-2}} \\ \cdots & \cdots & \cdots & \cdots \\ \mathbf{0}_{2^{f-2} \times 2^{f-2}} & \mathbf{0}_{2^{f-2} \times 2^{f-2}} & \cdots & \mathbf{J}_{2^{f-2} \times 2^{f-2}} \end{pmatrix}$$

and $\mathbf{X}_2 = (\mathbf{x}_1 \ \mathbf{x}_2 \ \cdots \ \mathbf{x}_{2^f - 1})$ is comprised of r repeated sections of 2^f rows, each of which can be written as:

$$\begin{pmatrix} \mathbf{M}_{2,1} \\ \mathbf{M}_{2,2} \\ \mathbf{M}_{2,3} \\ \mathbf{M}_{2,4} \end{pmatrix}$$

representing the treatments that appear in each of the four kinds of blocks. Within each submatrix $\mathbf{M}_{2,i}$, $i = 1, 2, 3, 4$, each column contains entries that sum to zero *except for* the three columns associated with the confounded effects. If we associate these three effects with the last three columns of $\mathbf{X}_2$, this leads to:

$$\mathbf{H}_1 \mathbf{X}_2 = (\mathbf{0} \ \mathbf{0} \ \mathbf{0} \ \cdots \mathbf{0} \ \mathbf{x}_{2^f - 3} \ \mathbf{x}_{2^f - 2} \ \mathbf{x}_{2^f - 1})$$

and

$$(\mathbf{I} - \mathbf{H}_1)\mathbf{X}_2 = (\mathbf{x}_1 \ \mathbf{x}_2 \ \mathbf{x}_3 \ \cdots \ \mathbf{x}_{2^f - 4} \ \mathbf{0} \ \mathbf{0} \ \mathbf{0})$$

indicating that the factorial effects associated with the last three columns of $\mathbf{X}_2$ are not estimable.

Letting m index blocks, for fixed block effects, an analysis of variance decomposition can be constructed as shown in Table 12.6. For random block effects, a split-plot ANOVA can be constructed as demonstrated in Table 12.7.

Partial confounding can also be used in this case. For example, suppose we need a 2^{4-2} blocking scheme, but can perform two complete replicates of four blocks each. We might choose to confound different sets of factorial effects in

Table 12.6 ANOVA Decomposition for 2^f in Regular Blocks of Size 2^{f-2}, Blocked by Confounding AB, AC, and BC

Source	Degrees of Freedom	Sum of Squares
blocks	$4r - 1$	$\sum_m 2^{f-2}(\bar{y}_{m\cdots} - \bar{y}_{\cdots})^2$
treatments	$2^f - 4$	$N \sum (\widehat{-})^2$ excluding $\widehat{(\alpha\beta)}^2, \widehat{(\alpha\gamma)}^2,$ and $\widehat{(\beta\gamma)}^2$
residual	$(2^f - 4)(r - 1)$	difference
corrected total	$r2^f - 1$	$\sum_{mijl}(y_{mijl} - \bar{y}_{\cdots})^2$

Table 12.7 ANOVA Decomposition for 2^f in Regular Blocks of Size 2^{f-2}; AB, AC, and BC Confounded with Random Blocks

Stratum	Source	Degrees of Freedom	Sum of Squares
whole-plot	AB,AC,BC	3	$N\widehat{(\alpha\beta)}^2 + N\widehat{(\alpha\gamma)}^2 + N\widehat{(\beta\gamma)}^2$
	blocks | AB,AC,BC	$4r - 4$	$\sum_m 2^{f-2}(\bar{y}_{m...} - \bar{y}_{....})^2 - N\widehat{(\alpha\beta)}^2 - \cdots$
	corrected total	$4r - 1$	"blocks" in Table 12.6
split-plot	other factorial effects	$2^f - 4$	"treatments" in Table 12.6
	residual	$(2^f - 4)(r - 1)$	"residual" in Table 12.6
	corrected total	$r2^f - 1$	"corrected total" in Table 12.6

each of the two replicates, e.g.:

$$\text{Replicate 1: } I = \pm \text{ ABC} = \pm \text{ BCD}(= \pm \text{ AD})$$
$$\text{Replicate 2: } I = \pm \text{ ABCD} = \pm \text{ BC}(= \pm \text{ AD}).$$

In this situation, $(\alpha\beta\gamma)$ and $(\beta\gamma\delta)$ estimates and sums of squares are computed using the data from replicate 2 only, and $(\alpha\beta\gamma\delta)$ and $(\beta\gamma)$ estimates and sums of squares are computed using the data from replicate 1 only. AD is confounded with blocks in both replicates, and so $(\alpha\delta)$ cannot be estimated. An ANOVA decomposition in this case could be constructed as shown in Table 12.8.

12.6 Regular blocks: general case

The ideas already introduced for blocks of size 2^{f-1} and 2^{f-2} can be generalized to construct designs with smaller blocks by sequentially re-splitting treatments into groups s times, to obtain 2^s blocks of 2^{f-s} units each per replicate. For example, suppose we need to construct a 2^{6-4} blocking system:

- $2^6 = 64$ treatments
- $2^4 = 16$ blocks
- $2^{6-4} = 4$ units per block

Table 12.8 ANOVA Decomposition for 2^4 in Regular Blocks of Size 2^2, Demonstrating Partial Confounding

Source	Degrees of Freedom	Sum of Squares
blocks	7	$\sum_m 4(\bar{y}_{m....} - \bar{y}_{....})^2$
unconfounded effects	10	$32 \sum (\hat{-})^2$ excluding confounded effects and $(\alpha\delta)$
confounded effects	4	$16[\widehat{(\alpha\beta\gamma)}^2 + \widehat{(\beta\gamma\delta)}^2 + \widehat{(\alpha\beta\gamma\delta)}^2 + \widehat{(\beta\gamma)}^2]$
residual	10	difference
corrected total	31	$\sum_{mijkl}(y_{mijkl} - \bar{y}_{.....})^2$

We can think in terms of the four factorial effects "independently" selected to confound with blocks, one corresponding to each of the sequence of "splits" of the 64 treatments, with the understanding that new generalized interactions are also confounded with blocks at each split after the first. For our example, we might consider the following confounding scheme:

"Split"	Selected Effect	Generalized Interactions						
1	ABF							
2	ACF	BC						
3	BDF	AD	ABCD	CDF				
4	DEF	ABDE	ACDE	BCDEF	BE	AEF	ABCEF	CE

With each split of the treatments except for the first, generalized interactions are added between the new "independent" effect and *each* of the previously implicated effects (whether labeled as independent effects or generalized interactions). In all, $2^s - 1$ effects or "words" are confounded with blocks (all possible combinations of the symbolic products of s independent effects). Again, any four of these $2^4 - 1 = 15$ "words" can be treated as "independent" when generating the sequence of treatment "splits" and the resulting pattern of blocks will be the same. In this example, we have constructed a blocking scheme for 2^6 treatments in 2^4 blocks of size 2^2 that allows fully efficient inference about all main effects and all but four 2-factor interactions and several effects of higher order. Effects lost to (or confounded with) blocks include six 3-factor interactions, three 4-factor interactions, and two 5-factor interactions.

In order to avoid confounding more and more factorial effects with blocks, it might be tempting to pick a previously identified generalized interaction as a new "independent" effect for identifying a new split. For example, we might have considered selecting CDF (a generalized interaction identified after the third split) as the "independent" effect for the fourth split, instead of DEF. But this does not work because CDF is *already* constant within blocks, and so cannot be used for further "splitting."

As blocking schemes become more complicated, it becomes less obvious which treatments are applied to units in each block, but the general principles introduced for $s = 1$ and 2 still apply. One block in our 2^{6-4} example consists of the set of treatments that simultaneously satisfy the four constraints represented by $I = +\text{ABF} = +\text{ACF} = +\text{BDF} = +\text{DEF}$. The four treatments to be applied in this block can be identified by constructing a small table of the elements of $\mathbf{M}_2$ that satisfy these requirements. Hence the A, B, and F columns can be filled in first to assure $\text{ABF} = +1$, then C can be added to assure $\text{ACF} = +1$, et cetera:

A	B	F	C	D	E
+	+	+	+	+	+
−	−	+	−	−	−
−	+	−	+	−	+
+	−	−	−	+	−

or

abcdef
f
bce
ad

.

We can similarly determine the treatments for any of the other 15 blocks by sequentially selecting factor levels so as to satisfy all constraints, e.g., I = +ABC = −ACF = −BDE = +DEF:

A	B	F	C	D	E
−	−	+	+	+	+
+	+	+	−	−	−
+	−	−	+	−	+
−	+	−	−	+	−

or

cdef
abf
ace
bd

Note that we only need to deal with the constraints associated with four "independent" effects; the constraints associated with the generalized interactions are automatically satisfied. The single block that includes the treatment defined by the low level for each factor is sometimes called the *principal block*:

A	B	F	C	D	E
−	−	−	−	−	−
+	+	−	+	+	+
+	−	+	−	+	−
−	+	+	+	−	+

or

(1)

abcde
adf
bcef

identified in this case by

$$I = -ABF = -ACF = -BDF = -DEF.$$

In general, the generating relation for the principal block assigns −/+ to words of odd/even length.

Once the treatments in one block have been identified, another approach to constructing the remaining blocks is by reversing the signs for one or more factors. For example, given the principal block above, we might reverse the signs for factors A and B to obtain:

A	B	F	C	D	E
+	+	−	−	−	−
−	−	−	+	+	+
−	+	+	−	+	−
+	−	+	+	−	+

or

ab
cde
bdf
acef

a block denoted by

$$I = -ABF = +ACF = +BDF = -DEF.$$

The signs attached to ACF and BDF are reversed relative to the principal block because they each contain only one of A or B. The sign attached to ABF is not reversed because it contains both A and B, and that for DEF is not reversed because it contains neither A nor B. Signs are also determined in this way for all 11 generalized interactions, depending on the involvement of an odd or even number of the symbols "A" and "B" in each.

Table 12.9 General form of ANOVA Decomposition for 2^f in Regular Fixed Blocks of Size 2^{f-s}

Source	Degrees of Freedom	Sums of Squares
blocks	$r2^s - 1$	$\sum 2^{f-s}(\bar{y}_{block} - \bar{y})^2$
treatments	$2^f - 2^s$	$\sum N\widehat{(-)}^2$ omitting confounded effects
residual	difference	difference
corrected total	$r2^f - 1$	$\sum (y - \bar{y})^2$

Following the more detailed patterns discussed for $s = 1$ and 2, a fixed-block ANOVA decomposition for r complete replicates of 2^f treatments, each split into 2^s blocks of size 2^{f-s} using the same confounding scheme in each replicate, can be constructed as shown in Table 12.9, and a random-block (split-plot) ANOVA can be written as given in Table 12.10.

As with $s = 1$ and 2, partial confounding can also be used when blocks should be regarded as having fixed effects. For example, in a design comprised of 5 replicates, if ABC is an element of the confounding pattern in replicates 1 and 2, $(\alpha\beta\gamma)$ is estimated from the data from replicates 3–5 only and the corresponding single-degree-of-freedom component of SST is $3 \times 2^f \widehat{(\alpha\beta\gamma)}^2$. Some effects may be confounded in more replicates than others, but no formal inferences can be made for any effect confounded with blocks in all replicates.

12.7 Conclusion

When 2^f factorial treatments are compared in a complete block design, each factorial effect is orthogonal to blocks, and the usual additive-treatments-and-blocks analysis is possible. When smaller blocks are needed, BIBDs may be used, or the experiment can be performed in regular blocks of size 2^{f-s}, $s < f$. BIBDs are not orthogonally blocked, but analysis described in Chapter 7

Table 12.10 General form of Split-Plot ANOVA Decomposition for 2^f in Regular Random Blocks of Size 2^{f-s}

Stratum	Source	Degrees of Freedom	Sum of Squares
whole-plot	confounded effects	$2^s - 1$	$N \times$ (sum of $2^s - 1$ squared effect estimates)
	blocks | confounded effects	$2^s(r - 1)$	difference
	corrected total	$r2^s - 1$	"blocks" in Table 12.9
split-plot	other factorial effects	$2^f - 2^s$	"treatments" in Table 12.9
	residual	difference	"residual" in Table 12.9
	corrected total	$r2^f - 1$	"corrected total" in Table 12.9

leads to equal information for each factorial effect. In designs based on regular blocks, blocks are orthogonal to some factorial effects, but are intentionally confounded with others. When multiple complete replicates are included in the experiment, inferences can be made about confounded effects via a split-plot analysis when block effects are regarded as random, or by using partial confounding when they are regarded as fixed.

12.8 Exercises

1. Construct a BIBD for a 2^3 factorial experiment using as few blocks of size 6 as you can. What is the variance of $\hat{\alpha}$ (or for that matter, any factorial effect) for your design?

The following table lists data for a 2^4 experiment, including three complete replicates. Use these data to work exercises 2–6.

Data for 2^f Factorial Exercises 2–6

Treatment	Replicate		
	1	2	3
(1)	48.87	53.78	54.37
d	31.52	35.04	34.88
c	68.32	66.98	72.80
cd	49.84	47.19	51.28
b	24.67	40.99	43.08
bd	37.28	46.81	44.87
bc	51.16	49.53	52.26
bcd	39.62	37.83	35.55
a	56.30	65.11	56.41
ad	50.14	44.70	53.37
ac	63.10	58.01	66.53
acd	46.01	49.11	57.13
ab	45.56	50.02	52.21
abd	48.90	50.08	47.64
abc	57.06	50.53	51.43
abcd	50.67	45.38	56.48

2. Assume the experiment was executed as a completely randomized design (i.e., no blocks, and a model with only an overall mean, treatment-related effects, and homogeneous, independent noise for each measurement). Compute the least-squares estimates for each of the 15 factorial effects and their standard errors, and sums of squares and degrees of freedom for an ANOVA partitioned with lines for treatments, residual, and corrected total.

3. Assume the experiment was executed as a randomized complete block design, with each replicate constituting one block. How will the effect estimates differ from those in exercise 2? Compute sums of squares and degrees of freedom for an ANOVA partitioned with lines for blocks, treatments, residual, and corrected total.

4. Assume that blocks of size 4 must be used. Select a generating relation to use (i.e., two factorial effects and their generalized interaction). Use the same generating relation in each replicate (i.e., don't use partial confounding), do not confound any main effects with blocks, and confound as few two-factor interactions as possible. Assuming blocks are fixed, how will the effect estimates differ from those in exercises 2 and 3? Compute sums of squares and degrees of freedom for an ANOVA partitioned with lines for blocks, treatments, residual, and corrected total.

5. Repeat exercise 4, but now under the assumption that the block effects can be regarded as random. With this form of analysis, all factorial effects can be represented as "treatments" in the ANOVA, but not all will be compared to the same denominator mean square. How will the effect estimates differ from those in previous parts? Compute sums of squares and degrees of freedom for an ANOVA partitioned with lines for confounded effects, unconfounded effects, and residual terms appropriate for each of them.

6. Continuing to use blocks of size 4, assume that block effects are to be regarded as fixed, and select a different generating relation for each replicate (i.e., use partial confounding) in such a way that no main effect is confounded in any replicate, and at least 2/3 information is available for each interaction (i.e., no interaction is confounded in more than one of the three replicates). Compute the least-squares estimates for each of the 15 factorial effects and their standard errors, and sums of squares and degrees of freedom for blocks, partially confounded effects, unconfounded effects, residual, and corrected total.

7. The following are the treatment combinations from one block of a (complete) blocked 2^5 experiment:

A	B	C	D	E
+	+	+	+	+
+	−	+	+	−
+	−	−	−	+
+	+	−	−	−
−	−	+	−	+
−	+	+	−	−
−	+	−	+	+
−	−	−	+	−

  (a) Determine the generating relation for this blocking system.

  (b) Find the treatments to be included in each of the other blocks in this system.

8. Construct a row-column blocking system for an unreplicated 2^4 experiment as follows. First, write a generating relation for defining four blocks of size 4 by selecting two factorial effects and their generalized interaction to define row-blocks. Next, write a second generating relation to define

column-blocks; do not include any factorial effects in both generating relations. Assume that both row- and column-blocks have additive, fixed effects.

(a) Make a 4×4 table showing how the 16 treatments are assigned in your design.

(b) What individual factorial effects would be estimable from an experiment run using your design?

## CHAPTER 13

Two-level factorial experiments: fractional factorials

13.1 Introduction

Because of the exponential relationship between the number of factors and number of treatments they define, even a moderate number of factors can generate more treatments than can practically be included in many experiments. Blocking can help reduce the number of trials than must be executed on any one day, or with any one batch of experimental material, but the total experiment size (N) of the two-level design comprised of complete replicates, whether blocked or not, must be a multiple of 2^f. In this chapter, we discuss *fractional factorial* designs that can be much smaller because they do not include all possible treatments. At first glance, this may seem quite strange; if an experiment is undertaken to compare a collection of treatments, how can a design that intentionally omits some of these treatments supply the desired information?

The honest answer to this question is that it cannot, unless additional assumptions are made about the relationships between the treatments of interest. Those assumptions, while sometimes not stated explicitly, are based on the idea that *most* of the differences between factorial treatments can *often* be described in terms of main effects and interactions of relatively low order, i.e., that interactions of higher order may be relatively negligible. On the other hand, since we usually do not *know* this to be true of any given experimental situation in advance, we must not depend too heavily on assumptions of this kind in analysis of the data. A pragmatic approach is to realize that fractional factorials provide partial but valuable information about treatment comparisons if further assumptions cannot be made, and that this information is more definitive if they can.

13.2 Regular fractional factorial designs

A two-level *regular fractional factorial design* consists of the treatments in one block of a regular blocked full 2^f experiment, as discussed in Chapter 12. For example, a "one quarter fraction of a complete 2^5 factorial experiment" can be formed by constructing a 2^{5-2} blocked design of the 32 treatments in four blocks of size 8, and then using only one of those blocks as the *entire* (unblocked) experimental design. We might specify such a design symbolically by writing:

$$I = +ABC = -ADE(= -BCDE),$$

a generating relation similar to the general form used in Chapter 12, specifying two independently selected "splitting" effects and noting their generalized interaction. But here, "+" and "−" are assigned to each "word," not "±" as in the expression of a blocking scheme for a complete factorial experiment, since we are identifying only one of the blocks:

A	B	C	D	E
+	+	+	+	−
+	+	+	−	+
−	−	+	+	+
−	−	+	−	−
−	+	−	+	+
−	+	−	−	−
+	−	−	+	−
+	−	−	−	+

or

abcd
abce
cde
c
bde
b
ad
ae

.

These are the eight treatments to be included in the unreplicated fractional factorial experiment we have selected. More generally, we might choose to replicate each treatment by applying it to r units to allow estimation of σ^2 and increase precision of estimates, leading to $N = r2^{5-2}$. For any amount of replication (value of r), the fractional factorial requires $1/4$ of the experimental units needed with a blocked or unblocked complete, or "full," factorial experiment.

We know that even with a full factorial design organized in regular blocks, we "lose" experimental information about the factorial effects involved in the confounding scheme (ABC, ADE, BCDE in this case). It should be intuitively clear that this information is also unavailable with the fractional factorial experiment – one of the blocks in the full replicate – since we can't hope to *increase* information by eliminating $3/4$ of the experiment. In fact, even more information is lost because fewer treatments are evaluated. Specifically, the 2^5 treatment structure implies that there are $2^5 - 1$ factorial effects of interest, or $2^5 - 4 = 28$ factorial effects after "sacrificing" ABC, ADE, and BCDE. But because the fractional factorial plan contains only eight treatments, we cannot separately estimate all 28 of these factorial effects; additional information has clearly been lost.

As an example of the additional information loss, look at the signs associated with the BC interaction in our unreplicated fraction:

A	B	C	D	E	BC
+	+	+	+	−	+
+	+	+	−	+	+
−	−	+	+	+	−
−	−	+	−	−	−
−	+	−	+	+	−
−	+	−	−	−	−
+	−	−	+	−	+
+	−	−	−	+	+

.

It is clear from this table that the BC interaction is completely *aliased* with the A main effect in this design; that is, the contrast in cell means associated with α is exactly the same as that associated with $(\beta\gamma)$. Neither A nor BC would be confounded with blocks in the full replicate design (as ABC, ADE, and BCDE are), and neither is aliased with I in this fraction since each is "+" in four runs and "−" in four runs. But A and BC *are* aliased with each other in this fraction; that is, we would not be able to separate the influence of these two effects using data from the fractional factorial experiment. Information about these two effects *is* available in the blocked full-replicate design because:

$$I = +ABC \quad \rightarrow \quad +A = +BC \text{ in two blocks}$$

$$I = -ABC \quad \rightarrow \quad +A = -BC \text{ in two blocks}$$

so that the aliasing between A and BC is "broken" because the relationship between the two effects, while constant *within* any one block, is different from block to block.

In fact, in a regular fractional factorial plan *all* factorial effects are aliased in groups. The most obvious group of aliased effects is ABC, ADE, BCDE, and I, since each is represented by a column of "+"s or "−"s throughout the design. But the remainder of the factorial effects are also aliased in groups of size 4. In comparison, a full 2^f factorial experiment executed in regular blocks sacrifices information about *just* those factorial effects confounded with blocks; the remainder are orthogonal to blocks and to each other, so they can be estimated or tested.

Again, the generating relation (or "defining relation", or "identifying relation") for this fractional factorial design is:

$$I = +ABC = -ADE = -BCDE$$

i.e., the relationship between the effects intentionally aliased with the intercept. Recall that I, ABC, ADE, and BCDE are "words" or "generators" that stand for columns in the model matrix. We can use element-wise multiplication of these columns to identify the groups of aliased effects. Continuing our example, we have identified BC as an effect aliased with A. But more completely, we find all aliases of A by symbolically "multiplying" all words in the generating relation by "A":

$$I = +ABC = -ADE = -BCDE \quad (2^s\text{words})$$

$$A = +AABC = -AADE = -ABCDE$$

$$A = +BC = -DE = -ABCDE$$

because "AA" $= x_1 \times x_1 = 1$, et cetera. Hence the A main effect is confounded with $3 = 2^s - 1$ factorial effects: positively with BC, and negatively with each of DE and ABCDE:

A	B	C	D	E	BC	DE	ABCDE
+	+	+	+	−	+	−	−
+	+	+	−	+	+	−	−
−	−	+	+	+	−	+	+
−	−	+	−	−	−	+	+
−	+	−	+	+	−	+	+
−	+	−	−	−	−	+	+
+	−	−	+	−	+	−	−
+	−	−	−	+	+	−	−

In the same way, the aliases of the other four main effects can be determined:

$$B = +AC = -ABDE = -CDE$$
$$C = +AB = -ACDE = -BDE$$
$$D = +ABCD = -AE = -BCE$$
$$E = +ABCE = -AD = -BCD.$$

There are two additional alias groups that do not contain main effects. Noting that neither BD nor BE is an element of any alias group identified to this point, these effects can be used to determine the remaining two groups:

$$BD = +ACD = -ABE = -CE$$
$$BE = +ACE = -ABD = -CD.$$

These alias groups form a partition of all the factorial effects; each effect is a member of exactly one group. The seven degrees of freedom available for among-treatment analysis represent the information that can be gained concerning the seven alias groups after removing the group containing the experimentally uninteresting "I". For the general 2^{f-s} factorial structure, where a single subset of 2^{f-s} treatments is used (one block from a 2^{f-s} blocking scheme), all 2^f factorial effects (including μ) are partitioned into 2^{f-s} groups of size 2^s, and the effects within each group are completely aliased.

13.3 Analysis

The result of the 2^{5-2} fractional factorial experiment is eight estimable *strings* of effects, i.e., the linear combination of each group of aliased effects in which the coefficients are $+1$ and -1 to represent positive and negative aliasing, seven of which don't include I (or μ). For example, we noted above that one alias group is

$$A = +BC = -DE = -ABCDE.$$

As a result, the data contrast that is the usual estimate of α is really an estimate of a "string" of effects:

$$E[\hat{\alpha}] = \alpha + (\beta\gamma) - (\delta\epsilon) - (\alpha\beta\gamma\delta\epsilon).$$

Table 13.1 Data from Bacteriocin Experiment of Leal-Sanchez et al. (2002)

Coded Factors					Responses ($\log_{10}$ AU/ml)	
Glucose	Inoculum Size	Aeration	Temperature	Sodium	Strain A	Strain B
−	−	−	+	+	0.00	2.44
+	−	−	−	+	2.90	5.05
−	+	−	+	−	2.44	4.10
+	+	−	−	−	3.35	7.03
−	−	+	−	−	3.35	5.28
+	−	+	+	−	2.14	3.95
−	+	+	−	+	2.60	4.82
+	+	+	+	+	1.30	2.74

Similar expressions can be written for the expected value of the other main effects; for each of them except $\hat{\alpha}$, their alias strings each include only 1 two-factor interaction for the fractional factorial design we have selected. Given significance of some collection of these strings in a replicated experiment, or apparent significance via a normal plot in an unreplicated study, the individual effects that are most likely "real" must be identified by other information – expert knowledge, hierarchy or heredity rules, and/or further experiments.

13.4 Example: bacteria and bacteriocin

Leal-Sanchez, Jimenez-Diaz, Maldonado-Barragan, Garrido-Fernandez, and Ruiz-Barba (2002) conducted experiments for the purpose of optimizing the production of bacteriocin from bacteria in controlled laboratory cultures. Bacteriocin is a natural food preservative that can potentially be useful in canned foods. The five factors studied in one experiment were the amount of of glucose (1% or 2% wt/vol), initial inoculum size (5 or 7 $\log_{10}$ CFU/ml), aeration (1 or 0 liter/min), temperature (25 or 35°C), and sodium (3% or 5% wt/vol). For each of eight selected combinations of the factor levels, an experimental trial was performed in which the bioreaction was allowed to take place under standardized conditions. For each trial, the responses collected were the maximum (over time) bacteriocin activity detected, expressed in units of $\log_{10}$ AU/ml, for each of two bacteria strains, *L. plantarum 128/2* (response A) and *L. fermentum ATCC 14933* (response B). The experimental design used was a 2^{5-2} single-replicate, regular fractional factorial plan, listed in coded factors along with the responses for each run in Table 13.1.

13.5 Comparison of fractions

13.5.1 Resolution

Just as some generating relations produce better (for most purposes) blocked, full-replicate designs than others, some fractional factorial designs are also better than others. *Resolution*, introduced by Box and Hunter (1961), is an index used to compare fractional factorial designs for overall quality of the inferences that can be drawn, and is defined as the length of the shortest word (or order of the lowest-order effect) aliased with "I" in the generating relation. Generally, designs with greater resolution are deemed better, and a design goal is often to find a fractional factorial design of greatest resolution for a given number of runs and number of factors. The resolution of a design is sometimes denoted by a subscripted roman numeral; for example, any fractional factorial formed as one block of the complete six-factor design denoted by "I = ±ABCD = ±ABEF (= ±CDEF)" would be denoted as a 2_{IV}^{6-2} fractional factorial plan.

To see why fractions with large resolution are generally desirable, consider some general cases:

- Suppose a design is of resolution II, with a generating relation including +AB. Then A = +B = ..., that is, at least some pairs of main effects are aliased.

- Suppose a design is of resolution III, with a generating relation including +ABC. Then A = +BC = ..., that is, main effects will not be aliased with each other, but at least some will be aliased with two-factor interactions. So, resolution III plans support complete estimation of all main effects if all interactions are absent.

- Suppose a design is of resolution IV, with a generating relation including +ABCD. Then A = +BCD = ..., that is, main effects will not be aliased with each other or with two-factor interactions, but at least some will be aliased with three-factor interactions. Also AB = +CD = ..., that is, at least some pairs of two-factor interactions will be aliased. So, resolution IV plans support unbiased estimation of all main effects even if two-factor interactions are present, but cannot be used to estimate all effects in a second-order model.

- Suppose a design is of resolution V, with a generating relation including +ABCDE. Then A = +BCDE = ..., and AB = +CDE = ..., that is, all main effects can be estimated without bias if interactions of order less than 5 are absent, and all two-factor interactions can be estimated if no effects of greater order are present. So, resolution V plans support estimation of a complete second-order model, and supply unbiased estimates when that model is correct.

Generally, the *worst cases* of effect aliasing in a regular fractional factorial design – that is, the aliasing relationships involving the effects of lowest order – are determined by the lowest-order effect aliased with I. A general characterization is that in a design of resolution R, no O-order effect is confounded with any effect of order less than $R - O$. The fractional factorial designs most often used in practice are of resolution III, IV, and V, since designs of resolution II cannot separate the influence of all main effects, and designs of resolution VI and larger often require more than a practical number of units.

As a final example, consider the following three fractions of a 2^5 factorial structure:

- Design 1, generated with I = +ABCDE, is a 2_V^{5-1} fraction.

- Design 2, generated with I = +ABCD, is a 2_{IV}^{5-1} fraction, and so would usually be considered less desirable than design 1.

- Design 3, generated with I = +ABC = −BDE (= −ACDE) is a 2_{III}^{5-2} fraction of less resolution than either design 1 or design 2, *but* is also a smaller design, and so comparison based on resolution is not appropriate.

13.5.2 Comparing fractions of equal resolution: aberration

Resolution is a valuable index for grouping designs by overall practical value, but in most cases it categorizes reasonable choices into only a few groups, and there may be many designs of maximum resolution for a given problem. Fries and Hunter (1980) introduced design *aberration* as an additional index that can be used to "break ties" among designs of equal resolution. The aberration of a design is the number of words of shortest length (or effects of lowest order) that are aliased with "I" in its generating relation. For example, consider two 2_{IV}^{7-2} fractional factorial designs specified by:

- I = +<u>ABCD</u> = +<u>DEFG</u> (= +ABCEFG)
- I = +<u>ABCD</u> = +CDEFG (= +ABEFG)

The second has *less aberration* because it aliases a single four-factor interaction with "I" while the first aliases two four-factor interactions. In general, the goal is to find a design of:

1. maximum resolution (maximum length of shortest word), and among these

2. minimum aberration (minimum number of shortest words)

or stated another way, a design of minimum aberration from among those that are of maximum resolution. Designs of minimum aberration are generally desirable because they lead to the smallest number of alias relationships between low-order effects. For example, the first plan denoted above leads to six pairs of aliased two-factor interactions, namely:

(AB,CD), (AC,BD), and (AD,BC) because ABCD is in the generating relation

(DE,FG), (DF,EG), and (DG,EF) because DEFG is in the generating relation

But the second plan aliases only the first three pairs of interactions, and so would be considered superior.

These two criteria can be combined by looking at an ordered list of the number of words of each length for each candidate design:

Design	Length of Words							Resolution
	1	2	3	4	5	6	7	
I = +ABCD = +DEFG = +ABCEFG	(0	0	0	2	0	1	0)	IV
I = +ABCD = +CDEFG = +ABEFG	(0	0	0	1	2	0	0)	IV, minimum aberration
I = +ABC = +DEFG = +ABCDEFG	(0	0	1	1	0	0	1)	III

In such a list, the designs of maximum resolution are those for which the list contains the greatest number of "leading zeros," and among these, the designs of minimum aberration are those for which the first nonzero entry is smallest. This representation suggests other auxiliary criteria that might be used (for example, based on the size of the *second* nonzero entry in each list), and is often a convenient representation for use in computer searches for good fractional factorial designs.

13.6 Blocking regular fractional factorial designs

The basic idea of regular blocking for fractional factorials takes the same form as in complete experiments; that is, we select one or more factorial contrasts to be confounded with blocks, with the understanding that we cannot make inferences about these contrasts (at least when blocks are modeled as fixed effects). The primary difference here is that the contrasts selected are not individual factorial effects, but the strings of effects aliased because the design is a fraction. Continuing our previous 2^{5-2} example based on the generating relation $I = +ABC = -ADE (= -BCDE)$, we had eight estimable strings:

$$\text{“I”} = I + ABC - ADE - BCDE \qquad \text{“D”} = D + ABCD - AE - BCE$$
$$\text{“A”} = A + BC - DE - ABCDE \qquad \text{“E”} = E + ABCE - AD - BCD$$
$$\text{“B”} = B + AC - ABDE - CDE \qquad \text{“BD”} = BD + ACD - ABE - CE$$
$$\text{“C”} = C + AB - ACDE - BDE \qquad \text{“BE”} = BE + ACE - ABD - CD.$$

Without blocking, the last seven of these are associated with the 7 degrees of freedom that would be available for comparing treatments. The fractional factorial design:

ad	ae	b	bde
c	cde	abcd	abce

can be divided into two blocks of size 4 by confounding one of the effect strings with blocks. Suppose we select "BD", that is, we split the treatments included in the fraction using the BD column or any other column associated with an effect in the "BD" alias group. The resulting blocked fractional factorial is then:

block 1		block 2	
ae	bde	ad	b
c	abcd	cde	abce

from which we could estimate the six effect strings "A" through "E" and "BE". The contrast associated with "BD" is now confounded with the block difference, and its associated single-degree-of-freedom sum of squares would be separated as "block" variation in a fixed-block ANOVA. As is the case with full replicates, we *could* split each of these blocks a second time by confounding a second effect string, say "BE". But, this would involve a generalized interaction also, BD × BE = DE, as well as its aliases, i.e., the aliased group "A" = A + BC − DE − ABCDE. But since the main effect A is included in this group, its effect cannot be assessed while simultaneously correcting for block differences, even if we are willing to assume the other three factorial effects in the group are zero. Note, however, that this is the best we can hope for in this case; with eight data values and four blocks there remain only four degrees of freedom within blocks for assessing treatment differences, so the design:

block 1	block 2	block 3	block 4
bde	ae	b	ad
c	abcd	cde	abce

will clearly not support simultaneous estimation of all five main effects.

13.7 Augmenting regular fractional factorial designs

13.7.1 Combining fractions

Because fractional factorials lead to more ambiguous analysis than full factorial experiments, experimenters often use them in sequence so as to intelligently "build up" the information needed about the joint effects of the factors. For example, a preliminary experiment based on a fractional factorial design may be performed. If the results are uninteresting or do not show promise in the context of the study, the experimental effort may be terminated. But if the results are interesting or surprising, the natural reaction may be to perform a follow-up experiment to gain more information, by reducing the length of the estimable strings of effects to provide a more complete picture of the factorial effects. This can be easily accomplished in the framework of regular fractional factorial designs.

Suppose, for example, that the 2^{5-2} fractional factorial design discussed in Section 13.6, generated with:

$$I = +ABC = -ADE(= -BCDE)$$

has been completed. Recall that this design includes eight experimental treatments, and provides estimates of eight effect strings previously listed:

$$\text{``I'', ``A'', ``B'', ``C'', ``D'', ``E'', ``BD'', ``BE''}$$

each containing 4 factorial effects. Let's say the resulting data are interesting, but indicate that the collection of factors have more (or more complex) effects on the response than was expected. A reasonable reaction is to expand the study – that is, to *augment* the design. The initial 1/4 fraction can be converted to a regular 1/2 fraction by adding any one of the other 1/4 fractions based on the same generating relation, that is, with different signs attached to ABC, ADE, and BCDE. For example:

	first 1/4 fraction:	$I = +ABC = -ADE = (-BCDE)$
+	second 1/4 fraction:	$I = -ABC = -ADE = (+BCDE)$.
=	combined 1/2 fraction :	$I =$ $-ADE$

Note that with a fractional factorial of twice the original size, there are now twice as many estimable strings (counting "I", now 16 of them), and that each of them contains half as many effects (now two). Selecting a different second 1/4 fraction results in a different augmented design. For example:

	1st 1/4 fraction:	$I = +ABC = -ADE = (-BCDE)$
+	2nd 1/4 fraction:	$I = -ABC = +ADE = (-BCDE)$.
=	combined 1/2 fraction:	$I =$ $-BCDE$

This half-fraction is of greater resolution, and so would ordinarily be preferred if the purpose of the follow-up runs is equally focused on the effects of all factors.

To state this in a more general form, we may start with a regular 2^{f-s} fraction:

$$I = W_1 = W_2 = \ldots W_s (= W_1 W_2 \ldots)$$

where each W or symbolic product represents a signed factorial effect, including $2^s - 1$ words (selected effects and generalized interactions other than I). We then add a second 2^{f-s} fraction:

$$I = -W_1 = W_2 = \ldots - W_s (= -W_1 W_2 \ldots).$$

In doing this, we may select any fraction from the same family (other than the fraction used in the first experimental stage) by reversing the signs on any combination of the independent generators. In all, 2^{s-1} of all words (independent generators and generalized interactions) will have different signs in the two fractions. The combined design is a regular 2^{f-s+1} fractional factorial for which the generating relation contains all words for which the sign *is the same* in both 2^{f-s} fractions.

For example, suppose we begin with a 2_{III}^{6-3} fraction, with generating relation:

$$I = +ABC$$
$$= +CDE \,(= +ABDE)$$
$$= -ADF \,(= -BCDF \ = -ACEF = -BEF).$$

If we choose to continue the study after seeing the results of this experiment, it would often be good to select a second-stage fraction from the same group so that the generating relation for the combined experiment contains no words of length 3; this would increase the resolution of the combined experiment to IV. Note that if the second set of runs is identified by reversing the signs of the first three factors, this will accomplish our purpose. The second 2^{6-3} fractional will then have generating relation:

$$I = -ABC$$
$$= -CDE \,(= +ABDE)$$
$$= +ADF \,(= -BCDF \ = -ACEF = +BEF)$$

and the resulting combined design is a 2^{6-2} regular fraction with generating relation:

$$I = +ABDE = -BCDF(= -ACEF).$$

We could subsequently expand this 1/4 fraction to a 1/2 fraction the same way, by adding (say) the 2^{6-2} fraction defined by:

$$I = -ABDE = -BCDF(= +ACEF).$$

The combined result is a 2^{6-1} fractional factorial with generating relation:

$$I = -BCDF.$$

Note that if we had *begun* our investigation with a 2^{6-1} fractional factorial design, we would have had the opportunity to use a half-fraction of greater resolution, e.g., the fraction generated by:

$$I = +ABCDEF.$$

Hence there can be an "information cost" associated with sequential experimentation; the half-fraction, here a 2^{6-1} fraction developed by two "doublings" of a 2^{6-3} fraction, is not a design of maximum resolution. But if the experimental questions of interest can be adequately answered using only the data collected in the first, or first and second stages of the experiment, the sequential approach will have reduced the cost of the experimental program.

Finally, regardless of the regular 2^{f-1} half-fraction we have, a full factorial plan can be achieved with a final doubling – in our example, by adding $I = +BCDF$.

13.7.2 Fold-over designs

Recall from discussion of blocked complete factorial designs (Section 12.6) that, given one block of runs, other blocks can be constructed by reversing the signs of a selected set of factors. This leads to two practical techniques for augmenting a resolution III design, based on the analysis of the data. For example, suppose we begin with a 2_{III}^{6-3} fraction:

$$I = +ABC = +CDE \ (= +ABDE) = -ADF \ (= -BCDF = -ACEF$$
$$= -BEF).$$

Suppose analysis of the eight data values, or $8r$ values if each treatment is replicated, suggests that factor A is potentially important, and we want more information about factorial effects that involve this factor. In selecting a second 2^{6-3} fraction, we can *reverse the sign for only factor* A in the augmenting fraction, i.e., add a second stage defined by:

$$I = -ABC = +CDE \ (= -ABDE) = +ADF \ (= -BCDF = +ACEF$$
$$= -BEF).$$

Together, the two 2^{6-3} fractions form a regular 2^{6-2} fraction:

$$I = +CDE = -BCDF \ (= -BEF \ ).$$

Note that the generating relation for the combined fraction contains no effects involving factor A, because each of these have opposite signs in the two one-eighth fractions. As a result, the aliases of the main effect for A:

$$A = +ACDE = -ABCDF = -ABEF$$

are each four-factor interactions. Further, aliases of two-factor interactions involving A, e.g., $AB = +ABCDE = -ACDF = -AEF$, are each interactions involving at least three factors. Since the A main effect and all two-factor interactions involving A are individually estimable if there are no interactions of order 3, this means that the resulting design is effectively of resolution V for factor A (although not for all factors). Hence this combined design is especially effective for gaining additional information about the nature of the effects related to factor A.

Alternatively, suppose analysis following the first stage of experimentation suggests that *all* factors are potentially interesting, and we want more information on the entire system. If we *reverse the signs of all factors* in the augmenting fraction, the generating relation for the second one-eighth fraction is:

$$I = -ABC = -CDE \ (= +ABDE) = +ADF \ (= -BCDF = -ACEF =$$
$$+BEF).$$

Again, the two 2^{6-3} fractions form a regular 2^{6-2} fraction, but with a different generating relation:

$$I = +ABDE = -BCDE \ (= -ACEF \ ).$$

Here the shortest word is of length 4, implying that the design has been improved to resolution IV (for all factors). Generally, it is easy to see that reversing the signs on all factors in a second-stage fraction results in a combined experiment with a generating relation containing only words of even length. Hence if the original fraction is of resolution III, the combined fraction constructed in this way will be of resolution at least IV.

Other design augmentation strategies involving the addition of fold-over runs have been investigated. For example, John (1966) and Mee and Peralta (2000) have discussed approaches that involve adding fewer fold-over runs than the experiment doubling techniques discussed here.

13.7.3 Blocking combined fractions

The discussion of subsections 13.7.1 and 13.7.2 assumes that experimentation is done in stages, i.e., that related fractions are combined in sequence to result in a larger design with better properties. In many applications, experimenting in stages raises the question of whether the result should be viewed as a blocked design. If the operating conditions and/or the available experimental material are not the same during each stage, these potential systematic differences may suggest that the analysis of the combined data should account for possible block effects.

Again, consider the 2^{5-2} example in which two related fractions are used in sequence, but now considered as blocks:

$$\text{Block 1: } I = +ABC = -ADE(= -BCDE)$$
$$\text{Block 2: } I = -ABC = +ADE(= -BCDE).$$

Each block individually provides estimates of the same eight effect strings, but with different signs assigned to the effects within groups. For example, A + BC − DE − ABCDE is estimable in block 1, while A − BC + DE − ABCDE is estimable in block 2. Taken together, the combined fractional factorial design is defined by

$$I = -BCDE$$

which yields 16 estimable strings of two effects each; one of them, for example, is "ABC" = ABC − ADE. Note that "ABC" is comprised of factorial effects that each have a constant sign within each block, i.e., that are elements of the original generating relation, but for which the signs are different in the two blocks. Put another way, these two effects are not only aliased with each other, but they are also confounded with (or "sacrificed to") blocks if we regard the combined experiment as blocked. The remaining alias groups of effects are orthogonal to blocks, and so the associated effect strings can be estimated.

Consider another example: begin with a 2^{6-3} fraction (block 1):

$$
\begin{aligned}
I &= +ABC \\
  &= +CDE \quad (= +ABDE) \\
  &= -ADF \quad (= -BCDF = -ACEF = -BEF).
\end{aligned}
$$

At this point, the seven estimable strings (excluding "I") contain eight effects each. Now add the 2^{6-3} fraction (block 2):

$$
\begin{aligned}
I &= -ABC \\
  &= -CDE \quad (= +ABDE) \\
  &= +ADF \quad (= -BCDF = -ACEF = +BEF).
\end{aligned}
$$

The result is a 2^{6-2} fraction:

$$I = +ABDE = -BCDF(= -ACEF)$$

in which the 15 estimable strings contain four effects each, and the specific string ABC + CDE − ADF − BEF is confounded with blocks because it is positive throughout block 1 and negative throughout block 2.

Subsequently, we could add two additional 2^{6-3} fractions (blocks 3 and 4):

$$
\begin{aligned}
I &= +ABC \\
  &= +CDE \quad (= +ABDE) \\
  &= +ADF \quad (= +BCDF = +ACEF = +BEF)
\end{aligned}
$$

and

$$
\begin{aligned}
I &= -ABC \\
  &= -CDE \quad (= +ABDE) \\
  &= -ADF \quad (= +BCDF = +ACEF = -BEF)
\end{aligned}
$$

which together form

$$I = +ABDE = +BCDF(= +ACEF).$$

When combined with blocks 1 and 2, the result is a 2^{6-1} fraction:

$$I = +ABDE.$$

This blocked half-fraction yields 31 estimable strings containing two effects each, and three of them:

$$BCDF + ACEF, ABC + CDE, \text{ and } ADF + BEF$$

are confounded with blocks. Finally, we might add the remaining four 2^{6-3} fractions from this group (blocks 5 through 8):

$$I = -ABDE$$

to complete a full 2^6 experiment. Estimable strings are now reduced to individual effects, seven of which (those included in the original generating relation) are confounded with blocks.

Finally, recall from subsection 12.4.1 that information about the factorial effects confounded with blocks can be "recovered" through inter-block analysis when block effects can be assumed to be random. Such split-plot analyses can be applied to blocked fractional factorial plans also, where effect strings confounded with blocks are analyzed at the whole-plot stratum. Bingham and Sitter (2001) and Loeppky and Sitter (2002) discuss in more depth the problem of designing fractional factorial split-plot experiments.

13.8 Irregular fractional factorial designs

A major constraint attached to the use of regular fractional factorial designs is the requirement that the number of treatments included be a power of 2. This can be a serious practical problem, especially when f is relatively large.

For example, the smallest resolution III regular fraction for 10 two-level factors contains 16 runs. A larger regular fraction in 32 runs could be chosen instead, but there are no regular alternatives available between these two sizes. In this section, we describe *Plackett-Burman* (1946) designs, another class of resolution III, two-level fractional factorial designs for which first-order effects are orthogonally estimable. (These are a particular collection of designs in a broader class called *orthogonal arrays*, e.g., Hedayat, Stufken, and Sloane (1999).) The number of treatments included in the smallest Plackett-Burman design that will accommodate f factors is the *smallest multiple of 4* that is at least $f+1$, and larger designs can be constructed for any larger value of N that is also a multiple of 4. So for 10-factor experiments, designs without replication can be constructed in 12,16,20, ... runs, and designs for which each treatment is equally replicated can be constructed in any multiple of these numbers.

Construction of the designs of Plackett and Burman is easy if one has access to their paper or the tables first published there. The table rows are indexed by values of N for an unreplicated design (multiples of 4), and for a given value of N the table entry is a sequence of $N-1$ "+" and "−" symbols, representing one of the combinations of factor levels to be included in the design. For a given value of N, the corresponding design in $f = N-1$ factors is generated by using:

- the sequence of symbols to specify factor levels in the first run,
- successive cyclical permutations of these symbols to specify runs 2 through $N-1$, and
- $- \ - \ - \ ... \ -$ (i.e., the treatment in which all factors are set to their respective low levels) in run N.

Plackett and Burman presented a table running through most values of N that are multiples of 4, through 100, and subsequent authors have extended their work. A small segment of this table follows:

N	First Row
8	+ + + − + − −
12	+ + − + + + − − − + −
16	+ + + + − + − + + − − + − − −
20	+ + − − + + + + − + − + − − − − + + −

In fact, reference to the "first" run, et cetera, and to low and high levels of each factor can (and usually should) be randomized in any application. This can be done by constructing an $N \times (N-1)$ design matrix via the rules above, randomly assigning physical factors to columns, within each column randomly assigning the two factor levels to the "+" and "−" symbols, and performing the sequence of N experimental runs in a randomized order.

Because Plackett-Burman designs are generally of greatest interest when the number of experimental runs must be limited, they are often implemented in unreplicated studies. However, using $r > 1$ "copies" of the basic design, when acceptable, maintains the balance of the design and allows formal inference based on a pure error mean square.

For example, suppose we wish to construct a Plackett-Burman design in eight runs for seven factors. Using the rules above, we determine that (ignoring run-order randomization) the model matrix for an unreplicated eight-run design, assuming a main-effects model, is:

$$
\mathbf{M} = \left(\begin{array}{c|ccccccc}
+ & + & + & + & - & + & - & - \\
+ & - & + & + & + & - & + & - \\
+ & - & - & + & + & + & - & + \\
+ & + & - & - & + & + & + & - \\
+ & - & + & - & - & + & + & + \\
+ & + & - & + & - & - & + & + \\
+ & + & + & - & + & - & - & + \\
+ & - & - & - & - & - & - & -
\end{array}\right).
$$

For 4, 5, or 6 factors, any subset of these columns can be used. Further examination shows that in any of these cases, the design just constructed is actually a *regular* fractional factorial design of resolution III; a generating relation for the 7-factor design is I = +ACD = +BDE = +CEF = +DFG (plus generalized interactions). This is true of all Plackett-Burman designs for which N is a power of 2. Plackett-Burman designs for which N is not a power of 2 are called *nongeometric* designs, and while they also provide orthogonal estimates of main effects, the alias structure between these estimates and interactions is more complex than in the regular case. Hamada and Wu (1992) show that this relatively complex aliasing structure can actually be an advantage when a primary goal of the experiment is to identify a few factorial effects that adequately describe the behavior of the response variable.

13.9 Conclusion

Regular fractional factorial designs allow inferences concerning 2^f treatments in experiments employing only 2^{f-s} treatments, but useful analysis of the resulting data generally requires additional assumptions about the nature of the factorial effects. These assumptions usually amount to at least a tentative statement that some factorial effects are absent, e.g., higher-order interactions (following hierarchical assumptions about the model) or all effects involving some or most factors ("factor sparsity"). Knowledge of the system being studied and selection of an appropriate fraction in the presence of that knowledge are important for successful use of fractional factorials.

Resolution and aberration are useful indices for selecting a good general-purpose fractional factorial design for a given number of factors in a given number of runs. Sequential experimentation can be organized around combining fractions, guided by interim analysis of data collected through each stage. Sequential experimental plans can be treated as blocked to account for unintended stage-to-stage variability.

The simplicity of analysis of data collected from regular fractional factorial designs stems from the fact that any two factorial effects are either orthogonal (when they are in different alias groups) or completely confounded (when they are in the same alias group). This structure imposes substantial constraints on the size of regular fractions; specifically, the number of treatments included must be a power of 2. When this restriction is not practical, irregular fractions, which do not preserve all of the structure of regular plans, may be used. One popular class of resolution III irregular fractions is the Plackett-Burman series, for which the number of included treatments is a multiple of 4.

13.10 Exercises

1. Recall that in Section 13.7, we began with a 2^{6-3} fractional factorial plan:

$$I = +ABC$$
$$= +CDE \, (= +ABDE)$$
$$= -ADF \, (= -BCDF \; = -ACEF = -BEF)$$

and after doubling it twice, came to the half-fraction:

$$I = -BCDF.$$

Is it possible to double this plan differently, or begin with a different 2^{6-3}_{III} fraction, so that a 2^{6-1} fraction of greater resolution results from two doublings?

2. In subsubsection 13.7.2, we discussed two forms of fold-over designs. In one, the initial runs are augmented by a new set in which the sign of only one factor is reversed; the main effect associated with the selected factor is then aliased with interactions of order at least 4. In the other, the signs of all factors are reversed in the second set of runs; every main effect is then aliased with interactions of order at least 3. If the signs of *two* $(< f)$ factors, say A and B, are reversed in the second set of runs, what can be said about the resulting aliases of the two corresponding main effects? (Montgomery and Runger (1996) address this and related experimental augmentation strategies.)

3. For the experiment of Leal-Sanchez et al. (2002) described in Section 13.4:

   (a) Determine the generating relation for the design used.

   (b) Determine the estimable strings (including all factorial effects) for the experiment. Tentatively assuming that all interactions are absent, compute estimates of the five strings that include main effects, for each of the two responses (strains). Which (if any) of these effect strings are significant? (Use 2-degree-of-freedom t-statistics to answer this.)

   (c) Based on your analysis in part (b), what 2^{5-2} fraction would you recommend to Leal-Sanchez et al. for the next stage of their investigation?

4. Suppose you begin a study with an N-run Plackett-Burman design (not necessarily a regular fractional factorial). You decide to augment this initial design with its complete fold-over, i.e., N more runs selected by reversing all factors in all the original runs. Prove that in the completed $2N$-run experiment, estimates of main effects are not aliased by any two-factor interactions. Assume that the two N-run fractions do not need to be treated as blocks.

5. A fractional factorial design of resolution V allows estimation of all parameters in a model containing an intercept, main effects, and two-factor interactions. Therefore the number of treatments included in the design must be at least as large as the number of parameters in this model. Using this information:

   (a) Find a lower bound on the number of treatments in a *regular* fractional factorial of resolution V for 8 factors.

   (b) Find a generating relation that can be used to construct a resolution V fraction of this size.

6. For each of the following, write a generating relation for a fractional factorial design of resolution III, for which the main effect for factor A is not confounded with 2- or 3-factor interactions.

   (a) 2^{4-1}

   (b) 2^{6-2}

   (c) 2^{8-3}

7. Addelman (1961) introduced the idea of *3/8 fractions*; combinations of three 1/8 fractions based on generating relations containing the same "words" but with different signs. Note that these are *not* regular fractional factorial designs, but sometimes require fewer treatments than regular fractions of the same resolution. For example, a 3/8 fraction of a 2^6 factorial experiment generated with:

$$I = +\text{ABC} = +\text{CDE} = -\text{ADF}$$
$$I = -\text{ABC} = -\text{CDE} = +\text{ADF}$$
$$I = +\text{ABC} = +\text{CDE} = +\text{ADF}$$

   contains 24 treatments and allows joint estimation of all main effects and 2-factor interactions if all higher-order interactions are zero, while a regular resolution V fraction would require at least 32 runs. Using a computer, determine which pairs of factorial effect estimates are *correlated* using this 3/8 fraction, and the size of these correlations.

8. Taguchi (as described by, e.g., Kackar, 1985) discussed the use of *product arrays* in industrial experiments. An example of a product array in six factors can be constructed by generating the 3-factor fraction associated

with I = +ABC and the 3-factor fraction associated with I = +DEF, and constructing the 16-treatment design comprised of every combination of the four treatments in the first design with the four treatments in the second. The result is a regular fractional factorial design in all 6 factors. What is the generating relation of this product (array) design?

## CHAPTER 14

Factorial group screening experiments

14.1 Introduction

The smallest regular two-level factorial designs we have studied for examining the main effects of f factors are the resolution III fractions, for which the number of included treatments must be at least the smallest power of 2 greater than f. In many cases, somewhat smaller Plackett-Burman plans can be used. However, there are occasionally situations in which preliminary *factor screening* experimentation must be carried out for a large number of factors, with the expectation that most will have little or no influence on the responses, and with the goal of identifying those few that do have nonnegligible effects for subsequent follow-up experimentation. In some screening situations, even orthogonal resolution III plans may require an unrealistically large number of experimental runs. Clearly, the use of smaller experimental designs will require even more assumptions. But such assumptions may be justified when effect sparsity is expected and follow-up experiments will be conducted to provide more precise estimation of effects found to be "active" in the screening study.

Factorial *group screening* designs rely on the seemingly unusual strategy of completely aliasing subsets of main effects, so that the overall size of the experiment will be small relative to the number of individual factors. We will describe factorial group screening plans that are comprised of orthogonal designs that are actually of resolution II – that is, with identifying relations including words such as AB – in the individual factors.

For example, suppose a two-level factorial experiment is needed to discover which few of 25 factors actually have some effect on the response. The first stage of a group screening study might be based on an experiment in which the factors are divided into five groups of five factors each. Within each group, the factors are intentionally aliased – that is, all five are applied either at their respective high levels or low levels in each run. The first experiment can then be thought of as being executed to examine the effects of five "group factors"; possible designs would then include the full 2^5 factorial design, or 2^{5-1}_V or 2^{5-2}_{III} fractions, any of which might be replicated. Suppose that a main effects model (in the five group factors) is used to analyze the data from the first stage experiment, and that only one of the groups appears to be "active," that is, has an effect judged to be nonzero. A tentative conclusion might then be that the factors in the remaining groups do not affect the response, and that at least one factor from the "active" group is important. A second stage experiment could then be constructed to obtain information about the effects of the five individual factors in this group, the other 20 factors being held at constant levels. Again, a full 2^5 factorial or fractional 2_V^{5-1} or 2_{III}^{5-2} plans, possibly replicated, could serve this purpose. If (say) resolution III fractions had been used in each stage, and each had been performed in two replicates to allow for estimation of variance, the total experimental program would have required 32 experimental runs, whereas a replicated regular fractional factorial of resolution III in all 25 individual factors would have required 64 runs, and a replicated Plackett-Burman design would have required 56 runs.

There are clearly risks involved in using such a strategy. Some apparently "inactive" groups removed in the first stage might actually include important factors for which the effects "cancel" in the group. Also, there is no guarantee that only one first-stage group is judged to be active; in the worst case, *all* group factors might appear "active" in the screening experiment, requiring a second-stage experiment large enough to reconsider *all* individual factors. We will cover these issues more carefully in Section 14.4, but note here that when the majority of individual factors actually do have little or no effect on the response, and appropriate care is taken in the factor grouping assignments, group screening can be a very effective and efficient experimental strategy.

14.2 Example: semiconductors and simulation

Ivanova, Malone, and Mollaghasemi (1999) used a group screening experiment to determine which of 17 inputs were most influential on the output of a whole-line semiconductor manufacturing stochastic simulation model. The inputs (or factors for our purposes) listed in Table 14.1 were each considered at two different levels, and the output (response) examined was the number of wafers produced in a certain period of time for one of the products. Because the semiconductor manufacturing process contains a large number of sequential steps (around 250 in this case), overall production characteristics are heavily dependent on the "queueing" patterns that develop between steps. An initial "queue size analysis" identified three general kinds of process steps as being most critical in plant throughput, specifically *stepper*, *implanter*, and *etcher* steps. Further, it was expected that for any kind of process machinery, *mean time to failure*, *mean time before repair*, *lot dispatch rules*, *number of machines*, and *operator ratio* could have an important impact on process performance. Guided by this general knowledge of the system, five group factors were defined associated with mean time before failure, et cetera, each containing three individual factors associated with steppers, implanters, and etchers, respectively. Within each group, "high" and "low" levels of each individual factor would be expected to have the same qualitative effect on the output variable, e.g., an increase in mean time to failure would likely increase throughput whether it is encountered in stepper, implanter, or etcher stages. Finally, two additional model inputs were considered, *lot release rules* and *hot*

Table 14.1 Factor Groups in Screening Study of Ivanova et al. (1999)

	Individual Factor	Group Factor
1	Mean Time Before Failure/steppers	A
2	Mean Time Before Failure/implanters	A
3	Mean Time Before Failure/etchers	A
4	Mean Time To Repair/steppers	B
5	Mean Time To Repair/implanters	B
6	Mean Time To Repair/etchers	B
7	Lot Dispatch Rule/steppers	C
8	Lot Dispatch Rule/implanters	C
9	Lot Dispatch Rule/etchers	C
10	Number of Machines/steppers	D
11	Number of Machines/implanters	D
12	Number of Machines/etchers	D
13	Operator Ratio/steppers	E
14	Operator Ratio/implanters	E
15	Operator Ratio/etchers	E
16	Lot Release Rules	F
17	Hot Lots Percentage	G

lots percentage; these inputs are not so clearly related to the first 15, and were isolated as individual factors (i.e., groups of only one individual factor each) in the screening experiment.

The initial screening experiment was designed as the regular 2^{7-3}_{IV} fractional factorial in the seven group factors, associated with the generating relation:

$$I = +ABCE = +BCDF = +ACDG$$

so that $t = 16$ unique treatments (model input parameter vectors) were included in the study. For each input vector, the model was run five times with a different random number seed each time, that is, $r = 5$ replications were included for each treatment for a total of $N = 80$ response values. When the seven (group) main effects were tested at the $\alpha = 0.15$ level, those for groups B, E, and F were significant. As a result, a second stage of experimentation was carried out focusing on the seven individual factors in these groups (4, 5, 6, 13, 14, 15, and 16). At this stage, an unreplicated regular 2^{7-2}_{IV} fraction in $N = 32$ runs was used; subsequent testing of effects at the $\alpha = 0.05$ level indicated that factors 14, 15, and 16, along with three two-factor interactions each involving one of these factors, were significant. Note that the entire experimental program required 112 runs; the smallest regular resolution IV fraction in the original 17 inputs would have contained 64 factor combinations, so that two replicates of this plan would have involved 128 simulation runs, and five complete replicates, as used by Ivanova et al. in their screening experiment, would have required 320 runs.

14.3 Factorial structure of group screening designs

The earliest forms of group screening designs were not developed for factorial experiments, but for situations in which multiple samples of material were intentionally combined and evaluated with one physical test to determine whether any of the samples in the pool contained a substance of interest. An application that received substantial attention was the testing of pooled blood specimens for a relatively rare antigen; if analysis of the pooled sample was negative, all individuals represented in the pool were classified as antigen-free, but if analysis of the pooled sample was positive, each individual had to be retested individually. Dorfman (1943) provided a statistical framework for such studies. Watson (1961) described how these ideas can be extended to factorial experiments. Watson's work and subsequent developments are reviewed by Morris (2006).

As noted in Section 14.1, factorial group screening may be a reasonable approach to experimentation when (1) it is reasonable to assume that most factors have no or negligible influence on the response, a situation often called *effect sparsity*, and (2) the immediate experimental goal is to determine which few of the factors actually do have nonnegligible effects. Suppose the f individual factors of interest are divided into g numbered groups containing $f_1, f_2, \ldots, f_g$ individual factors, respectively. For convenience, label the first *group factor* A, and the individual factors in this group A_1 through A_{f_1}, and use similar notation for the remaining individual factors and groups. In terms of the individual factors, a group screening design is a resolution II plan for which the identifying relation contains:

$$I = A_1 A_2 = A_2 A_3 = \ldots = A_{f_1-1} A_{f_1}$$

$$B_1 B_2 = B_2 B_3 = \ldots = B_{f_2-1} B_{f_2}$$

$$\ldots$$

$$G_1 G_2 = G_2 G_3 = \ldots = G_{f_g-1} G_{f_g} \ldots$$

plus generalized interactions implied by these words. There are $\sum_{i=1}^{g}(f_i-1) = f - g$ words (not counting generalized interactions) aliased with "I" in this expression, so if no additional words are included, the associated fractional factorial design includes $2^{f-(f-g)} = 2^g$ treatments – a full factorial arrangement in the g group factors. A little thought will show that the generalized interactions implied in this identifying relation include all words of *even* length that can be formed from factor labels associated with the same group; for example, $A_1 A_3 A_4 A_6$ is a generalized interaction because it can be formed as the symbolic product:

$$A_1 A_2 \times A_2 A_3 \times A_4 A_5 \times A_5 A_6.$$

Similarly, words that include an even number of factor levels from each of multiple groups, e.g., $A_1 A_2 B_1 B_2 B_3 B_4$, are also generalized interactions. But words that include an odd number of factor labels from any group are not included. To see this, note that the product of any $A_i A_j$ and $A_k A_l$ results in a two-letter word if one index from the first word matches one from the second, or a four-letter word if there are no common indices. By extension, further multiplication by words comprised of factor labels from the first group cannot result in a word of odd length.

If a smaller design is desired, additional words can be added to the generating relation containing one factor label (or equivalently, any odd number of factor labels) from each of several groups, for example $A_1B_1C_1$. At this point, we can simplify notation by realizing that since all factors in a group are confounded – i.e., are indistinguishable within the design – we can ignore the subscripts identifying particular individual factors and refer to this simply as ABC, associated with a *group* three-factor interaction. Addition of this new "independent" word to the generating relation results in a design of half the previous size, now 2^{g-1} treatments, and can be thought of as the half-fraction identified by I = ABC in the *group* factors. Hence while the design is of resolution II in the individual factors (because, for example, A_1A_2 is in the generating relation), the group factor design is of resolution III. Depending on the number of group factors and the required resolution of the group factor design, further words comprised of the unsubscripted group labels can be added to produce a smaller fraction; if s such independent words are selected, the resulting design will include 2^{g-s} treatments. In the example of Section 14.2, the investigators divided 17 factors into seven groups and used a resolution IV, one-eighth fraction, or 2^{7-3}_{IV}, of 16 treatments in the first stage of screening.

Data analysis following a group screening design focuses on the groups, since no information is available that allows separation of the influence of individual factors within a group. For example, if the main effects associated with the individual factors in group 1 are denoted by $\alpha_1, \alpha_2, \ldots \alpha_{f_1}$, then the expectation of each response from the screening experiment contains either

$$+\alpha_1 + \alpha_2 + \alpha_3 + \cdots + \alpha_{f_1}$$

for runs in which all factors in group 1 are set at their high levels, or

$$-\alpha_1 - \alpha_2 - \alpha_3 - \cdots - \alpha_{f_1}$$

for runs in which all factors in group 1 are set at their low levels. As a result, we may write a model for the data in the screening experiment substituting a group main effect, $\alpha = \sum_{i=1}^{f_1} \alpha_i$, which appears positively for runs in which the "group factor" is at the high level and negatively for runs in which it is at the low level. Similarly, all $f_1 f_2$ two-factor interactions associated with one individual factor from group 1 and one individual factor from group 2 must take a common sign in each run, so their sum can be replaced with a group two-factor interaction, $(\alpha\beta) = \sum_{i=1}^{f_1} \sum_{j=1}^{f_2} (\alpha_i \beta_j)$. But two-factor interactions for which both factors are members of the *same* group are aliased with μ, e.g., recall that A_1A_2 is aliased with I in the generating relation and so $+(\alpha_1\alpha_2)$ appears in the expectation of each response.

More generally, the mean structure for a group effects factorial model may be written:

$$E(y) = \mu_G + \alpha + \beta + \gamma + \cdots + (\alpha\beta) + \cdots + (\alpha\beta\gamma) + \cdots$$

with the understanding that each main effect or interaction term actually includes the sum of all main effects or interactions from the indicated group or groups. Interactions involving an even number of factors from each group are aliased with μ. Interactions involving an even number of factors from some groups, but an odd number of factors from one group, are aliased with the main effect for the last group. For example, $(\alpha_1\alpha_2\beta_1)$ is included in β because A_1A_2 is included in the generating relation, and so $B_1 = A_1A_2B_1$. Hence all 2^f individual effect factorial terms are included in one of the 2^g group effect terms if a full 2^g factorial design is used. If a fractional factorial design in the group factors is used instead (i.e., a 2^{g-s} fraction), further aliasing exists based on the fraction selected.

For a small example, suppose six factors are combined in three groups of size 2. Factorial effects in an individual effects model include μ; the six main effects α_1, α_2, β_1, β_2, γ_1, and γ_2, 15 two-factor interactions such as $(\alpha_2\gamma_1)$, et cetera, through the single six-factor interaction $(\alpha_1\alpha_2\beta_1\beta_2\gamma_1\gamma_2)$. The group effects model contains μ_G; the three main effects α, β, and γ; three two-factor interactions including $(\alpha\beta)$; and the single three-factor interaction $(\alpha\beta\gamma)$. Table 14.2 displays the relationship between group effects and individual effects. In general, each p-th order group interaction contains the sum of all p-th order individual interactions in which each factor is associated with a different group, *and* other individual interactions of order $p+2$, $p+4$, et cetera, for which some groups contribute an even number of individual factors. If the group screening design is one or more replicates of the complete 2^3 design in the group factors, all parameters in the group effect model are estimable, implying that all equivalent sums of parameters in the individual effects model are estimable.

Table 14.2 Relationship Between Group Model Parameters and Individual Factor Parameters in the Example of Section 14.3

Group Factor Model Terms	Individual Factor Model Terms
μ_G	$\mu + (\alpha_1\alpha_2) + (\beta_1\beta_2) + (\gamma_1\gamma_2) + (\alpha_1\alpha_2\beta_1\beta_2)$ $+ (\alpha_1\alpha_2\gamma_1\gamma_2) + (\beta_1\beta_2\gamma_1\gamma_2) + (\alpha_1\alpha_2\beta_1\beta_2\gamma_1\gamma_2)$
α	$\sum_{i=1}^2 \alpha_i + (\alpha_i\beta_1\beta_2) + (\alpha_i\gamma_1\gamma_2) + (\alpha_i\beta_1\beta_2\gamma_1\gamma_2)$
β	$\sum_{i=1}^2 \beta_i + (\alpha_1\alpha_2\beta_i) + (\beta_i\gamma_1\gamma_2) + (\alpha_1\alpha_2\beta_i\gamma_1\gamma_2)$
γ	$\sum_{i=1}^2 \gamma_i + (\alpha_1\alpha_2\gamma_i) + (\beta_1\beta_2\gamma_i) + (\alpha_1\alpha_2\beta_1\beta_2\gamma_i)$
$(\alpha\beta)$	$\sum_{i=1}^2 \sum_{j=1}^2 (\alpha_i\beta_j) + (\alpha_i\beta_j\gamma_1\gamma_2)$
$(\alpha\gamma)$	$\sum_{i=1}^2 \sum_{j=1}^2 (\alpha_i\gamma_j) + (\alpha_i\beta_1\beta_2\gamma_j)$
$(\beta\gamma)$	$\sum_{i=1}^2 \sum_{j=1}^2 (\beta_i\gamma_j) + (\alpha_1\alpha_2\beta_i\gamma_j)$
$(\alpha\beta\gamma)$	$\sum_{i=1}^2 \sum_{j=1}^2 \sum_{k=1}^2 (\alpha_i\beta_j\gamma_k)$

If a fraction is used for the group screening design, group factorial effects are also confounded. For example, using the 2^{3-1} half-fraction associated with I = ABC means that α is confounded with $(\beta\gamma)$, and so the sum of individual factorial effects associated with *both* of these group parameters is estimable, but the two individual sums can no longer be separately estimated.

14.4 Group screening design considerations

14.4.1 Effect canceling

As discussed in Section 14.3, two-factor interactions involving factors in the same group are confounded with μ. Hence factor screening experiments usually focus on the identification of factors with nonnegligible main effects, at least in the first experimental stage. (However, Lewis and Dean (2001) have discussed the use of group screening experiments to identify nonzero two-factor interactions as well.) Suppose for the moment that all interactions are actually zero, and recall that a resolution III or IV design in the group factors allows estimation of

$$\alpha = \alpha_1 + \alpha_2 + \cdots + \alpha_{f_1} \quad \beta = \beta_1 + \beta_2 + \cdots + \beta_{f_2} \quad \cdots \quad \gamma = \gamma_1 + \gamma_2 + \cdots + \gamma_{f_g}.$$

If $\hat{\alpha}$ is judged to be sufficiently different from zero relative to apparent background noise (through comparison to mean square error [*MSE*] or a normal or half-normal plot) group 1 will be judged to be "active"; if not, the individual factors in group 1 will be assessed to be unimportant.

Note, however, that some or all of $\alpha_1, \alpha_2, \ldots, \alpha_{f_1}$ can be nonzero, but their sum must be zero. For example, if $\alpha_1 = -\alpha_2$ and $\alpha_i = 0$, $i = 3, 4, \ldots, f_1$, then $\alpha = 0$ even through the main effects associated with factors 1 and 2 may be very large. This possibility is sometimes called *effect cancelling*, and constitutes one of the biggest risks in group screening.

If the *potential direction* of each main effect can be assumed, factor groups can be formed to minimize the risk of effect cancelling. That is, if the experimenter is willing to make statements of the form *"If factor A really is active, I would expect response values to be generally larger when it is set to level l,"* then factors can be grouped, and the levels designated "+" and "−" can be arranged so that the *anticipated* signs of individual factor main effects in each group are the same. If the experimenter is correct in all (or some) anticipated effect signs, this eliminates (or reduces) the risk of failing to identify active factors due to effect cancelling.

14.4.2 Screening failure

The *efficiency* of group screening is directly related to the number of individual factors that can be eliminated from consideration in the first (grouped) stage. As discussed in Section 14.1, if five groups of five factors are screened, and only one group is passed on to the second stage of experimentation, then

20 individual factors have been eliminated from further study, potentially resulting in substantial cost savings. But if all five groups appear to contain active factors, very little has been gained in the first stage of experimenting. Here, the second-stage experiment might well be the same (25-factor) study that would have been employed had the screening design not been executed. This phenomenon might be called *screening failure*.

Again, prior knowledge about the likely influence of individual factors can be used to reduce the risk of screening failure. Other things being equal, screening is relatively more efficient if the important individual factors are all assigned to one group or a relatively few groups. Conversely, screening will be ineffective (or relatively inefficient) if important individual factors are assigned to all (or relatively many) groups. If the experimenter is willing to classify individual factors by categories such as "likely important," "perhaps important," and "likely unimportant," or even assign subjective probabilities for the activity of each factor, this information can be used to isolate the factors thought to be most critical in one or a few groups. If a large number of factors are confidently labeled "almost surely inactive," one or a few relatively active large groups may be formed for these factors to minimize the value of g, and so also the size of the screening experiment. Note, however, that if such a large group is unexpectedly classified as "active" in the analysis of first-stage data, this will require a larger second-stage study.

14.4.3 Aliasing

Now suppose that the individual factor model actually includes some two-factor interactions. Recall (Chapter 13) that even if a resolution III fraction is used as a design for all (individual) factors, there are two-factor interactions aliased with at least some main effects. The extent of possible aliasing increases when a resolution III design in the group factors is used. For example, for three factor groups, each of size 3 factors, and the 2_{III}^{3-1} fraction generated by I = ABC, the group 1 main effect, $\alpha = \alpha_1 + \alpha_2 + \alpha_3$, is aliased with the group 2-by-group 3 interaction $(\beta\gamma) = \sum_{i=1}^{3} \sum_{j=1}^{3} (\beta_i \gamma_j)$, that is, *nine* individual two-factor interactions appear in the estimable function associated with the group 1 main effect. If some of these are actually nonzero and of opposite sign from $\alpha_1 + \alpha_2 + \alpha_3$, this could result in effect cancelling, even if the α_i's are all of the same sign. Conversely, even if α_1, α_2, and α_3 are all zero, nonzero elements of $(\beta\gamma)$ may lead to erroneous classification of group 1 as "active," increasing the number of individual factors that must be included in the second-stage experiment.

As with ordinary (individual factor) fractional factorial experiments, the best protection against aliasing main effects with two-factor interactions in a group screening context is to increase the resolution of the fraction to IV, thereby ensuring that group main effects (sums of individual main effects) are not aliased by any two-factor interactions. This may be even more important in group screening because so many (individual) two-factor interactions have the potential to alias group main effect estimates if a resolution III plan is used.

14.4.4 Screening efficiency

The primary reason for using a group screening approach to factorial experimentation is the reduction of experimental effort required to identify the active factors. However, the number of experimental runs that will be needed to screen f factors cannot be precisely known *a priori* because the procedure is inherently sequential; the experimental design for the follow-up stage (and even the number of factors that will be included in the experiment) cannot be determined until the data collected in the first stage are analyzed. However, if an investigator is willing to supply a probability that each factor is active, calculations of the expected number of experimental runs required can be made.

Watson (1961) presented an analysis of the expected number of runs required in both stages of a screening study, in which the probability that each factor is active is deemed to be p, and the status of each factor is regarded as being independent of that of all others. He showed that when the numbers of runs in each of the first- and second-stage designs is minimal (i.e., one more than the number of group factors in the first stage, and one more than the number of individual factors being tested in the second stage), and no mistakes are made in screening (i.e., the nonzero factorial effects are large enough that they can be easily differentiated from background noise), the number of group factors that minimizes the expected number of total runs required is approximately

$$g = f\sqrt{p}$$

and that when each group is of equal size, the expected number of runs required for this value of g is approximately

$$N = 2g + 2.$$

Hence the "optimal" number of groups is smaller, and the number of factors per group larger, as individual factors are given a smaller probability of being active. The rule is a useful guideline; however, experimenters are often unable or unwilling to offer a "firm" value of p. Further, in many experiments there are groups of factors that are known to be at least qualitatively similar in their action (e.g., as demonstrated in the example of Section 14.2), that are inconsistent with the idea of "independent activity" for individual factors. Finally, the assumptions that minimal designs are used at each stage, and that errors are not made in classifying factor groups at the first stage (each made to simplify the mathematics of the argument) are generally unrealistic in practice. It should be remembered that even factor groups of size 2 lead to initial screening designs in roughly half the number of runs that would have been required for single-stage experimentation, and that relatively few experimental runs will be required in the second stage if most factors actually are not active.

14.5 Case study

Hendrix (1979) conducted an experiment to determine which of 15 controlled factors, each with two levels, had substantial impact on the cold crack resistance of an industrial product. Experimental factors were characteristics of the production process, e.g., *solvent* either recycled or refined, and *dry roll temperature* either 75° or 80°. The experiment was actually designed as an unreplicated 2_{III}^{15-11} fraction in 16 runs; half-normal plots of the resulting main effects estimates suggest that two factors are active, with main effects of approximately 3 and 1. The remaining estimates are all less than 0.30 in absolute value; the average of their squared values is 0.035 (a "pseudo-MSE"), suggesting that σ may be about 0.187.

Suppose that the authors had decided to investigate these 15 factors via group screening instead, based on an *a priori* assumption that the probability of any factor being independent is approximately 0.10. This suggests a first-stage experiment of $15 \times \sqrt{0.1} = 4.7 \approx 5$ groups, and without further a priori information, the 15 factors might be randomly split into groups of size 3. If this group structure is accepted, the next choice to be made is the particular 5-factor design to be used in the first-stage experiment. If we are convinced (or are willing to assume) that interactions are negligible, an eight-run 2_{III}^{5-2} plan is an obvious choice. Without replication, and assuming a model containing only main effects, this plan will provide a MSE with only 2 degrees of freedom; still, if the active main effects are on the order of $10 \times \sigma$ (as they appear to be in the analysis of Hendrix) there is a good chance that the groups containing the main effects would be detected.

Suppose now that the two active factors were actually randomly assigned to two different groups; the probability that this will have happened under random group assembly is 0.80. If both of these groups, and no others, are flagged as "active," six individual factors will need to be tested in the second-stage experiment. Here, an eight-run 2_{III}^{6-3} seems reasonable. It will produce an MSE with only 1 degree of freedom, but assuming the nature of random variation is the same in both experiments, the two mean squares might be pooled to yield a 3-degrees-of-freedom estimate of σ^2. Again, while this may not be an especially precise estimate, the chance of correctly identifying the two active factors is substantial. If both active factors had actually been assigned to the same initial group, and only that group had been flagged as active, a four-run 2_{III}^{3-1} might have been adequate for a follow-up design, but note that the number of runs in this design is the same as the number of parameters in the first-order model, and so the MSE from the first-stage experiment would have also been used in the second stage.

If all had "gone well," the two active main effects might have been identified using group screening with a total of either 16 or 12 experimental runs (with probability 0.80 and 0.20, respectively). Recall that Hendrix accomplished the actual study as a single-stage, 16-run resolution III fraction, so little would have been gained in this case. Still, it should be noted that since their design provided *no* degrees of freedom for estimating σ^2, use of a half-normal plot was required. Group screening is *most* effective when the number of factors is quite large, and the number of active factors is (relatively) quite small. For example, continue to suppose that there actually are two active main effects with values of three and one, and that σ is about 0.187, but that the total number of factors being screened is $f = 60$ (four times as many as before). A single orthogonal resolution III design that allows estimation of all main effects would require at least 64 runs. If our *a priori* probability of any factor being independent is approximately 0.025 (1/4 as large as before), Watson's analysis suggests a first-stage experiment of $60 \times \sqrt{0.025} = 9.5 \approx 10$ groups. An initial resolution III group design from the Plackett-Burman series could be executed in 12 runs, and a follow-up design to identify the active elements of two groups could be carried out in 16 runs. Even if the first design is replicated to provide 12 degrees of freedom in pure error, the total experimental effort of 40 runs would represent a substantial savings over single-stage experimentation.

14.6 Conclusion

Group factor screening is a technique that is primarily useful when a small fraction of a large number of factors are expected to be active. It is most effective when at least some prior information is available concerning the potential influence of each factor on the response. Information that is especially helpful is the probability that each factor is active, the likely direction of each factor's main effect if active, and relationships between factors that are likely to have similar effects if active. When these conditions are met, group factor screening can reduce the experimental effort required to identify the active factors in an experimental system.

Watson's (1961) discussion of how group screening can be used in factorial experimentation includes a more general analysis than is described here. He considers practical issues of how many active factors (those with nonzero main effects) one can expect to discover, and how many inactive factors one should expect to "falsely discover," as functions of the design selected. His analysis is predicated on the generally unrealistic assumption that all nonzero main effects have the same value, but more general analytical expressions that are simple enough to be useful may not be achievable. Mauro and Smith (1982) described a numerical study of the performance of group screening plans under the somewhat more general assumption that all nonzero main effects have the same absolute value, but can have different signs (allowing for effect cancelling). Patel (1962) discussed construction of group screening strategies involving more than two sequentially constructed designs, where relatively large factor groups that appear active in one stage are split into relatively smaller groups in the next, with the final experiment designed in those individual factors that are elements of "active groups" in all previous stages.

14.7 Exercises

1. Consider a two-level factorial group experiment in which 12 individual factors are to be screened in six groups, each of size 2. The first-stage screening plan that will be used is the quarter fraction associated with the identifying relation I=ABCD=CDEF=ABEF. In the following, assume that all interactions of order 3 or more are zero, both in the group effects and individual effects models.

   (a) Identify the 16 strings of estimable effects in the notation of the group factor model (counting the string including "I," or the model intercept).

   (b) Rewrite these 16 strings in terms of the effects in the individual factor model.

2. Continuing with the group screening design described in exercise 1 ($g = 6$, $f = 12$, $N = 2^{6-2} = 16$), suppose now that no interactions are present, and that there are three nonzero individual main effects of value 10, 10, and -10. If factors are assigned to groups randomly, what is the probability that "effect cancelling" will occur, i.e., that two nonzero main effects in one group will have a zero sum?

3. Suppose 40 factors are to be screened in 10 groups of size 4. The first-stage group screening design will consist of two replicates of a 12-run Plackett-Burman design (i.e., $N = 24$). Suppose that in fact, one main effect is nonzero and has a value of one, and also that $\sigma = 2$. Suppose a two-tailed t-test (or equivalently, an F-test with one numerator degree of freedom) with type I error probability of $\alpha = 0.10$ is used to classify each group factor as active or nonactive.

   (a) What is the probability that the truly active factor will be passed on to the second-stage experiment (i.e., that its group will be identified as "active")?

   (b) What is the expected value of the number of factors that will need to be investigated individually in the second stage?

4. The success of group screening depends critically on effect sparsity, i.e., that most of the considered factors have no (or very little) influence on the responses. But such prior information is generally uncertain, and it should be understood what is likely to happen if the assumption turns out to be false. Suppose group screening is used to investigate 44 factors in groups of size 4, and that an unreplicated 12-run Plackett-Burman design is used to investigate the group effects in the first stage.

   (a) What will be the likely conclusion if, in fact, *all* main effects are nonzero and have values that are normally distributed with mean zero and variance σ_E^2? Assume that $\sigma_E^2 \gg \sigma^2$, where the latter is the variance of individual observations.

    (b) If two replicates of the Plackett-Burman design had been used as the initial design (i.e., $N = 24$), might the result be different? Why?

5. Suppose a group screening design is set up by grouping 21 factors into seven groups of size 3. A 2_{III}^{7-4} fraction generated by I = ABC = CDE = EFG is used. Suppose that the only real factorial effects are the three main effects, three two-factor interactions, and a single three-factor interaction associated with factors 1, 2, and 3, and suppose that all seven of these parameters are positive and large relative to σ. Ignoring the possibility of effect cancelling, how many factors will likely need to be individually investigated in the second stage if:

    (a) factors 1, 2, and 3 are all assigned to group A

    (b) factors 1 and 2 are assigned to group A, and factor 3 is assigned to group B

    (c) factors 1, 2, and 3 are assigned to groups A, B, and C, respectively.

6. In many cases, it is reasonable to regard the first-stage (grouped factor) and second-stage (individual factor) experiments of a screening study as two blocks of the same experiment. This view can yield an improved estimate of σ^2 for use in the final analysis. Suppose f individual factors are screened in g groups of size f_1 each in the first stage of a screening study using r replicates of a minimal orthogonal resolution III design in g factors (i.e., where the number of unique treatments t_1 is the smallest multiple of four that is larger than g). Suppose only group 1 appears to be active in the first stage, and that an unreplicated minimal orthogonal resolution III design in f_1 factors is used as a second-stage design, and that the remaining $f - f_1$ factors are each fixed at one of their two levels in the second stage. Write degrees of freedom for an analysis of variance for the combined experiment, with lines for block difference, main effects for both individual and group factors (remembering to account for linear dependencies), and residual, with the latter divided into lack of fit and pure error components. For simplicity, assume that both $g + 1$ and $f_1 + 1$ are multiples of four.

7. The "ultimate" group screening plan consists of only $g = 1$ group containing all f factors; it can be replicated r times for a total of $N = 2r$ runs. If the potential direction of each main effect is fairly certain, and there are no interactions, such an experiment can determine quickly whether *any* of the factors actually affect the response. Let $\alpha = \sum_{i=1}^{f} \alpha_i$ be the single group main effect parameter, the sum of all individual main effects parameters, and σ^2 be the error variance.

    (a) Write an expression for the probability that the group will be determined to be active if a t- (or equivalently, F-) test is used with type I error probability of 0.10.

(b) Suppose that, in fact, $\alpha_1 = \alpha_2 = \alpha_3 = 1$ and that all other individual main effects are zero. However, suppose also that the single three-factor interaction involving these three factors is also nonzero, specifically $(\alpha_1\alpha_2\alpha_3) = -3$. Write an expression for the probability that the group will be determined to be active, again with a type I error probability of 0.10.

## CHAPTER 15

Regression experiments: first-order polynomial models

15.1 Introduction

To this point, we have discussed experiments that are executed to understand differences among a finite set of treatments. In some cases, all treatments in this set are included in the experiment; this is true of all experiments for "unstructured" treatments (Chapters 3–8), and for full factorial experiments, either blocked or unblocked (Chapters 9–12). In contrast, fractional factorial designs (Chapter 13) do not include all treatments from the set. The consequence of incomplete experimentation is that treatment effects cannot be uniquely estimated unless additional assumptions can be made that effectively eliminate some model parameters. For example, regular fractional factorial experiments support estimation of effect strings – linear combinations of aliased effects – under the complete factorial model; they allow estimation of individual effects only under the assumption of a reduced model including no more than one effect in each aliased group. For treatment sets of finite size, the choice between an experiment that includes all treatments and an experiment that does not requires balancing of experimental costs and prior knowledge about the system under study. Experiments that include all treatments generally cost more than those that do not, but may be unnecessary if higher-order factorial effects are reasonably assumed to be negligible.

We shift attention in this chapter to experiments with *functional treatment structure* for which the set of possible treatments is not finite in size, and so experiments necessarily include only a subset of them. In the simplest case, regression experiments are carried out to compare treatments that are "indexed" by points in a continuous *experimental region*, denoted R. For example, a cellular biology experiment may be set up to investigate the growth rate of a certain kind of cell culture as a function of the relative proportion of two nutritional ingredients in the substrate on which it is grown. Even if all other aspects of the scenario are held constant, one ingredient might make up anywhere from 20% to 50% (by weight, say) of the two-ingredient blend, and the resulting (infinite) set of treatments correspond to an experimental region expressed as $R = [0.2, 0.5]$. Some descriptions of regression design differentiate between the experimental region indexing the treatments of interest, and the *design region* indexing the treatments that may actually be included in the experimental design. While this is a useful distinction in many contexts, we will not make it here.

While the "mechanics" of experimental design (e.g., randomization and blocking) and analysis (e.g., inference based on linear models) we have studied in previous chapters are also relevant for regression experiments, it is important to recognize that the balance between experimental cost and prior knowledge is fundamentally different for these studies. Since an experiment including all treatments of interest *cannot* be conducted, there is of necessity more reliance on modeling assumptions. This leads to increased interest in diagnostic procedures such as tests for lack of fit, that are not directly related to the experimental questions of greatest interest, but are important steps in validating the model on which interesting inferences can be made.

15.1.1 Example: bacteria and elastase

Chen, Ruan, Zhang, and He (2007) describe a series of laboratory experiments performed to study the production of elastase (an enzyme used in a number of industrial food processing applications) using cultures of *Bacillus licheni-formis* ZJUEL31410, a mutant bacterial strain that produces relatively large quantities of elastase. The first of the experiments reported was a "screening" exercise, to determine which of six continuous, controllable variables representing process conditions were most influential on yield. Table 15.1 lists data collected over 11 experimental runs, in which process temperature, time of reaction, volume of the culture, proportional volume of the inoculum, seed age, and shake speed of the flask were each varied over three evenly spaced values, and the resulting concentration of elastase recorded.

Note that *in principle*, the number of unique values of each control variable was not necessarily limited to three – the experimenters were interested in modeling responses across the indicated range of each variable, and there was (apparently) no operational constraint on the number of unique values that might have been used for each.

Table 15.1 Data from Elastase Experiment of Chen et al. (2007)

Run #	Temperature (°C)	Time (h)	Volume (ml)	Inoculating volume(%)	Seed age(h)	Shake speed(r/min)	Elastase (U/ml)
1	30	30	30	7	10	180	294
2	40	18	30	3	10	220	265
3	30	18	20	7	10	220	307
4	40	30	30	7	26	220	351
5	30	30	20	3	26	220	311
6	40	30	20	3	10	180	355
7	30	18	30	3	26	180	110
8	40	18	20	7	26	180	274
9	35	24	25	5	18	200	355
10	35	24	25	5	18	200	375
11	35	24	25	5	18	200	403

15.2 Polynomial models

While we shall continue to use linear statistical models as a basis for thinking about the structure of data, it is helpful to change notation slightly to emphasize the continuous nature of the experimental region. Let d denote the number of controlled variables used to define a treatment, i.e., the *dimension* of the experimental region, and let $\mathbf{x} \in R$ be a d-element vector or point corresponding to any particular treatment. For unblocked experiments, let y_{ij} denote the jth observation taken at the ith treatment included in the experiment, and say:

$$y_{ij} = \alpha + \mathbf{x}_i'\boldsymbol{\beta} + \epsilon_{ij},$$

$$i = 1 \ldots t, \quad j = 1 \ldots n_i,$$

$$\epsilon_{ij} \text{ iid with } E(\epsilon_{ij}) = 0 \text{ and } Var(\epsilon_{ij}) = \sigma^2 \qquad (15.1)$$

where $\mathbf{x}_i$ encodes the set of experimental conditions for the ith of t distinct treatments appearing in the design, and $\boldsymbol{\beta}$ is a d-element set of parameters to be estimated. Note that in this first-order polynomial model, the elements of the parameter vector $\boldsymbol{\beta}$ represent *slopes* of the expected response corresponding to each controlled variable. (Chapter 16 addresses second-order polynomial models.) Together with the intercept α, these parameters specify a hyperplane in $(d+1)$-dimensional space that represents the assumed structure of the expected response for all treatments corresponding to points in R. A matrix model for the entire N-run experiment, $N = \sum_{i=1}^{t} n_i$, can then be written as:

$$\mathbf{y} = \alpha\mathbf{1} + \mathbf{X}_2\boldsymbol{\beta} + \boldsymbol{\epsilon},$$

$$E(\boldsymbol{\epsilon}) = \mathbf{0}, \quad Var(\boldsymbol{\epsilon}) = \sigma^2\mathbf{I}. \qquad (15.2)$$

It is important to note that the elements of $\boldsymbol{\beta}$ actually represent treatment *differences*, corresponding to the usual concept of experiments being comparative studies. For example, under the first-order model, β_1 is the difference in expected response between any two treatments for which x_1 varies by one unit of value while the other independent variables are held constant. The intercept, α, is an experiment-wide effect, and so is regarded as a nuisance parameter in true experimental studies. Another very important practical aspect of regression models is the effect of the physical *units of measurement* associated with each variable. (Note that "units of measurement" are completely unrelated to "experimental units.") As an example, consider the experiment of Chen et al. (2007) described in subsection 15.1.1. The specified first-order polynomial model for these data, along with units of measurement, is:

$$\begin{aligned}
y \quad &= \alpha + x_1\,\beta_1 + x_2\,\beta_2 + x_3\,\beta_3 + x_4\,\beta_4 + x_5\,\beta_5 \; + x_6\,\beta_6 + \epsilon. \\
\text{U/ml} \quad & \qquad\quad °\text{C} \qquad\quad \text{h} \qquad\;\; \text{ml} \qquad\; \% \qquad\;\; \text{h} \quad \text{r/min}
\end{aligned}$$

Because the left and right sides of the model equation must agree in their units as well as in their numerical values, each term in the fitted model must be in units of U/ml, units of elastase produced per milliliter of culture. (A measurement "unit" of elastase was defined by Chen et al. to be the amount required to solubilize 20 mg of elastin-Congo red at standard conditions.) Hence β_1,

and its estimate and the standard error of that estimate, must take units of
U/(ml×°C) so that the term $x_1\beta_1$ will agree in units with the response variable. Note that because the six controlled variables in this model are expressed
in different physical units, their corresponding coefficients also have different
units; β_3, for example, carries units of U/ml^2. It is especially important to
understand this when comparing the results of different studies, where the
same physical units may not have been used in recording experimental conditions or data. For example, in comparable studies in which shake speed (x_6) is
recorded in revolutions per *second* and response data are recorded in U/ml, a
comparable estimate of β_6, in measurement units of (sec × U)/(ml × r), would
need to be roughly 60 times the estimate found from a regression performed
using the data recorded in Table 15.1.

In analysis of data from regression experiments, controlled variables are
often linearly rescaled so that the highest and lowest values used for each
(coded) controlled variable are +1 and −1. This would be done for x_2 in the
experiment of Chen et al. by defining:

$$x_2(\text{scaled}) = 2 \times (x_2(\text{h}) - 24\text{h})/(30\text{h} - 18\text{h}) \qquad (15.3)$$

where 30, 18, and 24 are the highest, lowest, and average physical values in
time units (h). In fact, the fraction on the right side of equation (15.3) is
"unitless" since both numerator and denominator are in the same physical
units. Since controlled variables rescaled in this way are all (strictly speaking)
unitless, the physical units attached to the regression coefficients are reported
on the scale of the response variable (U/ml). However, this can be misleading
if taken out of context. What is *really* happening here is that x_2 has been
coded to "units of half the range covered in the experiment," a scale on which
$[-1, +1]$ represents the range of time values selected by the experimenter. So,
for example, if the smallest and largest reaction times selected in a comparable
experiment had been 12 h and 36 h, scaled values of x_2 would still be "unitless"
and take values ranging from −1 to +1, and resulting estimates of β_2 would
still be in units of U/ml, but direct comparison to the estimates from this
experiment would require multiplication by a factor of $\frac{30-18}{36-12}$, the ratio of
ranges (of physical values) used in the two experiments.

Note finally that in models where x's are defined as (unitless) indicator
variables – as with most of the material in this book describing experiments
in which experimental conditions are unstructured, or are represented only by
the points on a factorial lattice – all model coefficients are given in the same
units as the response variable.

15.3 Designs for first-order models

15.3.1 Two-level designs

Where first-order models are assumed to be adequate representations of $E(y)$
as a function of **x**, designs that employ two appropriately selected values for
each controlled variable are often popular and effective. In fact, the two-level

factorial and fractional factorial designs discussed in Chapters 11–13, where the symbolic "+1" and "−1" values are used to represent relatively large and small values of each controlled variable, are very efficient designs for regression experiments when analysis is based on a first-order model. Where the controlled variables are scaled so that $R = [-1, +1]^d$, all of these designs lead to information matrices of form $\mathcal{I} = N\mathbf{I}_{d \times d}$ for the parameter vector $\boldsymbol{\beta}$, and are "optimal" designs in this context in the sense that no design for this R leads to unbiased estimates of linear functions of $\boldsymbol{\beta}$ that have smaller variance.

15.3.2 Simplex designs

Recall that the smallest two-level orthogonal factorial designs we could construct for first-order factorial models were the Plackett-Burman designs (Section 13.6), requiring N be at least the smallest multiple of four greater than the number of factors included in the experiment. In the present context we are estimating d slopes (one associated with each element of $\mathbf{x}$), and a Plackett-Burman design of 12 runs, for example, could be used to orthogonally estimate a first-order regression model with one intercept (α) and as many as 11 slope parameters ($\boldsymbol{\beta}$). Because regression experiments allow selection of design points from a continuous experimental region, it is possible in some cases to construct orthogonal designs for first-order models that require slightly fewer points than the Plackett-Burman designs by using more than two values to represent at least some controlled variables.

The *simplex design* introduced by Box (1952) is one such design, which contains $N = d + 1$ distinct design points for any value of d (that is, the Plackett-Burman restriction that N must be a multiple of four is not required). The name *simplex* is due to the fact that the $d + 1$ treatments used in such a design are the vertices of a simplex in $\mathcal{R}^d$ – a geometric figure for which each pair of vertices is separated by the same distance. For example, an equilateral triangle is a simplex in $\mathcal{R}^2$, and a tetrahedron is a simplex in $\mathcal{R}^3$. Mathematically, a simplex design is described by any $(d + 1) \times d$ model matrix $\mathbf{X}_2$ for which $(\mathbf{1}|\mathbf{X}_2)'(\mathbf{1}|\mathbf{X}_2)$ is a diagonal matrix with nonzero diagonal elements. Such a matrix can always be constructed when R is, for example, a cuboid or spheroid in d-dimensional space. For example, the matrix:

$$\mathbf{X}_2 = \sqrt{N} \begin{pmatrix} +\dfrac{1}{\sqrt{2}} & +\dfrac{1}{\sqrt{6}} & +\dfrac{1}{\sqrt{12}} & \cdots & +\dfrac{1}{\sqrt{d(d+1)}} \\ -\dfrac{1}{\sqrt{2}} & +\dfrac{1}{\sqrt{6}} & +\dfrac{1}{\sqrt{12}} & \cdots & +\dfrac{1}{\sqrt{d(d+1)}} \\ 0 & -\dfrac{2}{\sqrt{6}} & +\dfrac{1}{\sqrt{12}} & \cdots & +\dfrac{1}{\sqrt{d(d+1)}} \\ 0 & 0 & -\dfrac{3}{\sqrt{12}} & \cdots & +\dfrac{1}{\sqrt{d(d+1)}} \\ \cdots & \cdots & \cdots & \cdots & \cdots \\ 0 & 0 & 0 & \cdots & -\dfrac{d}{\sqrt{d(d+1)}} \end{pmatrix} \qquad (15.4)$$

Table 15.2 Simplex Experiment Factor Values for the Study of Chen et al. (2007)

Variable		Values			Physical Units
temperature (x_1)	physical	40	30	35	°C
	scaled	$+\frac{1}{\sqrt{2}}$	$-\frac{1}{\sqrt{2}}$	0	–
time (x_2)	physical	30	18	26	h
	scaled	$+\frac{1}{\sqrt{6}}$	$-\frac{2}{\sqrt{6}}$	0	–
volume (x_3)	physical	30	20	27.5	ml
	scaled	$+\frac{1}{\sqrt{12}}$	$-\frac{3}{\sqrt{12}}$	0	–
inoculating volume (x_4)	physical	7	3	6.2	%
	scaled	$+\frac{1}{\sqrt{20}}$	$-\frac{4}{\sqrt{20}}$	0	–
seed age (x_5)	physical	26	10	23.3	h
	scaled	$+\frac{1}{\sqrt{30}}$	$-\frac{5}{\sqrt{30}}$	0	–
shake speed (x_6)	physical	220	180	–	r/min
	scaled	$+\frac{1}{\sqrt{42}}$	$-\frac{6}{\sqrt{42}}$	–	–

satisfies the requirements for any value of d. Individual columns of $\mathbf{X}_2$ can be scaled to allow them to fit within the bounds for each controlled variable as set by the definition of R. (The columns are already "centered" since each must be orthogonal to $\mathbf{1}$.) Note that, for this selection of $\mathbf{X}_2$, the last controlled variable appears at two values in the experiment, while all others appear at three. For any such $\mathbf{X}_2$ the information matrix for $\boldsymbol{\beta}$ is $\mathcal{I} = \mathbf{X}_2'(\mathbf{I} - \frac{1}{N}\mathbf{1}(\mathbf{1}'\mathbf{1})^{-1}\mathbf{1}')\mathbf{X}_2 = \mathbf{X}_2'\mathbf{X}_2 = N\mathbf{I}$, again, because each column of $\mathbf{X}$ is orthogonal to $\mathbf{1}$.

Example

Suppose Chen et al. (2007) had chosen to use a simplex design in the elastase study described in subsection 15.1.1. Using the model matrix "template" given in equation (15.4), and requiring the upper and lower physical values of each independent variable to be as shown in Table 15.1, coded and physical values for the six experimental factors would have been as displayed in Table 15.2. Note that, except in the case of temperature(x_1), coded "0" does not correspond to the center of the experimental range. Simplex designs can sometimes be rotated in d-dimensional space to yield more evenly distributed values.

15.4 Blocking experiments for first-order models

Full and regular fractional factorial designs can be blocked for regression experiments in essentially the same way they are blocked when factors take only two levels. For example, Section 13.6 discusses blocking of a 2^{5-2} fractional factorial with the defining relation I = +ABC = −ADE (= −BCDE), into two blocks of size 4 by confounding BD = ACD = − ABE = − CE with the block difference. Likewise, we can design a regression experiment to estimate the slopes associated with five continuous independent variables for the model

$$y = \alpha + \sum_{i=1}^{5} x_i \beta_i + \epsilon$$

as:

$$\mathbf{X}_2 = \begin{pmatrix} +1 & -1 & -1 & -1 & +1 \\ -1 & +1 & -1 & +1 & +1 \\ -1 & -1 & +1 & -1 & -1 \\ +1 & +1 & +1 & +1 & -1 \\ \hline +1 & -1 & -1 & +1 & -1 \\ -1 & +1 & -1 & -1 & -1 \\ -1 & -1 & +1 & +1 & +1 \\ +1 & +1 & +1 & -1 & +1 \end{pmatrix} \begin{matrix} \text{block 1} \\ \\ \\ \\ \text{block 2} \\ \\ \\ \end{matrix} .$$

In factorial notation, the six estimable effect strings (other than the two including I and BD) are

$$\begin{aligned} A + BC - DE - ABCDE \\ B + AC - ABDE - CDE \\ C + AD - ACDE - BDE \\ D + ABCD - AE - BCE \\ E + ABCE - AD - BCD \\ BE + ACE - ABD - CD \end{aligned}$$

The five slopes in an assumed first-order polynomial model correspond to the factorial main effects A, B, C, D, and E. Because the design is an orthogonal resolution III fractional factorial:

• these are each aliased only with factorial terms of order greater than one,

• they are orthogonal to each other, and

• they are orthogonal to the factorial string aliased with the block difference.

As a result, the eight-run regression experiment in two blocks of size 4 is fully efficient, with design information matrix $\mathcal{I} = 8\mathbf{I}_{5\times 5}$ for β, and provides one degree of freedom (that which would be associated with the BE + ACE − ABD − CD string in a factorial model) for estimating σ^2.

More generally, a common aim in blocking an experiment is that blocks be arranged in a way that does not reduce the treatment information – in the present case, the information regarding the d slopes associated with the independent variables. *Orthogonally blocked* experiments accomplish this by yielding the same design information matrix $\mathcal{I}$ as their unblocked counterparts. For example, consider the nine-run experiment in $d = 2$ independent variables corresponding to the coded model matrix:

$$\mathbf{X}_2 = \begin{pmatrix} +1 & +1 \\ 0 & +1 \\ -1 & +1 \\ +1 & 0 \\ 0 & 0 \\ -1 & 0 \\ +1 & -1 \\ 0 & -1 \\ -1 & -1 \end{pmatrix}.$$

Under the model that also includes an intercept:

$$\mathbf{y} = \alpha\mathbf{1} + \mathbf{X}_2\boldsymbol{\beta} + \boldsymbol{\epsilon}$$

the design information matrix for $\boldsymbol{\beta}$ is

$$\mathcal{I} = \mathbf{X}_2'\left(\mathbf{I} - \frac{1}{N}\mathbf{J}\right)\mathbf{X}_2 = \mathbf{X}_2'\mathbf{X}_2 = 6\mathbf{I}_{2\times2}$$

because $\mathbf{1}'\mathbf{X}_2 = \mathbf{0}$. Now suppose the same nine treatments are to be included in an experiment that requires grouping the nine units in three blocks of size 3. We can extend the model to accommodate this as:

$$\mathbf{y} = \mathbf{X}_1\boldsymbol{\theta} + \mathbf{X}_2\boldsymbol{\beta} + \boldsymbol{\epsilon}.$$

where $\mathbf{X}_1$ is a 9-by-3 matrix of indicator variables – a single "1" in each row – and $\boldsymbol{\theta}$ is a three-element vector of block parameters. (Because the rows of $\mathbf{X}_1$ sum to $\mathbf{1}$ we do not need to include the parameter α.) In particular, suppose we assign treatments to blocks as follows:

$$\mathbf{X}_1 = \begin{pmatrix} 1 & 0 & 0 \\ 0 & 1 & 0 \\ 0 & 0 & 1 \\ 0 & 0 & 1 \\ 1 & 0 & 0 \\ 0 & 1 & 0 \\ 0 & 1 & 0 \\ 0 & 0 & 1 \\ 1 & 0 & 0 \end{pmatrix}.$$

A special feature of this arrangement is that within each block, the values of both x_1 and x_2 sum to zero, that is $\mathbf{X}_1'\mathbf{X}_2 = \mathbf{0}$. Letting $\mathbf{H}_1 = \mathbf{X}_1(\mathbf{X}_1'\mathbf{X}_1)^-\mathbf{X}_1$, note that this implies $\mathbf{H}_1\mathbf{X}_2 = \mathbf{0}$. But $this$ in turn implies that

$$\mathcal{I} = \mathbf{X}_2'(\mathbf{I} - \mathbf{H}_1)\mathbf{X}_2 = \mathbf{X}_2'\mathbf{X}_2 = 6\mathbf{I}_{2\times2}$$

just as in the unblocked design. The orthogonal blocking structure associated with $\mathbf{X}_1'\mathbf{X}_2 = \mathbf{0}$ implies that there is no information reduction associated with blocks in the second design, just as the balanced structure associated with $\mathbf{1}'\mathbf{X}_2 = \mathbf{0}$ implies that there is no information reduction associated with the intercept in the first design.

Of course, there is also a sense in which the blocked experiment *does* entail an "information loss," in that fewer degrees of freedom can be allocated to estimating σ^2; mean square error (*MSE*) has six degrees of freedom in the unblocked experiment, but only four in the blocked experiment. Still, as discussed in earlier chapters, if blocking is being used effectively, the value of σ^2 should be smaller in the blocked experiment, offsetting the loss of *MSE* degrees of freedom.

15.5 Split-plot regression experiments

As with experiments in which factors have discrete levels, the operational restrictions of regression experiments sometimes require that they be designed and analyzed as split-plot studies. This may be related to:

- some controlled variables being more difficult (expensive or time-consuming) to change than others, or

- the practical need to apply some controlled variables to larger quantities of experimental material (plots) and other controlled variables to smaller subquantities of material (split-plots).

Split-plot regression experiments can be organized in a manner similar to that described in Chapter 10 for factors with discrete levels.

15.5.1 Example: bacteria and elastase reprise

Consider again the experiment of Chen et al. (2007) described in subsection 15.1.1. In this study, an experimental unit is referred to as a "flask" containing a bacterial culture prepared as required by the selected levels of culture volume, inoculating volume, and seed age, and processed according to the selected levels of temperature, time, and shake speed. Suppose, however, that this experiment was performed in a laboratory in which three flasks are processed together as a "group," and that due to operational constraints, all flasks in a group must be processed at the same temperature and for the same length of time. Table 15.3 contains a data set for this hypothetical experiment, in which some values have been altered (as indicated) for purposes of demonstration. Included in the table are coded values of each independent variable (-1, 0, and $+1$ for variables represented at three values, and -1 and $+1$ for variables represented at two levels) to facilitate calculation of ANOVA components.

Because anything unique that occurs during the processing of a group of flasks has the potential to affect all three experimental units in the same way, it is sensible to think of "group" as an experimental block within which the levels of flask volume, inoculating volume, seed age, and shake speed are

Table 15.3 Altered Data from Chen et al. (2007). Entries Marked with (*) are Altered from or Added to the Original Data Set. The Last Run Listed Did Not Appear in the 11-Run Design of Chen et al. The Second Value Listed for Each Controlled Variable is the Coded (unitless) Value

"Group"	Temp. (°C)		Time (h)		Volume (ml)		Inoc. Vol. (%)		Seed Age (h)		Shake (r/min)		Elastase (U/ml)
1	30	−1	30	+1	30	+1	7	+1	10	−1	180	−1	344*
1	30	−1	30	+1	20	−1	3	−1	26	+1	220	+1	361*
1	30*	−1	30*	+1	25	0	5	0	18	0	200	0	325*
2	40	+1	18	−1	30	+1	3	−1	10	−1	220	+1	265
2	40	+1	18	−1	20	−1	7	+1	26	+1	180	−1	274
2	40*	+1	18*	−1	25	0	5	0	18	0	200	0	275*
3	30	−1	18	−1	20	−1	7	+1	10	−1	220	+1	307
3	30	−1	18	−1	30	+1	3	−1	26	+1	180	−1	110
3	30*	−1	18*	−1	25	0	5	0	18	0	200	0	225*
4	40	+1	30	+1	30	+1	7	+1	26	+1	220	+1	351
4	40	+1	30	+1	20	−1	3	−1	10	−1	180	−1	355
4	40*	+1	30*	+1	25*	0	5*	0	18*	0	200*	0	300*

varied from unit to unit. Because levels of temperature and time are actually assigned to groups of three flasks, rather than independently for each, these groups of flasks are the experimental units for the purpose of evaluating the effects of these two factors. Hence the experiment should be regarded as a split-plot arrangement, with temperature and time applied to flask groups (the whole-plots) and the levels of the remaining controlled variables applied to individual flasks (the split-plots).

An ANOVA decomposition of these data is accomplished along the same lines as described in subsection 12.4.1 for two-level factorial experiments, where in this case the estimated slopes $\hat{\beta}_i$ are analogous to main effects, and terms of higher order are excluded from the model. A result of the assumption of a first-order model is that additional degrees of freedom are assigned to the residual lines; one extra degree of freedom for whole-plot residuals corresponding to the AB interaction, and four degrees of freedom for split-plot residuals (all of which would have been associated with terms in a higher-order model). Table 15.4 demonstrates this for the data of Table 15.3; coefficient estimates are calculated for the coded controlled variables as if the design were not blocked:

$$\hat{\beta}_1 = 12.3333, \quad \hat{\beta}_2 = 48.3333, \quad \hat{\beta}_3 = -28.375, \quad \hat{\beta}_4 = 23.125,$$
$$\hat{\beta}_5 = -21.875, \quad \hat{\beta}_6 = 25.125,$$

and $\bar{y}_{1:3}$, et cetera, refer to block average values. The design information matrices are $\mathcal{I}_{inter} = 12\mathbf{I}_{2\times2}$ for β_1 and β_2, and $\mathcal{I}_{intra} = 8\mathbf{I}_{4\times4}$ for β_3 through β_6. Standard errors for $\hat{\beta}_1$ and $\hat{\beta}_2$ are each based on the whole-plot *MSE*:

$$\sqrt{\frac{3201.3576}{1} \Big/ 12} = 16.3333,$$

Table 15.4 Split-Plot ANOVA Decomposition for the Data of Table 15.3

Stratum	Source	Degrees of Freedom	Sum of Squares
whole-plot	β_1 and β_2	2	$12(\hat{\beta}_1^2 + \hat{\beta}_2^2) = 29858.6270$
	residual	1	difference $= 3201.3576$
	corrected total	3	$3[(\bar{y}_{1:3} - \bar{y})^2 + (\bar{y}_{4:6} - \bar{y})^2$
			$+ (\bar{y}_{7:9} - \bar{y})^2 + (\bar{y}_{10:12} - \bar{y})^2]$
			$= 33059.9846$
split-plot	β_3 through β_6	4	$8 \sum_{i=3}^{6} \hat{\beta}_i^2 = 19597.5000$
	residual	4	difference $= 2578.5154$
	corrected total	11	$\sum_{i=1}^{12} (y_i - \bar{y})^2 = 55236.0000$

while those for $\hat{\beta}_3$ through $\hat{\beta}_6$ are each based on the split-plot *MSE*:

$$\sqrt{\frac{2578.5154}{4} \Big/ 8} = 8.9766.$$

15.6 Diagnostics

Many of the general diagnostic techniques described in Chapter 6 are useful in the context of regression experiments. Residual plots and power transformation are especially useful in the formulation and checking of polynomial models. In the case of regression experiments, the more specific question of *adequacy of fit* or *lack of fit* of a selected model, relative to a polynomial of higher order, is often of central concern. (Both phrases are used in practice; "adequacy of fit" refers to the null hypothesis, while "lack of fit" suggests the alternative. In the linear models literature, the latter may be more commonly used.) In subsection 15.6.1, we describe a particular test for adequacy of the assumed first-order model based on the addition of runs made at the *center point* of the experimental region. In subsection 15.6.2, the more general *F*-test for adequacy of fit described in subsection 2.7.1 is discussed in the context of the first-order regression model.

15.6.1 Use of a center point

Suppose a two-level design has been selected, and for each factor half the runs are made at one level, and half at the other. Two-level designs with this property are sometimes called *balanced designs*. Suppose coding is such that each element of $\mathbf{x}$ is $+1$ for half the runs in the experiment and -1 for the other half. Since $\mathbf{x}$ is a point from continuous R, we can also select points that are not at corners of the experimental region, such as the *center point* $\mathbf{x} = \mathbf{0}$. Let $\bar{y}_f$ be the average of all n_f data values taken from the factorial portion of the design, and let $\bar{y}_c$ be the average of all n_c data values collected from the center point treatment. Under the assumed first-order linear model:

$$E(\bar{y}_f) = E(\bar{y}_c) = \alpha$$

and since $\bar{y}_f$ and $\bar{y}_c$ are independent, and each is independent of MSE,

$$\frac{\bar{y}_f - \bar{y}_c}{\sqrt{MSE(n_f^{-1} + n_c^{-1})}} \sim t(N - d - 1).$$

On the other hand, if the model actually contains monomials of form $x_i^2 \beta_{ii}$, then

$$E(\bar{y}_f) = \alpha + \sum_{i=1}^{d} \beta_{ii}$$

because each $(\pm 1)^2 \beta_{ii} = \beta_{ii}$ is a component of the expected response at all factorial points, while $E(\bar{y}_c)$ is still α. Hence the t-statistic shown above can be the basis for a test of the hypothesis:

$$\mathrm{Hyp}_0 : \sum_{i=1}^{d} \beta_{ii} = 0.$$

Note that addition of the center point does not allow individual estimation of all "pure quadratic" coefficients, β_{ii}, $i = 1, \ldots, d$ (except when $d = 1$), but only their sum. It is certainly possible that individual β_{ii} might be nonzero, but in such a way that their sum is zero. In such a case, the test described has power equal to its selected level of the test; that is, the lack of fit due to the pure quadratic terms would be undetectable (except by accident). However, if a preliminary assumption of the first-order model is reasonable, the test does offer partial confirmation if the hypothesis is not rejected, or useful evidence to the contrary if it is.

This t-test for lack of fit can also be performed using data from a blocked experiment, provided:

• each block contains the same number of factorial runs,

• the factorial points in each block are balanced for each factor, i.e., have the same number of coded "+1" and "−1" values, and

• each block contains the same number of center point runs.

Under these conditions, $E(\bar{y}_f) = E(\bar{y}_c)$ if the first-order model is correct, and $E(\bar{y}_f) = E(\bar{y}_c) + \sum_i \beta_{ii}$ if pure quadratic terms are present (because parameters representing additive block effects contribute the same component to the expectation of each average). The only adjustment needed is a reduction of degrees of freedom to $N - d - b$, where b is the number of blocks, to reflect the addition of the $b - 1$ degree-of-freedom block component to the ANOVA decomposition.

15.6.2 General test for lack-of-fit

The test described in the subsection 15.6.1 is a popular "one degree of freedom" test for the adequacy of a first-order model in a linear regression problem

Table 15.5 ANOVA for First-Order Polynomial Model

Source	Degrees of Freedom	Sum of Squares
$\beta\|\alpha$	d	$SST = \mathbf{y}'(\mathbf{H} - \frac{1}{N}\mathbf{J})\mathbf{y}$
residual	$N - d - 1$	$SSE = \mathbf{y}'(\mathbf{I} - \mathbf{H})\mathbf{y}$
lack of fit	$t - d - 1$	$SSLOF = \mathbf{y}'(\mathbf{H_Z} - \mathbf{H})\mathbf{y}$
pure error	$N - t$	$SSPE = \mathbf{y}'(\mathbf{I} - \mathbf{H_Z})\mathbf{y}$
corrected total	$N - 1$	$SSCT = \mathbf{y}'(\mathbf{I} - \frac{1}{N}\mathbf{J})\mathbf{y}$

that is useful when the design satisfies the necessary balance properties. The test for lack of fit described in subsection 2.7.1 is more general, sometimes more powerful, and can be applied to any design containing replicated points. Suppose we have data from a design at t distinct experimental conditions, with $t > d + 1$, and with complete model matrix $\mathbf{X} = (\mathbf{1}|\mathbf{X}_2)$ of full column rank. Because $t > d + 1$, the experimental data contain more information than is minimally necessary to estimate a first-order model. The specific polynomial forms of the models that *can* be fit depends upon the selection of the t distinct points in the design. Further, if $N > t$, the design contains one or more groups of replicate runs – those runs coded with identical rows in $\mathbf{X}$. As in subsection 2.7.1, define a more general model:

$$\mathbf{y} = \mathbf{Z}\phi + \epsilon, \quad E(\epsilon) = \mathbf{0}, \quad Var(\epsilon) = \sigma^2\mathbf{I}$$

where the t columns of $\mathbf{Z}$ contain indicator variables for each of the t unique rows in $\mathbf{X}$. Letting $\mathbf{H} = \mathbf{X}(\mathbf{X}'\mathbf{X})^{-1}\mathbf{X}'$ and $\mathbf{H_Z} = \mathbf{Z}(\mathbf{Z}'\mathbf{Z})^{-1}\mathbf{Z}'$, an ANOVA decomposition of variability associated with the model and residuals, with further decomposition of residual variation into lack of fit and pure error components, is shown in Table 15.5.

In this notation, a test for $\text{Hyp}_0 : E(\mathbf{y}) = \alpha\mathbf{1} + \mathbf{X}_2\beta$, or "adequacy" of the assumed model, can be based on an F-statistic comparing the second and last components of the ANOVA decomposition:

$$\frac{SSLOF/(t - d - 1)}{SSPE/(N - t)} : F_{1-\alpha}(t - d - 1, N - t).$$

If $\text{Hyp}_0 : E(\mathbf{y}) = \alpha\mathbf{1} + \mathbf{X}_2\beta$ is not rejected, a test for $\text{Hyp}_0 : \beta = \mathbf{0}$, or "effectiveness" of the assumed first-order model, can be based on an F-statistic comparing the first and last components of variation:

$$\frac{SST/d}{SSPE/(N - t)} : F_{1-\alpha}(d, N - t)$$

again, using $SSPE$ as a denominator sum of squares, or on an F-statistic comparing the first component of variation to the pooled lack of fit and pure error components – the usual "residual" sum of squares based on the assumption of a first-order model:

$$\frac{SST/d}{SSE/(N - d - 1)} : F_{1-\alpha}(d, N - d - 1).$$

Table 15.6 ANOVA for First-Order Polynomial Model with Blocking

Source	Degrees of Freedom	Sum of Squares
$\beta\|\theta$	d	$\mathbf{y}'(\mathbf{H} - \mathbf{H}_1)\mathbf{y}$
residual	$N - b - d$	$\mathbf{y}'(\mathbf{I} - \mathbf{H})\mathbf{y}$
lack of fit	$N^* - b - d$	$\mathbf{y}'(\mathbf{H_z} - \mathbf{H})\mathbf{y}$
pure error	$N - N^*$	$\mathbf{y}'(\mathbf{I} - \mathbf{H_z})\mathbf{y}$
total corrected for blocks	$N - b$	$\mathbf{y}'(\mathbf{I} - \mathbf{H}_1)\mathbf{y}$

The advantage of the first form is that the denominator mean square is a valid estimator of σ^2 even if the first-order model is incorrect. If the first-order model *is* correct, the denominator mean square of the second test statistic is also valid, and is based on more degrees of freedom than the first, leading to a more powerful test.

A similar analysis can be constructed if the experiment is blocked by augmenting the model to incorporate the additional structure:

$$\mathbf{y} = \mathbf{X}_1\boldsymbol{\theta} + \mathbf{X}_2\boldsymbol{\beta} + \boldsymbol{\epsilon}$$

where $\mathbf{X}_1$ is a matrix of indicator variables containing one column per block, with a single 1 in each row, and $\boldsymbol{\theta} = (\theta_1, \theta_2, \ldots, \theta_b)'$ is a set of nuisance parameters associated with the b blocks. Because the columns of $\mathbf{X}_1$ sum to $\mathbf{1}$, α is mathematically redundant and can be dropped from the model. Suppose there are N^* unique experimental conditions *determined by both treatment and block assignment*, i.e., unique rows in $\mathbf{X} = (\mathbf{X}_1|\mathbf{X}_2)$, that $\mathbf{Z}$ is an alternative model matrix with one column corresponding to each of the N^* distinct experimental conditions, and that $N > N^* > \text{rank}(\mathbf{X})$. Then, an ANOVA decomposition of the variability remaining after blocks are eliminated is shown in Table 15.6.

Example

Suppose a regression experiment in two controlled variables is constructed as indicated in the following model matrix (in coded variables):

$$\mathbf{X}_2 = \begin{pmatrix} +1 & +1 \\ +1 & -1 \\ -1 & +1 \\ -1 & -1 \\ 0 & -1 \\ 0 & +1 \\ -1 & 0 \\ +1 & 0 \\ 0 & 0 \\ 0 & 0 \\ 0 & 0 \\ 0 & 0 \end{pmatrix}.$$

Table 15.7 Example ANOVA for First-order Polynomial Model where X_2 has Zero Column Sums

Source	Degrees of Freedom	Sum of Squares
$\beta\vert\alpha$	2	$SST = 6(\hat{\beta}_1^2 + \hat{\beta}_2^2)$
residual	9	$SSE = SSCT - SST$
lack of fit	6	$SSLOF = SSCT - SST - SSPE$
pure error	3	$SSPE = \sum_{i=9}^{12}(y_i - \bar{y}_{9:12})^2$
corrected total	11	$SSCT = \sum_{i=1}^{12}(y_i - \bar{y})^2$

Due to the simple structure of this design, it is easily verified that:

$$\mathbf{H} - \frac{1}{N}\mathbf{J} = \mathbf{H}_2 = \mathbf{X}_2(\mathbf{X}_2'\mathbf{X}_2)^{-1}\mathbf{X}_2' = \frac{1}{6}\mathbf{X}_2\mathbf{X}_2'$$

because $\mathbf{X}_2$ has zero column sums, i.e., $\mathbf{1}'\mathbf{X}_2 = \mathbf{0}$, so

$$\mathbf{y}'\mathbf{H}_2\mathbf{y} = 6\left[\hat{\beta}_1^2 + \hat{\beta}_2^2\right]$$

and

$$\mathbf{Z}'\mathbf{Z} = \mathrm{diag}(1,1,1,1,1,1,1,1,4), \quad \mathbf{H}_\mathbf{Z} = \begin{pmatrix} \mathbf{I} & \mathbf{0} \\ \mathbf{0} & \frac{1}{4}\mathbf{J} \end{pmatrix}.$$

So in this case, Table 15.6 can be evaluated as shown in Table 15.7, where $\hat{\beta}_1$ and $\hat{\beta}_2$ are estimates based on the assumed model, and $\bar{y}_{9:12}$ is the average of response values at the replicated center point.

Now suppose the 12 runs are organized in two blocks as:

$$\mathbf{X}_2 = \begin{pmatrix} +1 & +1 \\ +1 & -1 \\ -1 & +1 \\ -1 & -1 \\ 0 & 0 \\ 0 & 0 \\ \hline 0 & -1 \\ 0 & +1 \\ -1 & 0 \\ +1 & 0 \\ 0 & 0 \\ 0 & 0 \end{pmatrix} \begin{matrix} \\ \\ \text{block 1} \\ \\ \\ \\ \\ \text{block 2} \\ \\ \\ \\ \end{matrix}.$$

Since this design is orthogonally blocked, the corrected treatment sum of squares is again $6(\hat{\beta}_1^2 + \hat{\beta}_2^2)$, based on estimates from the assumed model. This leads to the specific ANOVA decomposition shown in Table 15.8, where

Table 15.8 Example ANOVA for First-Order Polynomial Model, Orthogonal Blocking

Source	Degrees of Freedom	Sum of Squares
$\beta \mid \theta$	2	$SST = 6(\hat{\beta}_1^2 + \hat{\beta}_2^2)$
residual	8	$SSE = SSCT - SST$
lack of fit	6	$SSLOF = SSCT - SST - SSPE$
pure error	2	$SSPE = \sum_{i=5}^{6}(y_i - \bar{y}_{5:6})^2$ $+ \sum_{i=11}^{12}(y_i - \bar{y}_{11:12})^2$
total, corrected for blocks	10	$SSCT = \sum_{i=1}^{6}(y_i - \bar{y}_{1:6})^2$ $+ \sum_{i=7}^{12}(y_i - \bar{y}_{7:12})^2$

$\bar{y}_{1:6}$, $\bar{y}_{5:6}$, $\bar{y}_{7:12}$, and $\bar{y}_{11:12}$ each denote averages of the indicated collection of data values.

15.7 Conclusion

When experimental treatments are defined by the selected values of one or more continuous, controlled variables, statistical analysis based on a regression model is often appropriate. Unlike factorial experiments in which each factor is expressed through a finite set of levels, the number of possible treatments in a regression experiment is infinite. But this means that any regression experiment must include only an "infinitely small fraction" of the possible treatments, and so model selection is critical. For many experimental settings, a first-order polynomial regression model adequately represents the relationship between controlled variables and the expected response.

Two-level factorial and fractional factorial designs, with the design points located at or near the corners of the experimental region, are effective plans for first-order polynomial regression experiments. Simplex designs are alternatives of primary use when experimental runs are very expensive or time-consuming. Both factorial and simplex designs can be augmented with center point runs to provide a test for the adequacy of the first-order model.

15.8 Exercises

1. Note that the experimental design used by Chen et al. (2007)(section 15.1.1) is a two-level fractional factorial plan with $n_c = 3$ added center point runs.

   (a) Determine the defining relation and resolution of the eight-run fractional factorial portion of this design.

   (b) Compare the average of data taken at the center point to the average of the data collected over the factorial portion of the design, using a t-statistic as described in subsection 15.6.1.

   (c) Perform the F-test for lack of fit as described in subsection 15.6.2.

(d) Are these two tests equivalent in this case? If you think so, explain why. If you don't think so, explain the difference in the hypotheses being tested by the two procedures.

2. In the experiment of Chen et al. (subsection 15.1.1), one data value (run #7, $y = 101$) is substantially different from the others. Do you think it is plausible that this response is consistent with the rest of the data, or do you think it is more likely an erroneous value or the result of a faulty experimental run? Use diagnostic graphs or indices to support your argument.

3. Some investigators prefer "one factor at a time" (OFAT) experiments to standard factorial plans when performing experiments to fit first-order regression models. One such design, that might be called a "plus-and-minus" plan, results in a $(2d + 1) \times d$ model matrix (in coded variables) of form:

$$
\mathbf{X}_2 = \begin{pmatrix}
-1 & 0 & 0 & \cdots & 0 \\
+1 & 0 & 0 & \cdots & 0 \\
0 & -1 & 0 & \cdots & 0 \\
0 & +1 & 0 & \cdots & 0 \\
0 & 0 & -1 & \cdots & 0 \\
0 & 0 & +1 & \cdots & 0 \\
\cdots & \cdots & \cdots & \cdots & \cdots \\
0 & \cdots & \cdots & \cdots & -1 \\
0 & 0 & 0 & \cdots & +1 \\
0 & 0 & 0 & \cdots & 0
\end{pmatrix}.
$$

That is, for each controlled variable, the design contains a pair of runs carried out at a high level and a low level, respectively, with all other controlled variables held at their nominal (center) values, augmented by a center point at which all variables are set to their nominal values. Note that this design, with $N = 2d + 1$, requires approximately the same number of runs as the fold-over Plackett-Burman plan (with $N = 2d^+$, where d^+ is the smallest multiple of four larger than d).

(a) For general d, compare the precision of the OFAT design given above to that of the comparable fold-over Plackett-Burman design, by deriving formulae for the standard error of first-order coefficients for each plan.

(b) Despite their poor estimation properties, some investigators like OFAT designs, especially under conditions where they want to reserve the right to terminate the experiment early. Explain how an OFAT design might have some advantage in circumstances where early termination might be necessary, and what restrictions on experimental randomization would be necessary in order to use an OFAT design in this way.

4. Continuing the discussion of exercise 3, there are also experimenters who believe that the best experimental design that can be constructed for regression modeling in (say) $R = [-1, +1]^d$ is a set of N points selected randomly and uniformly from this space. Their intuitive argument is that this procedure should uniformly "cover" the experimental space (at least on average), and so the resulting design should be relatively effective regardless of the shape of the response surface. Examine the efficiency of such designs for the case $d = 3$, $N = 20$ using a statistical program that supports simulation:

   - Construct 100 such random designs.
   - For each design, compute the standard error (apart from the MSE multiplier) for each of $\hat{\beta}_1$, $\hat{\beta}_2$, and $\hat{\beta}_3$.
   - For each of the three regression coefficients, construct a histogram of the 100 values computed.

   Compare these computed values to the standard error (again, apart from MSE) that would result from a 20-run design comprised of two replicates of a 2^3 factorial in the corners of R, and four center points.

5. One practical decision that must generally be made in planning a regression experiment is the range of physical values to be used for each controllable variable. In many cases, ranges need to be limited if the expected response is to be well approximated by a first-order polynomial; that is, the response may be a more complex function if the range of experimental conditions is larger. However, the precision of estimates and power of tests for the regression parameters of a first-order regression model are improved as the width of the experimental region is increased.

   Suppose an experiment in d controlled variables will be designed as a full unreplicated 2^d factorial plan plus n_c center points. The design may be executed one of two ways; in design A the upper and lower values of each controlled variable are set at specific values (coded as $+1$ and -1 in each case), and in design B these are each made more extreme by 20% (so that, for comparison, the coded factorial values are $+1.2$ and -1.2). How much reduction (proportionally) is realized in the standard deviations of the coefficient estimates by using design B instead of design A?

6. In industrial applications, one major goal of regression experiments is to find conditions under which a process can be expected to yield greater (or smaller) values of the expected response than are realized under current or "nominal" conditions. Suppose that in an experiment involving d controlled variables, we express these variables in scaled $(-1, +1)$ form, with the center point (zero for each variable) representing nominal conditions. Suppose we are willing to assume that a first-order polynomial accurately expresses the expected response as a function of the controlled variables, and even more, that we actually know the values of the coefficients in the model – β_1, β_2, $\beta_3, \ldots, \beta_d$. What point in the experimental space, at Euclidean distance

of δ from the center point (i.e., for which $\sum_{i=1}^{d} x_i^2 = \delta^2$), yields the highest expected response? (You can use the Method of Lagrangian Multipliers described in subsection 3.4.1 to solve this constrained optimization problem. Note that in practice, we would have to use estimates of the model parameters instead, and would perhaps also want to think about the resulting uncertainty in our result. This result is key to the "path of steepest ascent," first described by Box and Wilson (1951).)

7. Consider an experimental situation in which five controlled variables (in coded form) vary over $R = [-1, +1]^5$. Three possible experimental designs are being considered:

   - Design A: A complete, unreplicated 2^5 factorial plan with $n_c = 4$ added center point runs.

   - Design B: A 2_V^{5-1} fractional factorial plan with $n_c = 4$ added center point runs.

   - Design C: A 2_{III}^{5-2} fractional factorial plan with $n_c = 4$ added center point runs.

   Suppose for planning purposes that the first-order model is correct, and that

   $$\beta_1 = \beta_2 = \beta_3 = 1, \qquad \beta_4 = \beta_5 = 0, \qquad \sigma = 3$$

   and consider the F-test for the effectiveness of the regression model, i.e., the test of:

   $$\text{Hyp}_0 : \beta_1 = \beta_2 = \beta_3 = \beta_4 = \beta_5 = 0.$$

   For each of the three proposed designs, compute:

   (a) the degrees of freedom for MSE, fitting the first-order model

   (b) the noncentrality parameter (given the assumed parameter values) associated with the test

   (c) the power of the test (given the assumed parameter values) for type I error probability 0.05.

8. Show that the regular fractional factorial in $d = 3$ dimensions based on the generating relation I = ABC is also a simplex design.

9. Consider a more general form of the split-plot regression analysis described in Section 15.5. Suppose a regression design in d controlled variables is planned as an N-run two-level factorial or regular fractional factorial of resolution at least III, perhaps replicated. The experiment is to be executed in b blocks, each of size n. The arrangement is made so that:

   - d_w variables are constant within each block (i.e., are controlled at the whole-plot stratum), and the whole-plot design is a full or regular fractional factorial design in these d_w variables when blocks are regarded as units, and

- the remaining d_s variables, $d = d_w + d_s$, each take their respective high and low values an equal number of times in each block.

Construct formulae for the degrees of freedom in a split-plot ANOVA decomposition for whole-plot variables and residual sums of squares, and split-plot variables and residual sums of squares.

## CHAPTER 16

Regression experiments: second-order polynomial models

16.1 Introduction

We continue the discussion of regression models from Chapter 15, but now consider designs for experiments in which the anticipated model is quadratic in the controlled variables. Quadratic regression models are often used in process engineering contexts, where controlled variables represent process characteristics that can be adjusted (e.g., temperatures and processing rates), especially when the response variable is some measure of quality or quantity of product for which the expectation may be optimized within the experimental region. (Note that linear polynomials cannot mimic this behavior.) *Response surface* optimization strategies (e.g., Myers and Montgomery, 2002) are often based on a series of experiments, focusing on first-order regression in the early iterations to establish paths of "steepest ascent" to new experimental regions, and second-order regression in the final step(s) when the conditions that will optimize the process are within the region of the design.

Quadratic models are used in other contexts as well, even when optimization is not of primary interest, simply because they can represent more complicated response-to-factor relationships than first-order models. Of course, regression is possible with even higher-order polynomial models, but unless there are very few factors, the number of parameters in polynomial models of order 3 or more is quite large, requiring experiments that are often impractically large. Quadratic regression polynomials are often used as a pragmatic compromise between the need to use a model that is flexible enough to adequately express the physical phenomenon of interest, and the operational need to limit the size of the experimental design.

16.1.1 Example: nasal sprays

Dayal, Pillay, Babu, and Singh (2005) described a regression experiment carried out to model effectiveness characteristics of a nasal spray as functions of percentage levels of five ingredients. Preparations were made of hydroxyurea (HU, the model drug under investigation), two surface-active polymers (labeled HEC and PEU), and two ionic excipients ($CaCl_2$ and NaCl), each at three different values (% weight of the mixture) as summarized in Table 16.1. (Note that the five percentage values sum to less than 100 in each case; the remaining material was the same for each mixture tested.) Aerosol spray was

Table 16.1 Three Values for Each of Five Controlled Variables Used in the Nasal Spray Study of Dayal et al. (2005)

Variable	Levels		
HEC	0%	2%	4%
PEO	0%	2%	4%
CaCl$_2$	0%	15%	30%
NaCl	0%	15%	30%
HU	0%	2%	4%

generated in a controlled manner, and the viscosity, mean drug diffusion time (MDT), and droplet size were determined for three different experimental runs for each of 44 preparations. Model reported for expected MDT, recorded in 8h units, is:

$$\widehat{E(y)} = 20.1038$$
$$+ 35.3040\ x_1 + 65.3821\ x_2 + 4.3340\ x_3 + 1.8093\ x_4 - 9.1356\ x_5$$
$$- 1.6799\ x_1^2 - 7.3861\ x_2^2 - 0.0483\ x_3^2 + 0.0100\ x_4^2 + 0.4909\ x_5^2$$
$$- 4.7250\ x_1x_2 - 0.8533\ x_1x_3 - 0.0200\ x_1x_4 - 0.6312\ x_1x_5$$
$$+ 0.2775\ x_2x_3 - 0.7267\ x_2x_4 + 3.0687\ x_2x_5$$
$$- 0.0071\ x_3x_4 + 0.0642\ x_3x_5$$
$$- 0.0183\ x_4x_5$$

where x_1 through x_5 are the five controlled variables summarized in Table 16.1, each in units of % weight of the mixture.

16.2 Quadratic polynomial models

Extending the form of the first-order polynomial models presented in Section 15.2, again define a d-dimensional vector $\mathbf{x} = (x_1, x_2, x_3, \ldots, x_d)'$, a point in the d-dimensional experimental region R specifying a particular experimental treatment. In scalar notation, a second-order, or quadratic, polynomial model for the expected response may be written as:

$$E(y) = \alpha + \sum_{i=1}^{d} x_i\beta_i + \sum_{i=1}^{d} x_i^2\beta_{ii} + \sum_{i=1}^{d-1} \sum_{j=i+1}^{d} x_ix_j\beta_{ij}.$$

An equivalent matrix notation that is sometimes more convenient is:

$$E(y) = \alpha + \mathbf{x}'\boldsymbol{\beta}_1 + \mathbf{x}'\mathbf{B}_2\mathbf{x}$$

where $\{\boldsymbol{\beta}_1\}_i = \beta_i$, and $\mathbf{B}_2$ is a $d \times d$ matrix with $\{\mathbf{B}_2\}_{ii} = \beta_{ii}$, and $\{\mathbf{B}_2\}_{ij} = \{\mathbf{B}_2\}_{ji} = \beta_{ij}/2, i < j$.

In many regression experiments, an important goal is to identify values of the independent variables that result in extreme (large or small) expected response. If we actually knew the coefficients of the quadratic model, the *stationary point or points* at which derivatives of $E(y)$ with respect to each x_i are zero can be easily identified as the solution set to the matrix equation:

$$\frac{\partial}{\partial \mathbf{x}'} E(y) = \boldsymbol{\beta}_1 + 2\mathbf{B}_2 \mathbf{x}_0 = \mathbf{0}.$$

When $\mathbf{B}_2$ is not singular, the solution is unique and can be written in closed form as:

$$\mathbf{x}_0 = -\mathbf{B}_2^{-1}\boldsymbol{\beta}_1/2.$$

In this case, $\mathbf{x}_0$ maximizes $E(y)$ if all eigenvalues of $\mathbf{B}_2$ are negative, minimizes $E(y)$ if all eigenvalues of $\mathbf{B}_2$ are positive, and does neither – i.e., is the location of a saddlepoint – if $\mathbf{B}_2$ has both positive and negative eigenvalues. If one or more eigenvalues of $\mathbf{B}_2$ are zero, $\mathbf{x}_0$ is not unique, but is a collection of points along a line, plane, or hyperplane in R.

While we regard the model coefficients as unknown quantities that can only be estimated with uncertainty given data, the (estimated) coefficients from the fitted model can be substituted in the above expressions to yield an estimated stationary point. For example, in the experiment of Dayal et al. described in subsection 16.1.1, $\hat{\boldsymbol{\beta}}_1$ and $\hat{\mathbf{B}}_2$ can be constructed based on the first- and second-order coefficients, respectively, of the fitted model for MDT:

$$\hat{\boldsymbol{\beta}}_1 = \begin{pmatrix} 35.3040 \\ 65.3821 \\ 4.3340 \\ 1.8093 \\ 9.1356 \end{pmatrix} \quad \hat{\mathbf{B}}_2 = \begin{pmatrix} -1.6799 & -2.3625 & -0.4266 & -0.0100 & -0.3156 \\ -2.3625 & -7.3861 & 0.1388 & -0.3634 & 1.5344 \\ -0.4266 & 0.1388 & -0.0483 & -0.0036 & 0.0321 \\ -0.0100 & -0.3634 & -0.0036 & 0.0100 & -0.0092 \\ -0.3156 & 1.5344 & 0.0321 & -0.0092 & 0.4904 \end{pmatrix}.$$

Because $\hat{\mathbf{B}}_2$ is of full rank, these yield a unique estimated stationary point:

$$\hat{\mathbf{x}}_0 = -\hat{\mathbf{B}}_2^{-1}\hat{\boldsymbol{\beta}}_1/2 = \begin{pmatrix} -0.7899 \\ 2.5078 \\ 46.9822 \\ -27.7185 \\ -21.2646 \end{pmatrix}.$$

Eigenvalues of $\mathbf{B}_2$ are mixed in sign, with three positive and two negative, so the stationary point neither maximizes nor minimizes the fitted response model. However, an even more important observation is that $\hat{\mathbf{x}}_0$ is not physically meaningful at all in the context of this problem, because the individual elements represent percent components in a mixture, and three of the five estimates are negative! More generally, estimated stationary points that lie outside the region in which an experiment is conducted should be regarded with suspicion, even if they represent physically meaningful conditions, because extrapolations based on polynomial models are often unreliable even when the model is an adequate approximation to the truth in a limited region. Graphical representations of the fitted model, especially in and near the experimental region, are often the most practical tools for understanding the general shape of the function, and (when applicable) the most promising subregions or directions for process improvement.

For an N-run experiment, we can write a matrix model for the full experiment as:

$$\mathbf{y} = \alpha\mathbf{1} + \mathbf{X}_2\boldsymbol{\beta} + \boldsymbol{\epsilon} = \alpha\mathbf{1} + \mathbf{X}_L\boldsymbol{\beta}_L + \mathbf{X}_{PQ}\boldsymbol{\beta}_{PQ} + \mathbf{X}_{MQ}\boldsymbol{\beta}_{MQ} + \boldsymbol{\epsilon}$$

where each row of $\mathbf{X}_L$ is the transpose of the vector $\mathbf{x}$ associated with the corresponding run, and $\boldsymbol{\beta}_L$ is the vector of "linear" polynomial coefficients. $\mathbf{X}_{PQ}$, like $\mathbf{X}_L$, is of dimension N-by-d, the (i, j) element of $\mathbf{X}_{PQ}$ is the square of the corresponding element of $\mathbf{X}_L$, and $\boldsymbol{\beta}_{PQ}$ is the set of "pure quadratic" coefficients – those associated with squared controlled variables. The rows of $\mathbf{X}_{MQ}$ each contain $d(d-1)$ elements, the pairwise products of all distinct pairs of elements from the vector $\mathbf{x}$, and $\boldsymbol{\beta}_{MQ}$ contains the "mixed quadratic" coefficients – those associated with products of two controlled variables.

16.3 Designs for second-order models

In order to support estimation of the coefficients in a quadratic polynomial model, an experimental design must include at least three distinct values for each controlled variable. Thinking about the simplest case of a single variable ($d = 1$) should make this clear; here the model contains three parameters, and so a minimum of three distinct design points (values of x) are necessary to provide unique estimates.

A number of classes of designs have been developed for quadratic polynomial regression; we will briefly describe four of them. In each case, we will assume that the controlled variables have been scaled so that the experimental region R is $[-1, +1]^d$.

16.3.1 Complete three-level factorial designs

Perhaps the most obvious class of experimental designs for quadratic polynomial regression is the complete factorial design (CFD) where each controlled variable takes on three coded levels within R. Denoting those levels as $\{-f, 0, f\}$ for each variable, and adding $n_c - 1$ additional runs at the center point of R, $\mathbf{x} = \mathbf{0}$, to provide replicate information from which σ^2 can be estimated, such a design contains a total of $N = 3^d + n_c - 1$ runs. For example, the model matrices for a three-level factorial plan in $d = 3$ controlled variables, with $n_c = 3$ center points, can be written as:

$$
X_L = \begin{pmatrix}
f & f & f \\
f & f & -f \\
f & -f & f \\
f & -f & -f \\
-f & f & f \\
-f & f & -f \\
-f & -f & f \\
-f & -f & -f \\
0 & f & f \\
0 & f & -f \\
0 & -f & f \\
0 & -f & -f \\
f & 0 & f \\
f & 0 & -f \\
-f & 0 & f \\
-f & 0 & -f \\
f & f & 0 \\
f & -f & 0 \\
-f & f & 0 \\
-f & -f & 0 \\
0 & 0 & f \\
0 & 0 & -f \\
0 & f & 0 \\
0 & -f & 0 \\
f & 0 & 0 \\
-f & 0 & 0 \\
0 & 0 & 0 \\
0 & 0 & 0 \\
0 & 0 & 0
\end{pmatrix}
\quad
X_{PQ} = \begin{pmatrix}
f^2 & f^2 & f^2 \\
f^2 & f^2 & f^2 \\
f^2 & f^2 & f^2 \\
f^2 & f^2 & f^2 \\
f^2 & f^2 & f^2 \\
f^2 & f^2 & f^2 \\
f^2 & f^2 & f^2 \\
f^2 & f^2 & f^2 \\
0 & f^2 & f^2 \\
0 & f^2 & f^2 \\
0 & f^2 & f^2 \\
0 & f^2 & f^2 \\
f^2 & 0 & f^2 \\
f^2 & 0 & f^2 \\
f^2 & 0 & f^2 \\
f^2 & 0 & f^2 \\
f^2 & f^2 & 0 \\
f^2 & f^2 & 0 \\
f^2 & f^2 & 0 \\
f^2 & f^2 & 0 \\
0 & 0 & f^2 \\
0 & 0 & f^2 \\
0 & f^2 & 0 \\
0 & f^2 & 0 \\
f^2 & 0 & 0 \\
f^2 & 0 & 0 \\
0 & 0 & 0 \\
0 & 0 & 0 \\
0 & 0 & 0
\end{pmatrix}
\quad
X_{MQ} = \begin{pmatrix}
f^2 & f^2 & f^2 \\
f^2 & -f^2 & -f^2 \\
-f^2 & f^2 & -f^2 \\
-f^2 & -f^2 & f^2 \\
-f^2 & -f^2 & f^2 \\
-f^2 & f^2 & -f^2 \\
f^2 & -f^2 & -f^2 \\
f^2 & f^2 & f^2 \\
0 & 0 & f^2 \\
0 & 0 & -f^2 \\
0 & 0 & -f^2 \\
0 & 0 & f^2 \\
0 & f^2 & 0 \\
0 & -f^2 & 0 \\
0 & -f^2 & 0 \\
0 & f^2 & 0 \\
f^2 & 0 & 0 \\
-f^2 & 0 & 0 \\
-f^2 & 0 & 0 \\
f^2 & 0 & 0 \\
0 & 0 & 0 \\
0 & 0 & 0 \\
0 & 0 & 0 \\
0 & 0 & 0 \\
0 & 0 & 0 \\
0 & 0 & 0 \\
0 & 0 & 0 \\
0 & 0 & 0 \\
0 & 0 & 0
\end{pmatrix}.
$$

Complete three-level factorial designs have a number of attractive statistical features, and are popular plans when d is relatively small. However, for larger values of d, the number of experimental runs required is impractically large in many settings. The designs described in the next three subsections also have relatively good statistical properties when quadratic polynomial models are used, and can be constructed in substantially fewer runs for $d > 3$.

16.3.2 Central composite designs

Box and Wilson (1951) described a general-purpose design that is perhaps the most widely used experimental plan in situations where a quadratic polynomial model is anticipated. The central composite design (CCD) can be thought of as being comprised of three "subdesigns," each located so that its "center of gravity" corresponds to the center point of R, $\mathbf{x} = \mathbf{0}$:

- an n_f-run orthogonal two-level design; either a full factorial plan in the d controlled variables, or a regular fractional factorial plan of resolution at least V,

- a collection of n_c runs taken at the center point, and

- an "axial" subdesign of $2d$ treatments, each of which is defined by setting one of the controlled variables to a standard nonzero value, either positive or negative, and all other controlled variables to zero.

The overall size of the design is then $N = n_f + 2d + n_c$. Where the controlled variables have been coded so that the experimental range of each is the same, the absolute value of each variable in the factorial portion of the design might be labeled f in all runs, while the absolute value of the nonzero controlled variable in each axial run might be labeled a. In this notation, model matrices for a central composite design in $d = 3$ controlled variables with three center point runs can be written as:

$$
\mathbf{X}_L =
\begin{pmatrix}
f & f & f \\
f & f & -f \\
f & -f & f \\
f & -f & -f \\
-f & f & f \\
-f & f & -f \\
-f & -f & f \\
-f & -f & -f \\
a & 0 & 0 \\
-a & 0 & 0 \\
0 & a & 0 \\
0 & -a & 0 \\
0 & 0 & a \\
0 & 0 & -a \\
0 & 0 & 0 \\
0 & 0 & 0 \\
0 & 0 & 0
\end{pmatrix}
\quad
\mathbf{X}_{PQ} =
\begin{pmatrix}
f^2 & f^2 & f^2 \\
f^2 & f^2 & f^2 \\
f^2 & f^2 & f^2 \\
f^2 & f^2 & f^2 \\
f^2 & f^2 & f^2 \\
f^2 & f^2 & f^2 \\
f^2 & f^2 & f^2 \\
f^2 & f^2 & f^2 \\
a^2 & 0 & 0 \\
a^2 & 0 & 0 \\
0 & a^2 & 0 \\
0 & a^2 & 0 \\
0 & 0 & a^2 \\
0 & 0 & a^2 \\
0 & 0 & 0 \\
0 & 0 & 0 \\
0 & 0 & 0
\end{pmatrix}
\quad
\mathbf{X}_{MQ} =
\begin{pmatrix}
f^2 & f^2 & f^2 \\
f^2 & -f^2 & -f^2 \\
-f^2 & f^2 & -f^2 \\
-f^2 & -f^2 & f^2 \\
-f^2 & -f^2 & f^2 \\
-f^2 & f^2 & -f^2 \\
f^2 & -f^2 & -f^2 \\
f^2 & f^2 & f^2 \\
0 & 0 & 0 \\
0 & 0 & 0 \\
0 & 0 & 0 \\
0 & 0 & 0 \\
0 & 0 & 0 \\
0 & 0 & 0 \\
0 & 0 & 0 \\
0 & 0 & 0 \\
0 & 0 & 0
\end{pmatrix}.
$$

16.3.3 Box-Behnken designs

Box and Behnken (1960) introduced a series of experimental designs for quadratic regression, with structure based on patterns found in balanced incomplete block designs. Box-Behnken designs (BBD) are symmetric three-level designs, consisting of a combination of two-level factorial plans, each constructed using only a subset of the controlled variables. To design an experiment in d controlled variables, one first selects a balanced incomplete block design (BIBD) for d treatments in b blocks of size $k < d$. (This BIBD is not the design to be implemented in the regression experiment, but is used in the construction process.) Each treatment in the BIBD is associated with one of the regression variables. Then, for each block in the BIBD, a two-level factorial design or regular fractional factorial design of resolution at least V is selected for only those variables associated with BIBD treatments included in the block; all other controlled variables take the value of zero in these runs. The resulting designs are then combined, along with n_c center point runs, to form a Box-Behnken regression design of $N = bn_f + n_c$ runs. For example, a quadratic regression design for $d = 3$ controlled variables can be formed using the BIBD for three treatments in three blocks of two units each:

1	2
1	3
2	3

Using a complete 2^2 factorial design in the pairs of factors associated with each block, a Box-Behnken design with $n_c = 3$ center point runs can then be constructed with model matrices:

$$
\mathbf{X}_L = \begin{pmatrix}
f & f & 0 \\
f & -f & 0 \\
-f & f & 0 \\
-f & -f & 0 \\
f & 0 & f \\
f & 0 & -f \\
-f & 0 & f \\
-f & 0 & -f \\
0 & f & f \\
0 & f & -f \\
0 & -f & f \\
0 & -f & -f \\
0 & 0 & 0 \\
0 & 0 & 0 \\
0 & 0 & 0
\end{pmatrix}
\quad
\mathbf{X}_{PQ} = \begin{pmatrix}
f^2 & f^2 & 0 \\
f^2 & f^2 & 0 \\
f^2 & f^2 & 0 \\
f^2 & f^2 & 0 \\
f^2 & 0 & f^2 \\
f^2 & 0 & f^2 \\
f^2 & 0 & f^2 \\
f^2 & 0 & f^2 \\
0 & f^2 & f^2 \\
0 & f^2 & f^2 \\
0 & f^2 & f^2 \\
0 & f^2 & f^2 \\
0 & 0 & 0 \\
0 & 0 & 0 \\
0 & 0 & 0
\end{pmatrix}
\quad
\mathbf{X}_{MQ} = \begin{pmatrix}
f^2 & 0 & 0 \\
-f^2 & 0 & 0 \\
-f^2 & 0 & 0 \\
f^2 & 0 & 0 \\
0 & f^2 & 0 \\
0 & -f^2 & 0 \\
0 & -f^2 & 0 \\
0 & f^2 & 0 \\
0 & 0 & f^2 \\
0 & 0 & -f^2 \\
0 & 0 & -f^2 \\
0 & 0 & f^2 \\
0 & 0 & 0 \\
0 & 0 & 0 \\
0 & 0 & 0
\end{pmatrix} .
$$

16.3.4 Augmented pairs designs

Like central composite designs and Box-Behnken designs, the augmented pairs designs (APD) introduced by Morris (2000) are constructed by combining sets of design points. In this case, the first set of points is a two-level fractional factorial plan of size n_f and of resolution at least III (or a full two-level factorial design for $d = 2$); a Plackett-Burman design is generally recommended to minimize design size. For notational convenience, say the n_f points in this two-level design are specified by the vectors of coded controlled variables $x_1, x_2, x_3, \ldots, x_{n_f}$. The second set of points contains one experimental run determined by each distinct *pair* of runs in the first set, specified as:

$$x_{ij} = -(x_i + x_j)/2, \quad i > j$$

that is, each controlled variable in the new run takes a value that is the negative of the average value that variable takes in the two runs from the first set. n_c replicated runs at the design region center point comprise the final point set. The augmented pairs design contains a total of $N = n_f + \binom{n_f}{2} + n_c$ runs. Hence for $d = 3$, the model matrices for an augmented pairs design containing $n_c = 3$ center point runs are:

$$
\mathbf{X}_L =
\begin{pmatrix}
f & f & f \\
f & -f & -f \\
-f & f & -f \\
-f & -f & f \\
-f & 0 & 0 \\
0 & -f & 0 \\
0 & 0 & -f \\
0 & 0 & f \\
0 & f & 0 \\
f & 0 & 0 \\
0 & 0 & 0 \\
0 & 0 & 0 \\
0 & 0 & 0
\end{pmatrix}
\quad
\mathbf{X}_{PQ} =
\begin{pmatrix}
f^2 & f^2 & f^2 \\
f^2 & f^2 & f^2 \\
f^2 & f^2 & f^2 \\
f^2 & f^2 & f^2 \\
f^2 & 0 & 0 \\
0 & f^2 & 0 \\
0 & 0 & f^2 \\
0 & 0 & f^2 \\
0 & f^2 & 0 \\
f^2 & 0 & 0 \\
0 & 0 & 0 \\
0 & 0 & 0 \\
0 & 0 & 0
\end{pmatrix}
\quad
\mathbf{X}_{MQ} =
\begin{pmatrix}
f^2 & f^2 & f^2 \\
-f^2 & -f^2 & f^2 \\
-f^2 & f^2 & -f^2 \\
f^2 & -f^2 & -f^2 \\
0 & 0 & 0 \\
0 & 0 & 0 \\
0 & 0 & 0 \\
0 & 0 & 0 \\
0 & 0 & 0 \\
0 & 0 & 0 \\
0 & 0 & 0 \\
0 & 0 & 0 \\
0 & 0 & 0
\end{pmatrix}.
$$

Note that in this case ($d = 3$), the augmented pairs design has structure very similar to that of the central composite design with $a = f$, but is based on a 2^{3-1}_{III} fraction rather than a full 2^3 factorial plan. This similarity does not exist for larger values of d, where the second group of runs in the augmented pairs plan do not resemble the axial runs in the central composite design.

16.4 Design scaling and information

The information available through experiments arranged according to any of these designs, as reflected in their respective information matrices, is directly dependent on the "corrected" model matrix, $(\mathbf{I} - \mathbf{H}_1)(\mathbf{X}_L | \mathbf{X}_{PQ} | \mathbf{X}_{MQ})$, where $\mathbf{H}_1$ is the hat matrix associated with the model containing only nuisance parameters. All designs in each of the four classes described in Section 16.3 have two important "balance" properties:

- The elements in each column of $\mathbf{X}_L$ have a zero sum (and average).

- The elements in each column of $\mathbf{X}_{MQ}$ have a zero sum (and average).

These facts have immediate implications for the form of the information matrix. Specifically, the model containing only the "nuisance parameter" α, leads to a hat matrix of form $\mathbf{H}_1 = \frac{1}{N}\mathbf{J}$. As a result, $(\mathbf{I} - \mathbf{H}_1)\mathbf{X}_L = \mathbf{X}_L$ and $(\mathbf{I} - \mathbf{H}_1)\mathbf{X}_{MQ} = \mathbf{X}_{MQ}$. The zero-sum property cannot hold for the columns of $\mathbf{X}_2$, however, because all nonzero elements of this matrix must be positive. However, due to the symmetry of each of these designs,

- The elements in each column of $\mathbf{X}_{PQ}$ have the same average, say a_{PQ}.

This means that $(\mathbf{I} - \mathbf{H}_1)\mathbf{X}_{PQ} = \mathbf{X}_{PQ} - a_{PQ}\mathbf{J}$ for each of these designs. Thinking about the form of each design (and examining the model matrices given in Section 16.3 as examples) we can derive the form of a_{PQ} for each design:

Design	a_{PQ}
CFD	$(2 \times 3^{d-1})f^2)/N$
CCD	$(2^d f^2 + 2a^2)/N$
BBD	$r2^k f^2/N$
ADP	$(n_f + \frac{n_f}{2}(\frac{n_f}{2} - 1))f^2/N$

where r is the replication factor for the BIBD on which the BBD is based, i.e., the number of blocks in which each BIBD treatment is applied. Also, because the columns of $\mathbf{X}_L$ are orthogonal to each other in these designs, $\mathbf{X}_L'\mathbf{X}_L = Na_{PQ}\mathbf{I}$, and since the columns of $\mathbf{X}_L$ are orthogonal to those in $\mathbf{X}_{MQ}$, $\mathbf{X}_L'\mathbf{X}_{MQ} = \mathbf{0}$. As a result of this structure, the design information matrix for any of these designs may be written as:

$$\mathcal{I} = \mathbf{X}_2'(\mathbf{I} - \mathbf{H}_1)\mathbf{X}_2$$

$$= \begin{pmatrix} \mathbf{X}_L'\mathbf{X}_L & \mathbf{X}_L'(\mathbf{X}_{PQ} - a_{PQ}\mathbf{J}) & \mathbf{X}_L'\mathbf{X}_{MQ} \\ & (\mathbf{X}_{PQ} - a_{PQ}\mathbf{J})'(\mathbf{X}_{PQ} - a_{PQ}\mathbf{J})' & (\mathbf{X}_{PQ} - a_{PQ}\mathbf{J})'\mathbf{X}_{MQ} \\ & & \mathbf{X}_{MQ}'\mathbf{X}_{MQ} \end{pmatrix}$$

$$= \begin{pmatrix} Na_{PQ}\mathbf{I} & \mathbf{X}_L'\mathbf{X}_{PQ} & \mathbf{0} \\ & (\mathbf{X}_{PQ} - a_{PQ}\mathbf{J})'(\mathbf{X}_{PQ} - a_{PQ}\mathbf{J}) & \mathbf{X}_{PQ}'\mathbf{X}_{MQ} \\ & & \mathbf{X}_{MQ}'\mathbf{X}_{MQ} \end{pmatrix}. \quad (16.1)$$

Some additional simplifications of this form are also possible, but the details are not the same for each class of designs. For example, for any of the designs considered here,

- the elements of each column of $\mathbf{X}_{MQ}$ have the same sum of squared values, s_{MQ},

and for each design class, the values of s_{MQ} are:

Design	s_{MQ}
CFD	$4 \times 3^{d-2} f^4$
CCD	$2^d f^4$
BBD	$\lambda 2^k f^4$
ADP	$\frac{1}{8} n_f (n_f + 4) f^4$

where for the BBD, λ is the number of blocks in the associated BIBD containing any pair of treatments applied together (Chapter 7). Hence, s_{MQ} is the common diagonal elements of $\mathbf{X}'_{MQ}\mathbf{X}_{MQ}$. Further, $\mathbf{X}'_{MQ}\mathbf{X}_{MQ} = s_{MQ}\mathbf{I}$ for the CFD, CCD, and BBD design classes, but not (in general) for ADP designs because not all collections of four columns in the initial resolution III design segment have the same internal relationship. Information matrices for the four example designs described in Section 16.3 are:

$$\mathcal{I}^{CFD} = \begin{pmatrix} 18f^2\mathbf{I} & 0 & 0 \\ & 6f^4\mathbf{I} + 12f^4\mathbf{J} & 0 \\ & & 12f^4\mathbf{I} \end{pmatrix}$$

$$\mathcal{I}^{CCD}(a = f) = \begin{pmatrix} 10f^2\mathbf{I} & 0 & 0 \\ & 2f^4\mathbf{I} + 1.75f^4\mathbf{J} & 0 \\ & & 8f^4\mathbf{I} \end{pmatrix}$$

$$\mathcal{I}^{BBD} = \begin{pmatrix} 8f^2\mathbf{I} & 0 & 0 \\ & 4f^4\mathbf{I} + 4f^4\mathbf{J} & 0 \\ & & 4f^4\mathbf{I} \end{pmatrix}$$

$$\mathcal{I}^{APD} = \begin{pmatrix} 6f^2\mathbf{I} & 0 & 0 \\ & 2f^4\mathbf{I} + 4f^4\mathbf{J} & 0 \\ & & 4f^4\mathbf{I} \end{pmatrix}.$$

Note that because designs in the CFD, BBD, and APD classes are three-level plans with coded controlled variable values restricted to $-f$, 0, and f, the overall *scale* of $\mathcal{I}$ for these designs is governed by the value of f. In particular, consider two designs from any of these classes that are identical except that design 1 is scaled by f_1 and design 2 is scaled by f_2. It should be clear that if the information matrix for design 1 is $\mathcal{I}^1$, then the information matrix for design 2 is:

$$
\mathcal{I}^2 = \begin{pmatrix} \dfrac{f_2}{f_1}\mathbf{I}_d & 0 \\ 0 & \dfrac{f_2^2}{f_1^2}\mathbf{I}_{d(d+1)/2} \end{pmatrix} \mathcal{I}^1 \begin{pmatrix} \dfrac{f_2}{f_1}\mathbf{I}_d & 0 \\ 0 & \dfrac{f_2^2}{f_1^2}\mathbf{I}_{d(d+1)/2} \end{pmatrix}.
$$

Hence, all functions of the information matrix that reflect statistical performance (noncentrality and estimation precision) are superior for the design that has larger "span" as measured by f. For these designs, "f" is usually simply coded as "1," since it is understood that "larger is better," at least within the experimental range in which the model is assumed to be accurate.

The scaling of designs in the CCD class is a bit more complicated since they are five-level designs. Statistical indices of performance are generally improved (or at least, not made worse) by increasing either f or a, but their relative values are also important. Designs are often tabulated with "f" coded as "1" so that a represents the ratio of an axial deviation to a factorial deviation in the design. Performance measures can then be investigated with respect to this relative value of a, with the understanding that for a given ratio of a/f, "larger" is again "better" within the region of interest and model adequacy.

16.5 Orthogonal blocking

When an experiment is executed in blocks, an important question is whether this can be arranged so that efficiency is not lost. Equivalently, we can ask whether $\mathcal{I}$ is the same when block parameters are included in the model as when α is the only nuisance parameter, that is, whether the design is blocked orthogonally.

As in earlier chapters, when each experimental unit is associated with exactly one of b blocks, the hat matrix for the model containing only the block parameters (or the block parameters and intercept) is $\mathbf{H}_1 = \mathrm{diag}(\frac{1}{n_1}\mathbf{J}_{n_1 \times n_1}, \frac{1}{n_2}\mathbf{J}_{n_2 \times n_2}, \ldots, \frac{1}{n_b}\mathbf{J}_{n_b \times n_b})$, where rows are ordered by blocks, and n_i is the number of units/runs in block i. Now, we can take the observations made about the structure of $(\mathbf{I} - \mathbf{H}_1)\mathbf{X}_2$ for the unblocked case in Section 16.4, and ask what must be true of a blocking structure to yield the same result. First, recall that for unblocked designs, the four design classes we have considered are each such that:

- The elements of each column of $\mathbf{X}_L$ have a zero sum (and average).
- The elements of each column of $\mathbf{X}_{MQ}$ have a zero sum (and average).
- The elements of each column of $\mathbf{X}_{PQ}$ have the same average, a_{PQ}.

Hence for the blocked case, $(\mathbf{I} - \mathbf{H}_1)\mathbf{X}_L$ and $(\mathbf{I} - \mathbf{H}_1)\mathbf{X}_{MQ}$ will be as they are in the unblocked case if the elements of each column of $\mathbf{X}_L$ and $\mathbf{X}_{MQ}$ *sum to zero within each block*. Similarly, $(\mathbf{I} - \mathbf{H}_1)\mathbf{X}_{PQ}$ will be as it is in the unblocked case if the elements of each column of $\mathbf{X}_{PQ}$ *have an average value of a_{PQ} within each block*.

For example, consider the CFD in $d = 3$ controlled variables described in subsection 16.3.1. There are several ways the design can be divided into two blocks, each preserving zero sums within blocks for the columns of $\mathbf{X}_L$ and $\mathbf{X}_{MQ}$ within blocks. But constructing blocks in such a way that the within-block averages of elements within columns of $\mathbf{X}_{PQ}$ are the same is less obvious. One way to modify the design to meet this requirement is to add additional center points. If the first eight runs (as tabulated in the display in subsection 16.3.1) are grouped with eight center point replicates, while runs 9–26 are grouped with two center point replicates, the within-block average of elements from any column of $\mathbf{X}_{PQ}$ will be $f^2/2$, and the restrictions concerning the columns of $\mathbf{X}_L$ and $\mathbf{X}_{MQ}$ are also met. Note that the block sizes are not equal in this case (16 and 20), but the difference is small enough that this would not present problems in many applications.

CCDs are often blocked by grouping factorial points into one or more blocks, and axial points in another. As with CFDs, the number of center points in each block can be adjusted to meet the conditions for orthogonal blocking, but for these designs, the value of a (relative to f) can also be adjusted to help meet the condition related to the columns of $\mathbf{X}_{PQ}$.

16.6 Split-plot designs

The general structure for split-plot regression experiments is essentially the same regardless of the order of model used in analysis: the values of some controlled variables remain constant within blocks/plots, while the values of other controlled variables differ across units/split-plots within a block/plot. Model monomials (x_i, x_i^2, and $x_i x_j$) that have constant values within each block are assessed in the whole-plot section of the ANOVA decomposition, relative to a MSE reflecting block-to-block variation. Other model terms for which values change from unit to unit within blocks are assessed in the split-plot section of the ANOVA decomposition, relative to a MSE reflecting unit-to-unit (within-block) variation.

16.6.1 Example

Consider an experiment in $d = 2$ controlled variables designed as two complete replicates of a full 3^2 factorial. Suppose the experimental material can be supplied in batches sufficient to prepare three experimental runs, and that the three runs associated with one batch can be most easily made if only x_2 is varied among runs and x_1 remains fixed. So the envisioned experimental layout, with each (x_1, x_2) denoting the variable setting for a single unit, is:

block 1	$(-1,-1)$	$(-1,0)$	$(-1,+1)$
block 2	$(-1,-1)$	$(-1,0)$	$(-1,+1)$
block 3	$(0,-1)$	$(0,0)$	$(0,+1)$
block 4	$(0,-1)$	$(0,0)$	$(0,+1)$
block 5	$(+1,-1)$	$(+1,0)$	$(+1,+1)$
block 6	$(+1,-1)$	$(+1,0)$	$(+1,+1)$

If the differences among blocks (due, for instance, to variation between material batches) can be regarded as random, a split-plot analysis can be carried out. Let $\mathbf{X}_1$ be the 18-by-6 model matrix of indicator variables representing blocks, $\mathbf{X}_{wp}$ be comprised of the x_1 column from $\mathbf{X}_L$ and the x_1^2 column from $\mathbf{X}_{pq}$, and $\mathbf{X}_{sp}$ be comprised of the x_2 column from $\mathbf{X}_L$, the x_2^2 column from $\mathbf{X}_{PQ}$, and the single column from $\mathbf{X}_{MQ}$ representing $x_1 x_2$. The information matrices are then:

$$\mathcal{I}_{inter} = \frac{1}{3}\mathbf{X}_{wp}'\mathbf{X}_1\left(\mathbf{I} - \frac{1}{6}\mathbf{J}\right)\mathbf{X}_1'\mathbf{X}_{wp} = \begin{pmatrix} 12 & 0 \\ 0 & 4 \end{pmatrix}$$

$$\mathcal{I}_{intra} = \mathbf{X}_{sp}'(\mathbf{I} - \mathbf{H}_1)\mathbf{X}_{sp} = \begin{pmatrix} 12 & 0 & 0 \\ 0 & 4 & 0 \\ 0 & 0 & 8 \end{pmatrix}.$$

Standard errors of $\hat{\beta}_2$, $\hat{\beta}_{22}$, and $\hat{\beta}_{12}$ are computed as the square root of MSE_{sp} divided by their respective diagonal elements of $\mathcal{I}_{intra}$, while those for $\hat{\beta}_1$ and $\hat{\beta}_{11}$ are based on MSE_{wp} and the elements of $\mathcal{I}_{inter}$.

16.7 Bias due to omitted model terms

In most applications, quadratic regression models do not exactly represent the underlying relationship between $E(y)$ and $\mathbf{x}$. They are useful and popular because they can approximate many (but not all) "curved" functions reasonably well over a limited domain. And in truth, regression experiments are often performed in contexts where there is insufficient information to stipulate a more appropriate model form. But where there is some possibility that higher-order (i.e., greater than two) monomial terms would be needed to accurately reflect the underlying function, it is important to understand how much estimation bias might result from their omission from the quadratic model.

Recall our discussion of effect *aliasing* in Chapter 13, in which a relatively modest linear model (e.g., one containing only an intercept and main effects) is fit in conjunction with a relatively small fractional factorial design (e.g., of resolution III), and is subject to estimation bias if terms of higher order (e.g., interactions) are present but are not included in the model. Chapter 13 describes how "alias groups" of effects can be easily identified for regular fractional factorial experiments based on inspection of the identifying relation. The phenomenon of coefficient estimate bias is really much more general, and can occur with any linear model when some important terms are omitted (or cannot be estimated based on the selected design).

Generally, suppose we fit a model of matrix form $\mathbf{y} = \mathbf{X}\boldsymbol{\theta} + \boldsymbol{\epsilon}$ to data acquired in an experiment, but that the data were actually generated by a model of form $\mathbf{y} = \mathbf{X}\boldsymbol{\theta} + \mathbf{W}\boldsymbol{\phi} + \boldsymbol{\epsilon}$. In the present case, $\mathbf{X}$ would include the collection of columns from $\mathbf{1}$, $\mathbf{X}_L$, $\mathbf{X}_{PQ}$, and $\mathbf{X}_{MQ}$, while $\mathbf{W}$ might contain columns associated with third-order monomials (e.g., $x_i^2 x_j$). Assuming $\mathbf{X}$ is of full column rank, the unique least-squares estimate of $\boldsymbol{\theta}$ is $\hat{\boldsymbol{\theta}} = (\mathbf{X}'\mathbf{X})^{-1}\mathbf{X}'\mathbf{y}$. But, substitution of the "true" expectation of $\mathbf{y}$ into this linear form results in:

$$E(\hat{\boldsymbol{\theta}}) = \boldsymbol{\theta} + (\mathbf{X}'\mathbf{X})^{-1}\mathbf{X}'\mathbf{W}\boldsymbol{\phi}.$$

That is, the bias of each estimated coefficient is determined by the value of the omitted coefficients $\boldsymbol{\phi}$, *and* the form of $(\mathbf{X}'\mathbf{X})^{-1}\mathbf{X}'\mathbf{W}$, often called the *alias matrix*, which in turn is determined by the model (both the fitted model and the potentially omitted terms) and the experimental design. In practice, we don't know the value of $\boldsymbol{\phi}$; in fact, we are effectively *assuming* it is zero by our choice of model. However, other things being equal, an experimental design for which the alias matrix contains elements of relatively small absolute value is often preferred to one containing larger elements, because this offers relatively more protection against coefficient bias *if* it turns out that $\boldsymbol{\phi}$ is nonzero.

For example, consider once again the central composite design for $d = 3$ independent variables described in subsection 16.3.2, scaled so that $f = 1$ and $a = 2$. For this design:

$$\mathbf{X}'\mathbf{X} = \begin{pmatrix} 17 & 0 & 0 & 0 & 16 & 16 & 16 & 0 & 0 & 0 \\ 0 & 16 & 0 & 0 & 0 & 0 & 0 & 0 & 0 & 0 \\ 0 & 0 & 16 & 0 & 0 & 0 & 0 & 0 & 0 & 0 \\ 0 & 0 & 0 & 16 & 0 & 0 & 0 & 0 & 0 & 0 \\ 16 & 0 & 0 & 0 & 40 & 8 & 8 & 0 & 0 & 0 \\ 16 & 0 & 0 & 0 & 8 & 40 & 8 & 0 & 0 & 0 \\ 16 & 0 & 0 & 0 & 8 & 8 & 40 & 0 & 0 & 0 \\ 0 & 0 & 0 & 0 & 0 & 0 & 0 & 8 & 0 & 0 \\ 0 & 0 & 0 & 0 & 0 & 0 & 0 & 0 & 8 & 0 \\ 0 & 0 & 0 & 0 & 0 & 0 & 0 & 0 & 0 & 8 \end{pmatrix}$$

$(\mathbf{X}'\mathbf{X})^{-1} =$

$$
\begin{pmatrix}
0.3043 & 0.0000 & 0.0000 & 0.0000 & -0.0870 & -0.0870 & -0.0870 & 0.000 & 0.000 & 0.000 \\
0.0000 & 0.0625 & 0.0000 & 0.0000 & 0.0000 & 0.0000 & 0.0000 & 0.000 & 0.000 & 0.000 \\
0.0000 & 0.0000 & 0.0625 & 0.0000 & 0.0000 & 0.0000 & 0.0000 & 0.000 & 0.000 & 0.000 \\
0.0000 & 0.0000 & 0.0000 & 0.0625 & 0.0000 & 0.0000 & 0.0000 & 0.000 & 0.000 & 0.000 \\
-0.0870 & 0.0000 & 0.0000 & 0.0000 & 0.0516 & 0.0204 & 0.0204 & 0.000 & 0.000 & 0.000 \\
-0.0870 & 0.0000 & 0.0000 & 0.0000 & 0.0204 & 0.0516 & 0.0204 & 0.000 & 0.000 & 0.000 \\
-0.0870 & 0.0000 & 0.0000 & 0.0000 & 0.0204 & 0.0204 & 0.0516 & 0.000 & 0.000 & 0.000 \\
0.0000 & 0.0000 & 0.0000 & 0.0000 & 0.0000 & 0.0000 & 0.0000 & 0.125 & 0.000 & 0.000 \\
0.0000 & 0.0000 & 0.0000 & 0.0000 & 0.0000 & 0.0000 & 0.0000 & 0.000 & 0.125 & 0.000 \\
0.0000 & 0.0000 & 0.0000 & 0.0000 & 0.0000 & 0.0000 & 0.0000 & 0.000 & 0.000 & 0.125
\end{pmatrix}.
$$

But now suppose that the data-generation process actually involved three additional model terms beyond those in the quadratic polynomial, namely $x_1^3\beta_{111}$, $x_1^2x_2\beta_{112}$, and $x_1^2x_2^2\beta_{1122}$. In this case:

$$
\mathbf{W} =
\begin{pmatrix}
1 & 1 & 1 \\
1 & 1 & 1 \\
1 & -1 & 1 \\
1 & -1 & 1 \\
-1 & 1 & 1 \\
-1 & 1 & 1 \\
-1 & -1 & 1 \\
-1 & -1 & 1 \\
8 & 0 & 0 \\
-8 & 0 & 0 \\
0 & 0 & 0 \\
0 & 0 & 0 \\
0 & 0 & 0 \\
0 & 0 & 0 \\
0 & 0 & 0 \\
0 & 0 & 0 \\
0 & 0 & 0
\end{pmatrix},
\quad
\mathbf{X}'\mathbf{W} =
\begin{pmatrix}
0 & 0 & 8 \\
40 & 0 & 0 \\
0 & 8 & 0 \\
0 & 0 & 0 \\
0 & 0 & 8 \\
0 & 0 & 8 \\
0 & 0 & 8 \\
0 & 0 & 0 \\
0 & 0 & 0 \\
0 & 0 & 0
\end{pmatrix},
\quad
(\mathbf{X}'\mathbf{X})^{-1}\mathbf{X}'\mathbf{W} =
\begin{pmatrix}
0.0 & 0.0 & 0.3478 \\
2.5 & 0.0 & 0.0000 \\
0.0 & 0.5 & 0.0000 \\
0.0 & 0.0 & 0.0000 \\
0.0 & 0.0 & 0.0435 \\
0.0 & 0.0 & 0.0435 \\
0.0 & 0.0 & 0.0435 \\
0.0 & 0.0 & 0.0000 \\
0.0 & 0.0 & 0.0000 \\
0.0 & 0.0 & 0.0000
\end{pmatrix}.
$$

Hence, $E(\hat{\beta}_1) = \beta_1 + 2.5\beta_{111}$, $E(\hat{\beta}_2) = \beta_2 + 0.5\beta_{112}$, and $E(\hat{\beta}_{ii}) = \beta_{ii} + 0.0435\beta_{1122}$, $i = 1, 2, 3$. The estimated intercept is also biased by β_{1122}, but this is not of concern in a comparative experiment; the remaining coefficients in the estimated quadratic model remain unbiased.

Finally, it should be noted that in most real physical processes, $E(y)$ is not exactly related to $\mathbf{x}$ through a fourth-order polynomial either. The idea is that, when the addition of higher-order polynomial terms would make a much better approximation to the truth, but are omitted from the fitted model, this analysis provides a good approximation to the estimation bias that can be expected.

16.8 Conclusion

Regression experiments designed for quadratic polynomials provide more information about the nature of the expected response-versus-controlled variable relationship than those designed for first-order polynomial models, at the cost of a larger number of required experimental runs. Complete three-level factorial designs, central composite designs, Box-Behnken designs, and augmented pairs designs are four classes of experimental plans that are used for such experiments, each of which has specific strengths, but all of which have good overall properties. The conditions for orthogonal blocking are somewhat more extensive for quadratic models than for first-order models, but can often be achieved by selecting an appropriate number of center point runs in each block, and for CCDs, by adjusting the the value of a relative to f. Because quadratic polynomial models are often only approximate representations of real systems, designs should be considered both for their precision of estimation, and for the potential bias they imply for coefficient estimates in case higher-order monomials should also be present in the "true model."

16.9 Exercises

1. Construct orthogonal blocking schemes for each of the following designs in $d = 4$ factors, that will allow them to be executed in the indicated number of blocks of equal size:

   (a) A BBD based on the BIBD with six blocks of size 2 with $n_c = 6$

   (b) A CCD with $n_c = 8$ center points in two blocks; for this, also determine the required value of a (relative to f).

2. Designs in the BBD class require $n_c > 0$ center points in order to provide unique least-squares estimates of all model parameters; if $n_c = 0$, the model matrix is singular. Explain the nature of the singularity when $n_c = 0$, i.e., show what group of columns is linearly dependent.

3. Suppose an unblocked CCD in two variables with three center points is planned for a situation in which σ is expected to be three. The design is scaled to values $f = 1$ and $a = 1.5$.

   (a) If $\beta_1 = \beta_2 = 2$, $\beta_{11} = \beta_{22} = -1$, and $\beta_{12} = 0$, what is the power of the test for model effectiveness?

   (b) What are the degrees of freedom for the test for lack of fit?

4. An experiment is planned to model the effect of three controlled variables on responses using three replicates of a full three-level factorial experiment with coded values $(-1, 0, +1)$ for each variable. The experiment must be

run as a split-plot study, with variables 1 and 2 varying between plots (of size 3), and variable 3 (only) changing within plots. Assuming that a full quadratic polynomial model will be used for analysis, determine:

- degrees of freedom for each line of a split-plot ANOVA,
- standard errors for each coefficient estimate.

5. Rechtschaffner (1967) described a class of three-level designs that are *saturated* for full quadratic models, i.e., for which N is the number of model coefficients.

   For $d = 3$, the design matrix is:

   $$\begin{pmatrix} -1 & -1 & -1 \\ -1 & +1 & +1 \\ +1 & -1 & +1 \\ +1 & +1 & -1 \\ +1 & 0 & 0 \\ 0 & +1 & 0 \\ 0 & 0 & +1 \\ -1 & -1 & +1 \\ -1 & +1 & -1 \\ +1 & -1 & -1 \end{pmatrix}.$$

   For any value of $d > 3$, the design contains four groups of runs:

   - a single run in which all $x_i = -1$,
   - d runs in which $x_i = -1$, and all $x_j = +1$ for $j \neq i$, $i = 1, 2, 3, \ldots, d$,
   - d runs in which $x_i = +1$, and all $x_j = 0$ for $j \neq i$, $i = 1, 2, 3, \ldots, d$, and
   - $d(d-1)/2$ runs in which $x_i = x_{i'} = +1$, and all $x_j = -1$ for $j \neq i$ and $j \neq i'$, $i, i' = 1, 2, 3, \ldots, d$ and $i \neq i'$.

   Using a computer, calculate the information matrix $\mathcal{I}$ for Rechtschaffner designs, for $d = 3$, 4, and 5.

6. Suppose a regression experiment in $d = 3$ controlled variables is carried out with the anticipation that a second-order quadratic polynomial model will be adequate. However, suppose that the "true" functional relationship also includes three cubic monomials:

   $$\beta_{111} x_1^3, \quad \beta_{112} x_1^2 x_2, \quad \text{and} \quad \beta_{123} x_1 x_2 x_3$$

   and that these are not included in the fitted model. What is the resulting bias in the estimates of the quadratic polynomial model coefficients, if the design used is:

   - the complete three-level factorial design described in subsection 16.3.1 with $f = 1$?
   - the central composite design described in subsection 16.3.2 with $f = 1$ and $a = 1.5$?

- the Box-Behnken design described in subsection 16.3.3 with $f = 1$?
- the augmented pairs design described in subsection 16.3.4 with $f = 1$?

7. In some cases, information about a physical system being studied can be used in suggesting appropriate modifications to a regression model, and the experimental design used to fit it. Suppose, for example, that for a $d = 1$ problem, we know (or firmly assume) that $E(y(x))$ is an *even function* of x in the experimental region $R = [-1, +1]$, that is, that $E(y(x)) = E(y(-x))$ for any $x \in R$. Under this assumption, a polynomial model should contain no terms of odd order; suppose that we select a model of form:

$$y = \alpha + \beta_{11}x_1^2 + \beta_{1111}x_1^4 + \epsilon.$$

Suggest an experimental design in $N = 10$ observations, containing enough replicate points to provide at least four degrees of freedom for "pure error." Defend your suggested design based on calculated standard errors, the noncentrality of tests, or other relevant statistical indices.

8. A complete three-level factorial experiment is planned to provide data to fit a quadratic model with $d = 2$ controllable variables. The experimental budget allows for a total of $N = 18$ runs to be included, and blocking will not be needed. The investigator is considering four different experimental designs:

- Design 1: two complete replicates of the 3^2 factorial plan.
- Design 2: one complete replicate of the 3^2 factorial plan, plus two additional runs at each corner point (i.e., where each $|x_i| = 1$) and one additional run at the center point.
- Design 3: one complete replicate of the 3^2 factorial plan, plus two additional runs at each side point (i.e., where exactly one $x_i = 0$) and one additional run at the center point.
- Design 4: one complete replicate of the 3^2 factorial plan, plus nine additional runs at the center point.

Compare these designs relative to the sampling variance for each of:

- $\hat{\beta}_1$
- $\hat{\beta}_{11}$
- $\hat{\beta}_{12}$

9. Continuing exercise 8:

(a) If it turns out that $\beta_1 = 1$, $\beta_{11} = -1$, $\beta_2 = 1$, $\beta_{12} = 1/2$, and $\beta_{22} = 0$, which design offers the best power for testing

$$\text{Hyp}_0 : \beta_1 = \beta_2 = \beta_{11} = \beta_{22} = \beta_{12} = 0?$$

(b) Which design results in the least bias to the estimate of β_1 (under the assumed quadratic model) if the "true model" also includes the monomial $\beta_{122}x_1x_2^2$?

## CHAPTER 17

Introduction to optimal design

17.1 Introduction

The experimental designs described to this point have been presented as "templates" that, under general sets of conditions, lead to statistical analyses with good general properties. For example, the complete block designs discussed in Chapter 4 account for additive block differences without sacrificing precision in estimates of treatment contrasts, and the factorial and fractional factorial designs of Chapters 11–13 yield estimates of each main effect that are as efficient as those provided by one-factor designs of the same size, each of which supports estimation of *only* that main effect. In many cases, the designs presented are *optimal*, in the sense that no other experimental plan in the same number of runs can provide more precise estimates or powerful tests against broad collections of alternative hypotheses for the parameters of interest, given the assumed model.

Optimal design provides a more direct connection between experimental design and statistical performance by framing design selection as an optimization problem, in which standard errors are minimized or noncentrality parameters are maximized. An example is the *allocation problem* discussed in Chapter 3, where sample sizes were directly determined for a completely randomized design (CRD) using constrained optimization so as to obtain the best estimation precision possible given the available resources. In this chapter, we further develop the concept of design optimality.

17.2 Optimal design fundamentals

In order to intelligently discuss a formulation of optimal design, at least three questions must be answered. First, what experimental treatments are available for use? Second, what parametric assumptions are we willing to make about how treatments affect responses? Finally, what questions are we trying to answer by performing the experiment, i.e., what sort of information are we trying to "optimize"? Corresponding to these questions, our framework for optimal design construction requires specification of three entities:

1. *The experimental region:* In this book, the experimental region is denoted by R. It is the finite or infinite set of values for $\mathbf{x}$, the scalar or vector of independent variables that defines a treatment. For example, $R = \{1, 2, 3, \ldots, t\}$ is appropriate for an experiment in t unstructured treatments, $R = \{0, 1\}^f$ can be used in two-level factorial experiments, and

$R = [-1, 1]^d$ may be appropriate for regression problems when the d independent variables have each been scaled to $[-1, 1]$.

2. *The linear model:*
$$M : y = t'_x \beta + \epsilon$$
where each element of t_x is a function of the elements of x. For example, a main effects model for an experiment in f two-level factors yields $t'_x = (1, x')$ of $f+1$ elements, while a quadratic regression model in d independent variables leads to a vector t_x of $1 + 2d + d(d-1)/2$ elements.

3. *The criterion function:* For any specified design, $D = \{x_1, x_2, \ldots, x_N\}$, the concept of optimal design requires a function, $\phi_M(D)$, that can be used as a measure of quality of the inference that can be expected from the resulting data. Examples include the power of a test (in which case we hope to maximize the criterion) or the standard deviation of an estimate (to be minimized). Note that we subscript ϕ with M because the form of the model is generally a necessary component in formulating a specific criterion function; designs that are very good for factorial models containing only main effects can be very poor when interactions of higher order are included.

Within this context, the general idea is to identify the design or designs, comprised of runs from the experimental region, that maximize or minimize (whichever is appropriate) the criterion function, e.g.:

$$D_{opt} = \text{argmax}_D \phi_M(D) \text{ such that } D = \{x_1, x_2, \ldots, x_N\} \text{ and } x_i \in R.$$

This formalism relies on the assumption that the overall statistical quality of a proposed experiment, in the context of the questions being asked, can be expressed as a single scalar-valued function, ϕ_M. This assumption may not be entirely accurate, especially in more complex settings, but it does serve to make more specific the question of how "good" an experiment can be relative to its competitors.

While optimal designs can be useful in many contexts, there are settings in which they may be especially valuable:

1. *Nonstandard experimental regions.* "Standard designs" are usually constructed for relatively simple, or "generic" experimental regions, e.g., fractional factorial designs for $\{0, 1\}^f$, and central composite designs (CCDs) for regression experiments in $[-1, 1]^d$. But practical settings often involve restrictions on the experimental region. For example, if x_1 and x_2 represent temperature and pressure, respectively, in a regression experiment involving a chemical reactor, operating constraints may require $x_1 + 2x_2 \leq 2$, say; that is, $R = \{x : -1 \leq x_i \leq +1, x_1 + 2x_2 \leq 2\}$. This would make application of a CCD difficult. Because optimal designs are region-specific, they can be constructed for any proposed experimental region.

2. *Nonstandard models.* Standard designs such as balanced incomplete block designs (BIBDs), regular fractional factorials, and CCDs are widely used because they perform well in the wide variety of circumstances where

"standard models" are appropriate. However, there are also situations in which theory or previous experimentation suggests a different model form. For example, in a factorial setting, certain low-order interactions may be known or assumed to be absent, while other higher-order interactions are known or assumed to be present. Because optimal designs are model-specific, they can be constructed for any proposed linear model, not just the "standard" forms.

3. *Nonstandard experiment size.* For many of the classes of designs we have examined, there are restrictions on the value of N, the total number of experimental runs. For example, regular fractional factorial designs with two-level factors require the number of unique treatments to be a power of 2. Central composite designs for regression include a two-level factorial or fractional factorial component, plus $2d$ axial runs, plus center points. Because optimal designs are the solution to optimization problems for which N can be specified to be any desired integer value, attention need not be limited to the values that are required or convenient for any particular class of designs.

In addition, investigators sometimes use optimal designs as "benchmarks" against which to compare their standard designs. For example, if a certain CCD can be shown to be almost as good, with respect to an appropriate criterion ϕ_M, as the optimal design of the same size constructed for the same experimental region, this is a useful argument for justifying the (usually simpler and sometimes better-accepted) CCD. A major *disadvantage* of optimal design is that construction of a design requires solution of a mathematical optimization problem that is usually of high dimension, can often be approached only numerically, and for which "true" optimal solutions sometimes cannot be practically verified. Still, for many important applications, optimal design construction is an important and useful alternative to the use of standard experimental plans.

17.3 Optimality criteria

Much of the development of optimal design has been set in the context of unpartitioned models for which the model matrix is of full rank. Accommodation can usually be made for partitioned models, e.g., those containing nuisance parameters, but less-than-full rank extensions generally require more care. Our presentation will be given first for unpartitioned full-rank models (the simplest case), followed by the partitioned form (often applicable in regression design problems), and finally less-than-full rank forms (more relevant to experiments with unstructured treatments).

17.3.1 A-optimality

"A" optimality refers to designs for which the *average* variance of estimates of interest is minimized. With this in mind, suppose we consider the basic linear

model:
$$y = \mathbf{X}\boldsymbol{\theta} + \boldsymbol{\epsilon}$$
with full rank $\mathbf{X}$ and $\mathcal{I} = \mathbf{X}'\mathbf{X}$. If the estimates of interest are the (entire) set of elements of $\hat{\boldsymbol{\theta}}$ we note that their individual variances can be written as:
$$Var(\{\hat{\boldsymbol{\theta}}\}_i) = \sigma^2\{\mathcal{I}^{-1}\}_{ii}.$$
So an A-optimal design minimizes
$$\phi_M(D) = \text{trace}(\mathcal{I}^{-1})$$
over possible designs.

Now consider the partitioned linear model:
$$y = \mathbf{X}_1\boldsymbol{\theta}_1 + \mathbf{X}_2\boldsymbol{\theta}_2 + \boldsymbol{\epsilon}$$
in which $\mathbf{X}_{2|1} = (\mathbf{I} - \mathbf{H}_1)\mathbf{X}_2$ is of full rank. This is often true, for example, for regression experiments where $\boldsymbol{\theta}_1$ contains parameters of no experimental interest, and $\boldsymbol{\theta}_2$ contains first- and/or higher-order polynomial coefficients. In this case $\mathcal{I}$ is of full rank, a unique inverse of $\mathcal{I}$ exists,
$$Var(\{\hat{\boldsymbol{\theta}}_2\}_i) = \sigma^2\{\mathcal{I}^{-1}\}_{ii}$$
and an A-optimal design minimizes
$$\phi_M(D) = \text{trace}(\mathcal{I}^{-1}).$$

Finally, if $(\mathbf{I} - \mathbf{H}_1)\mathbf{X}_2$ cannot be of full rank for the model and design under consideration, $\mathcal{I}$ cannot be of full rank, and an alternative formulation is required. For example, individual treatment parameters are not estimable in an "effects" model, regardless of the design used. Let the rows of $\mathbf{C}$ be weights of estimable linear combinations of the elements of $\boldsymbol{\theta}_2$. Then these functions can be A-optimally estimated by a design for which the average variance of the elements of $\widetilde{\mathbf{C}\boldsymbol{\theta}}$ is minimized, so a criterion function for A-optimality is:
$$\phi_M(D) = \text{trace}(\mathbf{C}\mathcal{I}^-\mathbf{C}') = \text{trace}(\mathbf{C}'\mathbf{C}\mathcal{I}^-).$$

It is important to note that in this case, the optimal design is completely dependent on the selection of the matrix $\mathbf{C}$. For example, in a problem involving four unstructured treatments, A-optimal designs for
$$\mathbf{C} = \begin{pmatrix} -1 & 1 & 0 & 0 \\ -1 & 0 & 1 & 0 \\ -1 & 0 & 0 & 1 \end{pmatrix} \text{ and } \mathbf{C} = \begin{pmatrix} -1 & 1 & 0 & 0 \\ 0 & -1 & 1 & 0 \\ 0 & 0 & -1 & 1 \end{pmatrix}$$

are generally not the same. So useful formulation of an A-optimal design problem requires that $\mathbf{C}$ actually reflect the (estimable) treatment comparisons of primary interest.

17.3.2 D-optimality

"D"-optimality refers to designs for which the determinant of the covariance matrix of estimates of interest, sometimes called the *generalized variance*, is minimized. Statistical motivation comes from the fact that this determinant is monotonically related to the volume of the normal-theory simultaneous confidence ellipsoid for the parameters. For the standard linear model:

$$\mathbf{y} = \mathbf{X}\boldsymbol{\theta} + \boldsymbol{\epsilon}$$

with full rank $\mathbf{X}$ and $\mathcal{I} = \mathbf{X}'\mathbf{X}$, such designs minimize

$$\phi_M(D) = |\mathcal{I}^{-1}|$$

or equivalently, *maximize*

$$\phi_M(D) = |\mathcal{I}|$$

over possible designs. The latter form is often preferred in practice since it avoids the unnecessary numerical step of inverting $\mathcal{I}$.

For the partitioned linear model

$$\mathbf{y} = \mathbf{X}_1\boldsymbol{\theta}_1 + \mathbf{X}_2\boldsymbol{\theta}_2 + \boldsymbol{\epsilon}$$

for which $\mathbf{X}_{2|1}$ is of full rank and the elements of $\boldsymbol{\theta}_2$ are the parameters of interest, the D-optimal designs are those that maximize

$$\phi_M(D) = |\mathcal{I}|.$$

If $\mathbf{X}_{2|1}$ cannot be of full rank for the model and design, the criterion must be modified as with A-optimality. Again, let $\mathbf{C}$ be a matrix for which the rows are linearly independent estimable combinations of the elements of $\boldsymbol{\theta}_2$. $\mathbf{C}\mathcal{I}^-\mathbf{C}'$ is then unique and of full rank, and

$$\phi_M(D) = |\mathbf{C}\mathcal{I}^-\mathbf{C}'|$$

can be minimized over choice of design to minimize the generalized variance of $\widehat{\mathbf{C}\boldsymbol{\theta}_2}$.

In contrast to A-optimality, designs that are D-optimal for estimating $\mathbf{C}\boldsymbol{\theta}_2$ and $\mathbf{B}\boldsymbol{\theta}_2$ are *the same* if

- $\mathbf{C}$ and $\mathbf{B}$ contain the same number of rows, and
- there exists a square matrix $\mathbf{T}$ of full rank for which

$$\mathbf{C} = \mathbf{TB}, \text{ and so } \mathbf{B} = \mathbf{T}^{-1}\mathbf{C}.$$

This is true because the criterion we would adopt for $\mathbf{C}$ is:

$$\begin{aligned}
\phi_M(D) &= |\mathbf{C}\mathcal{I}^{-1}\mathbf{C}'| \\
&= |\mathbf{TB}\mathcal{I}^{-1}\mathbf{B}'\mathbf{T}'| \\
&= |\mathbf{T}|^2|\mathbf{B}\mathcal{I}^{-1}\mathbf{B}'|
\end{aligned}$$

But since $|\mathbf{T}|$ is a constant, designs that minimize $|\mathbf{C}\mathcal{I}^{-1}\mathbf{C}'|$ are also those which minimize $|\mathbf{B}\mathcal{I}^{-1}\mathbf{B}'|$. For the example given at the end of subsection 17.3.1,

$$\begin{pmatrix} -1 & 1 & 0 & 0 \\ -1 & 0 & 1 & 0 \\ -1 & 0 & 0 & 1 \end{pmatrix} = \begin{pmatrix} 1 & 0 & 0 \\ 1 & 1 & 0 \\ 1 & 1 & 1 \end{pmatrix} \begin{pmatrix} -1 & 1 & 0 & 0 \\ 0 & -1 & 1 & 0 \\ 0 & 0 & -1 & 1 \end{pmatrix}.$$

So either 3×4 matrix of parameter contrasts leads to the same D-optimal designs. Hence D-optimality, unlike A-optimality, is *invariant to nonsingular linear reparameterization.*

17.3.3 Other criteria

D- and A-optimality are highlighted in this chapter, but there are a number of other popular criteria in use. Two additional criteria that have particular appeal when model predictions, rather than parameters, are of primary interest are G-optimality (for "global") and I-optimality (for "integrated"). G-optimal designs are those for which the largest (with respect to $\mathbf{x} \in R$) value of $Var(\hat{y}(\mathbf{x}))$ is minimized, and so can be implemented by minimizing

$$\phi_M(D) = \max_{\mathbf{x} \in R} \mathbf{t}'_{\mathbf{x}} \mathcal{I}^{-1} \mathbf{t}_{\mathbf{x}}.$$

Rather than minimizing the largest predictive variance, I-optimality is defined so as to minimize the average (over R) response variance, and so can be implemented by minimizing

$$\phi_M(D) = \int_{\mathbf{x}} \mathbf{t}'_{\mathbf{x}} \mathcal{I}^{-1} \mathbf{t}_{\mathbf{x}} w(\mathbf{x}) dx$$

for an appropriate weight function w (which may be omitted if all regions of equal volume in R should receive the same weight). Note that as defined here, neither G- nor I-optimality is particularly well motivated for *comparative* experiments, but alterations to these criteria can be made to focus on, say, the variances of *differences* between predictions of the response at $\mathbf{x} \in R$ and at a specified reference point, say in the center of R. Although the connection is beyond the scope of this book, there is an interesting relationship between D- and G-optimality following from the *Equivalence Theorem* of Kiefer and Wolfowitz (1960).

17.3.4 Examples

A factorial example

Suppose a two-level factorial design is required for fitting the model:

$$y = \mu + \alpha x_1 + \beta x_2 + \gamma x_3 + (\alpha\beta)x_1 x_2 + \epsilon$$

where the three independent variables have been coded so that $R = \{-1, +1\}^3$. Note that the model is asymmetric in the factors, including only one two-factor interaction (perhaps due to previous experimental experience, or relevant

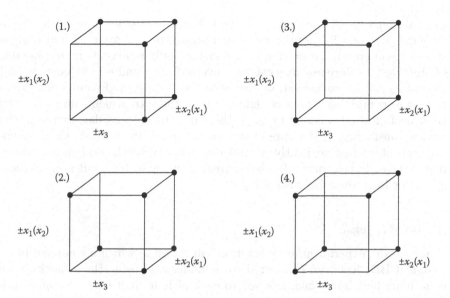

Figure 17.1 A-optimal ((1.) and (2.)) and D-optimal (all) designs for the factorial example.

theory about the system under investigation). Further, suppose that each experimental run is relatively expensive or time-consuming, and it has been determined that an experimental budget of $N = 6$ unblocked units will be required. So in this case, while R is "standard" for two-level factorial studies, the form of the model and the required number of experimental runs make it difficult to apply the designs we've discussed in earlier chapters.

Because this is a relatively small problem (in terms of f and N), complete enumeration and evaluation of all possible designs is not especially difficult (R17.1). The number of distinct experimental designs that can be constructed containing a single replicate assigned to each of six (out of eight) treatments is $8!/(6!2!) = 28$, and the number of designs that contain five distinct treatments, with one of them applied to two units, is $5 \times 8!/(5!3!) = 280$. These 308 designs are all that need be investigated, because others contain fewer than five distinct treatments, and so cannot support estimation of all five model parameters.

In fact, there are substantial groups of designs (as counted above) that are equivalent from the standpoint of most optimality criteria. For example, most criteria evaluate two two-level factorial designs as equivalent if one is constructed by reversing the application of "high" and "low" levels (coded $+$ and $-$) for some or all factors in each run of the other. Also, for this example, x_1 and x_2 play symmetric roles in the proposed model; for any design, a second design generated by exchanging corresponding values of x_1 and x_2 in each run would be evaluated as equivalent by most reasonable criteria.

Figure 17.1 displays the results of a complete search for the A- and D-optimal designs (for estimating α, β, γ, and $(\alpha\beta)$, given the nuisance parameter μ) for this problem. The two designs in the left half of the figure (labeled (1.) and (2.)) are the A-optimal arrangements, while all four designs shown are D-optimal. Note that each panel actually represents more than one distinct design, since any axis can be "reversed" ($+1$ and -1 exchanged), and x_1 and x_2 can be exchanged, as discussed above. The single remaining group of designs containing six out of eight of the factorial treatments omits the two treatments for any value of (x_1, x_2); these designs are singular for the model we are considering, and so are clearly not optimal. In contrast, the four parameters of interest *are* jointly estimable under many of the designs containing five of the eight treatments (with one treatment replicated), but none of these are A- or D-optimal for this setting.

A blocked example

Consider an "experiment augmentation" situation in which a complete block design (CBD) has been completed (or is underway) using three blocks, each containing four units, one assigned to each of four treatments. Suppose primary interest involves estimation of $\tau_2 - \tau_1$, $\tau_3 - \tau_1$, and $\tau_4 - \tau_1$, or in matrix form:

$$\mathbf{C}\tau = \begin{pmatrix} -1 & +1 & 0 & 0 \\ -1 & 0 & +1 & 0 \\ -1 & 0 & 0 & +1 \end{pmatrix} \begin{pmatrix} \tau_1 \\ \tau_2 \\ \tau_3 \\ \tau_4 \end{pmatrix}.$$

Now, suppose that unforeseen but fortunate circumstances make three additional blocks available, but that each of these blocks contains only *three* units, so that the CBD pattern cannot be continued. The question is, what treatments should be applied to the units in these extra blocks so that the complete six-block design is optimal for estimating the treatment contrasts of interest? Units to be allocated are shown graphically in Figure 17.2.

Thinking of each block individually, there are:

- four ways of applying three treatments to the (unordered) units
- 12 ways of applying two treatments to the units, applying one of these to two units and the other to one unit
- four ways of applying one treatment to all three units

or 20 different treatment assignment patterns in all. Hence there are $20^3 = 8000$ ways of assigning treatments to the identifiable blocks, and fewer if blocks are not viewed as identifiable. While this number is somewhat larger than the number of candidate designs in our factorial example, it is still small enough to make complete enumeration feasible.

Evaluation of designs requires:

$$\mathbf{H}_1 = \text{diag}\left(\frac{1}{4}\mathbf{J}_{4\times 4}, \frac{1}{4}\mathbf{J}_{4\times 4}, \frac{1}{4}\mathbf{J}_{4\times 4}, \frac{1}{3}\mathbf{J}_{3\times 3}, \frac{1}{3}\mathbf{J}_{3\times 3}, \frac{1}{3}\mathbf{J}_{3\times 3}\right)$$

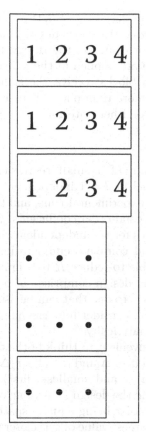

Figure 17.2 Choices in the augmented irregular block design example.

and

$$
\mathbf{X}_2 = \begin{pmatrix} \mathbf{I}_{4 \times 4} \\ \mathbf{I}_{4 \times 4} \\ \mathbf{I}_{4 \times 4} \\ \hline 9 \text{ rows} \end{pmatrix}
$$

where the last nine rows of $\mathbf{X}_2$ are used to code the treatment assignments in the last three blocks, as described above, leading to $\mathcal{I} = \mathbf{X}_2'(\mathbf{I} - \mathbf{H}_1)\mathbf{X}_2$ for the overall design. For the contrasts of interest, A- and D-optimal designs are those that minimize

$$\text{trace}(\mathbf{C}\mathcal{I}^-\mathbf{C}') \quad \text{and} \quad |\mathbf{C}\mathcal{I}^-\mathbf{C}'|,$$

respectively. (These forms are invariant to the selection of $\mathcal{I}^-$ since each row of $\mathbf{C}$ represents an estimable combination of the model coefficients.)

As in the factorial example, A-optimal designs are a subset of D-optimal designs for this problem. A-optimal designs are those that omit treatments 2, 3, and 4 each in one of the three added blocks, i.e., that add blocks in which treatments (1,2,3), (1,2,4), and (1,3,4) are applied. D-optimal designs are those that omit *any three* of treatments 1, 2, 3, and 4 each in one of the three additional blocks. For example, the design in which treatment groups (1,2,3), (1,3,4), and (2,3,4) are applied in the added blocks is also D-optimal. Designs in which any two blocks receive the same set of treatments, or in which any block contains more than one unit assigned to any treatment, are not optimal with respect to either criterion.

A regression example

Finally, consider the design of a small regression experiment in which a quadratic polynomial model in $d = 1$ independent variable is to be fit based on a design of only $N = 3$ experimental runs, and that interest is primarily in estimating the first- and second-order coefficients, β_1 and β_{11}. If $R = [-1, +1]$ is the set of allowable values for the independent variable, the number of possible designs is infinite and complete enumeration is not possible. In some cases, it might be acceptable to reduce R to a finite grid, say $\{-1.00, -0.95, -0.90, \ldots, 1.00\}$, since any design comprised of points in the continuous design region is at least "close" to one that can be assembled from points on the grid. However, here we shall consider how the question of optimal design can sometimes be addressed analytically.

For our purpose, it is convenient to think of the three values of x to be used as c (for *center* value), and $c - r\Delta$ and $c + (1 - r)\Delta$, where Δ is the difference between the largest (third) x and smallest (first) x. In this reexpression, r determines how asymmetric the design is: a value of $r = \frac{1}{2}$ corresponds to the largest and smallest values of x being equally spaced from c, while $r < / > \frac{1}{2}$ implies that the smallest/largest value of x is closer to c. The full model matrix is then expressed as:

$$\mathbf{X} = \begin{pmatrix} 1 & c - r\Delta & (c - r\Delta)^2 \\ 1 & c & c^2 \\ 1 & c + (1 - r)\Delta & (c + (1 - r)\Delta)^2 \end{pmatrix}.$$

Partitioning this into one- and two-column submatrices leads to:

$$\mathbf{X}_{2|1} = \frac{1}{3}\Delta \begin{pmatrix} -r - 1 & \Delta(r^2 + 2r - 1) - 2c(r + 1) \\ 2r - 1 & -\Delta(2r^2 - 2r + 1) + 2c(2r - 1) \\ 2 - r & \Delta(r^2 - 4r + 2) - 2c(r - 2) \end{pmatrix}.$$

Detailed but simple algebra can be used to show that the determinant of the 2×2 design information matrix is:

$$|\mathbf{X}'_{2|1}\mathbf{X}_{2|1}| = \frac{1}{3}\Delta^2[r(1 - r)]^2.$$

The first thing to be observed of this expression is that it is not a function of c, i.e., that the determinant, and therefore D-optimality, is "location invariant" along the real line. Next, note that $|\mathbf{X}'_{2|1}\mathbf{X}_{2|1}|$ is strictly increasing in the "spread" of the design, Δ, i.e., for any value of r, the design is made "best" with respect to the D-optimality criterion by making the range of x values as large as possible. Finally, for any value of Δ, the determinant is maximized when $r = \frac{1}{2}$, i.e., when the design is symmetric. Hence, given $R = [-1, +1]$, the design specified by $c = 0$, $\Delta = 2$, and $r = \frac{1}{2}$, or $x = \{-1, 0, 1\}$, is D-optimal.

17.4 Algorithms

While it is possible to mathematically derive or verify optimal experimental designs in some cases, they are most often constructed numerically. Numerical design optimization problems are generally difficult because:

1. The optimization is over many variables – often all of the elements of each **x** in D.

2. The objective function is often optimized for several designs, that is, there are often many optimal designs of a given size for a given experimental region, model, and criterion.

3. Many *near-optimal* designs that are dissimilar to optimal plans may also exist.

Further, since there are a very large number (sometimes infinite) of designs that *can* be constructed for a given situation, complete enumeration and evaluation is not a practical option except for very small or highly restricted experiments. As a result, designs are often constructed using algorithms that begin with an arbitrary or random starting design, and make a series of iterative changes with the purpose of improving the quality of the design at each step. When no further iterative improvements can be made, the search ends and the resulting design is reported. In most cases, this "converged" product of the search cannot be guaranteed to actually be optimal, so the calculation is often repeated with different starting designs. If the designs found via repeated executions of the algorithm are not equivalent, the best is reported with the understanding that it may very well be "good" with respect to ϕ_M, but "suboptimal."

The most widely used approach to the numerical construction of optimum experimental designs is undoubtedly through the use of various *point-exchange algorithms*. For simplicity, we shall describe this family of algorithms for problems in which R is finite. For this case, a simple point-exchange algorithm, described for a criterion function we wish to maximize, is:

1. Specify the N^*-point experimental region R, the model M, the experiment size N, and the criterion function ϕ_M to be used.

2. Specify a "starting design" D_0 by randomly selecting N points from R, and evaluate it by computing $\phi_M(D_0)$.

3. Construct the N^* designs, each of $N + 1$ points, that consist of the points of D_0 *and* one additional point from R. Evaluate each of these by computing ϕ_M, identify the best (the design that yields the highest value of

the criterion function), and denote this design D_0^+. If more than one $N + 1$ point design is best, select one of these randomly.

4. Construct the $N + 1$ designs, each of N points, that consist of all but one of the points in D_0^+. Evaluate each of these by computing ϕ_M, identify the best, and denote this design D_1. If more than one N-point design is best, select one of these randomly.

5. If $\phi_M(D_0) < \phi_M(D_1)$, that is, if the add-and-delete process has improved the initial design, repeat steps (3) and (4), substituting D_1 for D_0 and D_1^+ for D_0^+, to generate D_2. Continue iterations, generating D_{i+1} from D_i, $i = 2, 3, 4, \ldots$.

6. If $\phi_M(D_i) \geq \phi_M(D_{i+1})$, stop the search, reporting D_i as the optimal design.

For criterion functions that should be minimized, "highest" can be replaced by "lowest" in step 3, and the direction of the inequalities reversed in steps 5 and 6.

More elaborate point-exchange algorithms have also been developed. Mitchell's DETMAX (1972) extends the basic point-exchange algorithm for D-optimality by allowing iterative addition and deletion of more than one design point.

17.5 Conclusion

Many of the experimental designs used in scientific and other inquiries are selected from design *classes* that are appropriate for the specific circumstances at hand; examples include BIBDs and CRDs. A particular design is then picked from the identified class so as to meet the specific demands of the experimental conditions. In contrast, *optimal* designs are generated as the solution to a mathematical optimization problem defined by the experimental region, the form of the model, and an appropriate criterion function. In this chapter, we've focused on D- and A-optimal designs for experiments performed to compare treatments.

It should be immediately added that many of the "standard" experimental designs discussed in the first part of this book *are* optimal with respect to standard design regions and models, and popular optimality criteria. However, thinking about the experimental design process from the optimality perspective eliminates "structural" restrictions imposed by the standard patterns, for example, on allowable values of N in regular fractional factorials, or restrictions on the number of blocks and units per block for BIBDs. Treating design construction as an optimization problem allows more freedom in specifying the design region, model, and measure of goodness to actually reflect the "reality" of the circumstances in which the experiment is to be performed, but sometimes at the cost of nontrivial computational effort.

The theory of optimal design is a fascinating topic with substantial contributions by many researchers. This chapter is at best a very brief introduction to these ideas; the interested reader might be directed to the book of Atkinson and Donev (1992) for a more thorough treatment of this topic, to that of Silvey (1980) for a concise mathematical treatment of some of the central theory, or to Fedorov (1972) for an early, but still authoritative and widely referenced, text on optimal design theory.

17.6 Exercises

1. As noted in Section 17.1, "standard" designs are often optimal with respect to common criteria when appropriate experimental regions, values of N, and models are employed. In particular, regular two-level factorial and fractional factorial designs are D- and A-optimal for $R = \{-1, +1\}^f$ for any model for which they support estimation of all coefficients (main effects and interactions). Prove that a 2^2 factorial design ($N = 4$ runs) is D-optimal for estimating the two coefficients of a main effects model, given that the intercept must be included as a nuisance parameter. (Hint: Note that the elements of $\mathcal{I}$ are proportional to "sample variances" and "sample covariances" of the values of x_1 and x_2 used in the design.)

2. Continuing exercise 1, what if N had been six, rather than four? Write a short computer program to construct and evaluate all designs composed of six runs taken from the four possible unique treatments. Which of these are D-optimal for estimating the two main effects given that the intercept must be included as a nuisance parameter? Which are A-optimal?

3. For designs that enable estimation of all possible *contrasts* of unstructured treatment parameters but not the parameters individually, E-optimal designs are those that maximize the next-to-smallest eigenvalue of $\mathcal{I}$. (For all such designs, the *smallest* eigenvalue of $\mathcal{I}$ is zero.) Write a statistical motivation for this criterion. (Hint: Note that one generalized inverse for $\mathcal{I}$ in this case is $\sum_{i=2}^{t} \frac{1}{e_i} \mathbf{v}_i \mathbf{v}_i'$ where $e_i, i = 2, 3, 4, \ldots, t$ are the nonzero eigenvalues of $\mathcal{I}$ and $\mathbf{v}_i, i = 2, 3, 4, \ldots, t$ are their corresponding eigenvectors.)

4. Some critics of the use of optimal design note (correctly) that optimal designs can only be "optimal" with respect to one measure of statistical quality, while real experiments often serve to answer many questions. Consider a first-order linear regression experiment in one controlled variable performed within $R = [-1, +1]$ using an even number of units. The single design that is A- and D-optimal for estimating the slope parameter is the one that places half of the runs at $x = -1$ and the other half at $x = +1$. Comment on at least one aspect of this design that would be very *unappealing* in most realistic situations. How could the design be modified to overcome this weakness?

5. Sometimes the search for an optimal design is limited to a certain class or subclass of designs. Consider the design of a two-level factorial experiment, $R = \{-1, +1\}^f$, where a main effects model will be used for inference. Suppose the search is to be limited to orthogonal designs that are balanced – those for which each factor appears at level -1 for half the

runs, and at level $+1$ for the other half – but that it will not necessarily be restricted to regular fractional factorial designs. Direct results are that

$$\mathbf{X}_{2|1} = \mathbf{X}_2 \quad \text{and} \quad \mathbf{X}_2'\mathbf{X}_2 = N\mathbf{I}$$

where $\mathbf{X}_2$ is of dimension N-by-f. Now suppose that while a main effects (and intercept) model is to be fitted:

$$y = \mathbf{1}\mu + \mathbf{X}_2\boldsymbol{\theta}_{me} + \epsilon,$$

the data are actually generated by a model containing two-factor interactions:

$$y = \mathbf{1}\mu + \mathbf{X}_2\boldsymbol{\theta}_{me} + \mathbf{X}_3\boldsymbol{\theta}_{2fi} + \epsilon.$$

In this case, the least-squares estimate of $\boldsymbol{\theta}_{me}$ is biased:

$$E(\widehat{\boldsymbol{\theta}}_{me}) = \boldsymbol{\theta}_{me} + \mathbf{A}\boldsymbol{\theta}_{2fi}$$

(a) Define, in terms of model matrices, a criterion for which the optimal design minimizes the sum of squared alias coefficients, the elements of $\mathbf{A}$.

(b) If attention is further limited to regular fractional factorial designs, what designs are optimal with respect to your criterion?

Appendix A: Calculations using R

The calculations presented as examples in this book were performed using the statistical software package R; the software and documentation can be downloaded at the Web site www.r-project.org. This appendix should not be viewed as an introduction to R, and the package supports a much wider variety of calculations than those demonstrated here. However, the linear algebra capability and statistical modules available in R (and in some other software packages) provide a relatively easy way to produce numerical results needed to apply the ideas and methods discussed in this book.

The first section of this appendix provides a very brief description of some of the commands used in the examples. More extensive documentation can be easily found on the Web and in a number of books. The second section contains the R scripts used to generate the examples in the book, labeled here and in the text with the notation "(R#.#.)."

Some R Commands

- `matrix`: Construct a matrix with elements supplied in a list or from a file, e.g.,

  `A <- matrix(c(-1,2,3,1,-8,5),nrow=2,ncol=3)`

  produces a matrix of form

  $$\mathbf{A} = \begin{pmatrix} -1 & 3 & -8 \\ 2 & 1 & 5 \end{pmatrix},$$

  while

  `A <- matrix(c(-1,2,3,1,-8,5),nrow=2,ncol=3,byrow=T)`

  produces a matrix of form

  $$\mathbf{A} = \begin{pmatrix} -1 & 2 & 3 \\ 1 & -8 & 5 \end{pmatrix}.$$

- `%*%`: Compute a matrix multiplication; `A %*% B` will produce the matrix product of matrices $\mathbf{A}$ and $\mathbf{B}$, provided they have been defined appropriately.

- `t`: Transpose its matrix argument.

- `solve`: Solve linear systems. Here we use this command to invert full-rank matrices, i.e., `T <- solve(t(X)%*%X)` produces $(\mathbf{X}'\mathbf{X})^{-1}$.

- `boxplot`: Construct parallel boxplots. For example, if **y** and **x** are two vectors of the same length, and **x** has relatively few distinct values, `boxplot (split(y,x))` sorts the elements of **y** into groups by the corresponding values found in **x** and produces a boxplot of the data in each group.

- `sum`: Compute the sum of elements in a matrix or list, e.g., for the matrix defined above, `sum(A)` has the value 2.

- `length`: Count the number of elements in a matrix or list, e.g., `length(A)` has the value 6.

- `factor`: Reduce a column of data to distinct values that are regarded as ordinal "levels" of a factor; useful in converting real-valued data to categories for ANOVA.

- `aov`: Compute an analysis of variance and related quantities, primarily for balanced data sets. For example, `aov(y ~ A)` computes a one-way ANOVA for a response named y relative to a factor named A, while `aov(y ~ A+B)` computes a main-effects (no interaction) two-way ANOVA relative to factors A and B, and `aov(y ~ A*B)` also includes the interaction in the two-way ANOVA. The result is an "object" containing elements such as estimated coefficients (`coef`) and residuals (`resid`).

- `glht`: Can be used to perform some multiple comparisons procedures, including Tukey and Dunnett simultaneous confidence intervals. The package `multicomp` must be loaded before this command can be used. Simultaneous intervals can be displayed by applying the command `confint` to the resulting object.

- `lsfit`: Fit a linear model to the indicated data; for example `lsfit(X,y)` fits a linear model defined by the N-row model matrix **X** to the N-element column vector **y**. A summary of the fit, including fitted coefficients and the residual sum of squares, can be displayed by applying the command `ls.print` to the resulting object.

- `sort`: Sort the elements in a list or matrix, e.g., for the matrix defined above, `sort(A)` produces the ordered list $-8, -1, 1, 2, 3, 5$.

- `abs`: Compute absolute value of a scalar, or of the individual elements of a matrix, e.g., with the second version of **A** defined above, `abs(A)` produces

$$\mathbf{A} = \begin{pmatrix} 1 & 2 & 3 \\ 1 & 8 & 5 \end{pmatrix}.$$

- `plot`: Construct a two-dimensional scatterplot. For example, if **x** and **y** are vectors of the same length, `plot(x,y)` produces a plot of the corresponding (x, y) pairs.

- `qnorm`: Calculate the indicated quantile of a standard normal distribution, e.g., `qnorm(0.95)` has the value 1.644854.

Example Calculations

(R2.1): *Hat matrix when* **X** *is not of full column rank.*

```
X<-matrix(nrow=9,byrow=T,
c(1,0,1,0,0,
  1,0,1,0,0,
  1,0,0,1,0,
  1,0,0,1,0,
  1,0,0,0,1,
  0,1,1,0,0,
  0,1,1,0,0,
  0,1,0,1,0,
  0,1,0,0,1))
XpX <- t(X)%*%X
XpXstar <- XpX[2:5,2:5]
XpXginv <- matrix(0,nrow=5,ncol=5)
XpXginv[2:5,2:5] <- solve(XpXstar)
H <- X%*%XpXginv%*%t(X)
```

Alternatively, a generalized inverse can be computed numerically using the "ginv()" command after loading the MASS package:

```
XpX <- t(X)%*%X
XpXginv <- ginv(XpX)
H <- X%*%XpXginv%*%t(X)
```

avoiding the need to identify a maximal set of linearly independent columns in **X**.

(R2.2): *Design information matrix for a partitioned model.*

Partitioning the above **X** matrix into two sets of columns:

```
X1 <- X[,1:2]
X2 <- X[,3:5]
H1 <- X1%*%solve(t(X1)%*%X1)%*%t(X1)
X2bar1 <- X2 - H1%*%X2
Info <- t(X2bar1)%*%X2bar1
```

As in (R2.1), "ginv()" can be used to compute a generalized matrix when X_1 is not of full column rank.

(R2.3): *Central F-quantile and noncentral F probability.*

```
theta<-matrix(c(-2,1,1),nrow=3,ncol=1)
sigma<-1.25
lambda<-t(theta)%*%Info%*%theta/sigma^2
qf(.95,2,5)
1-pf(5.786,2,5,lambda)
```

or in compound form,

```
1-pf(qf(.95,2,5),2,5,t(theta)%*%Info%*%theta/sigma^2)
```

(R3.1): *Boxplots of the data of Matsuu et al., Figure 3.1.*

```
# Input of data from local directory
datamatrix <- matrix(scan("matsuu.data"),byrow=T,ncol=2)
trt        <- datamatrix[,1]
y          <- datamatrix[,2]
# Boxplot of data from Table 3.1
boxplot(split(y,trt),
    sub="Treatment Group Number",
    ylab="Epinephrine Levels",
    main="Boxplots of Data from Matsuu et al.")
```

(R3.2): *Power calculation example.*

```
n <- c(5,5,5,5)
tau <- c(0.,0.,-1.,-2.)
taubar <- sum(tau*n)/sum(n)
sig <- .75
lambda <- sum(n*(tau-taubar)^2 /sig^2)
df1 <- length(n)-1
df2 <- sum(n)-length(n)
critval <- qf(.99,df1,df2)
1-pf(critval,df1,df2,lambda)
```

(R4.1): *Boxplots of the block-corrected data of Kocaoz et al., Figure 4.1.*

```
# Input of data from local directory
datamatrix <- matrix(scan("kocaoz.data"),byrow=T,ncol=3)
blk        <- datamatrix[,1]
trt        <- datamatrix[,2]
y          <- datamatrix[,3]
# Boxplot of data from Table 4.1.
aovb.obj  <- aov(y ~ blk)
boxplot(split(resid(aovb.obj),trt),
    sub="Treatment Group Number",
    ylab="Tensile Strength",
    main="Block-Corrected Data from Kocaoz et al.")
```

(R6.1): *Analysis of variance in Table 6.3; data from Table 6.1.*

```
# Input of data from local directory
datamatrix <- matrix(scan("example.data"),byrow=T,ncol=4)
row        <- datamatrix[,1]
col        <- datamatrix[,2]
trt        <- datamatrix[,3]
y          <- datamatrix[,4]
# Analysis of variance Table 6.3
Rows       <- factor(row)
Columns    <- factor(col)
```

```
Treatments<- factor(trt)
aov.obj   <- aov(y ~ Rows + Columns + Treatments)
summary(aov.obj)
```

(R6.2): Simultaneous confidence intervals, following R6.1.

```
# Tukey intervals for all pairs of treatments
fit<-glht(aov.obj,linfct=mcp(Treatments="Tukey"))
confint(fit)
```

(R7.1): Analysis of variance; data from Table 7.1.

Beginning with a file named "file" containing:

19	1	0	0	0	0	0	0	0	0	0	0	0		1	0	0	0	0	0	0	0	0	0
17	1	0	0	0	0	0	0	0	0	0	0	0		0	1	0	0	0	0	0	0	0	0
11	1	0	0	0	0	0	0	0	0	0	0	0		0	0	1	0	0	0	0	0	0	0
												...											
14	0	0	0	0	0	0	0	0	0	0	0	1		0	0	1	0	0	0	0	0	0	0
24	0	0	0	0	0	0	0	0	0	0	0	1		0	0	0	0	1	0	0	0	0	0
21	0	0	0	0	0	0	0	0	0	0	0	1		0	0	0	0	0	0	1	0	0	0

the BIBD ANOVA decomposition can be assembled from:

```
data <- matrix(scan("file"),byrow=T,ncol=22)
y<-data[,1]
X1<-data[,2:13]
X2<-data[,14:22]
one<-lsfit(X1,y)
ls.print(one)
onetwo<-lsfit(cbind(X1,X2),y)
ls.print(onetwo)
```

(R9.1): Boxplots of Figure 9.1.

```
# Input of data from local directory
datamatrix <- matrix(scan("soudki.data"),byrow=T,ncol=4)
ratio1     <- datamatrix[,1]
ratio2     <- datamatrix[,2]
ratio3     <- datamatrix[,3]
y          <- datamatrix[,4]
Treatment  <- ratio1
r2         <- factor(ratio2)
r3         <- factor(ratio3)
aovb.obj   <- aov(y ~ r2*r3)
boxplot(split(resid(aovb.obj),Treatment),
        sub="TA/C",
        ylab="Compressive Strength",
        main="Main-Effect Boxplots for TA/C, Data from Soudki et al.")
```

(R11.1): Half-normal plot of Figure 11.4.

```
thetahat<- t(X)%*%y/16
absthetahat<-sort(abs(thetahat))
```

```
qvec<-qnorm(ppoints(2*length(absthetahat)))
plot(qvec[qvec>0],absthetahat,xlab="nonnegative normal quantiles",
      ylab="|estimates|")
pt.lab <- c(" "," "," "," "," "," "," "," "," "," "," "," ","ABC","AB","B","A")
text(qvec[qvec>0],absthetahat,paste(" ",pt.lab,sep=" "),adj=0)
```

(R17.1): *Complete search for optimal irregular fractional factorials.*
The following script performs a complete evaluation of all six-run fractions,
containing at least five distinct design points, of a 2^3 factorial, evaluating each
by the D- and A-criteria as described in Section 17.3.4. A warning: while this
script is written to make the search technique clear, it is not a computationally efficient way to perform this calculation because R is not especially well
suited to programs with nested loops. There are, however, more advanced R
programming techniques that allow this kind of calculation without resorting
to explicit looping.

```
R = matrix(
c(+1,+1,+1,+1,-1,-1,-1,-1,
  +1,+1,-1,-1,+1,+1,-1,-1,
  +1,-1,+1,-1,+1,-1,+1,-1),
ncol=3)
Rplus = cbind(R,R[,1]*R[,2])
Answers = matrix(0,308,8)
P = diag(1,6)-matrix(c(1/6),6,6)
X = matrix(0,6,4)
j=0

# Designs with 6 distinct points
for (i1 in c(1:7)) {
for (i2 in c((i1+1):8)) {
ilist1=c(1:8)
ilist2=ilist1[ilist1!=i1&ilist1!=i2]
ic=0
for(i in ilist2){
ic=ic+1
X[ic,]=Rplus[i,]}
j=j+1
Info=t(X)%*%P%*%X
a=sum(diag(ginv(Info)))
d=det(Info)
Answers[j,1]=ilist2[1]
Answers[j,2]=ilist2[2]
Answers[j,3]=ilist2[3]
Answers[j,4]=ilist2[4]
Answers[j,5]=ilist2[5]
Answers[j,6]=ilist2[6]
Answers[j,7]=a
```

```
Answers[j,8]=d
}}

# Designs with 5 distinct points, one rep
for (i1 in c(1:6)) {
for (i2 in c((i1+1):7)) {
for (i3 in c((i2+1):8)) {
ilist1=c(1:8)
ilist2=ilist1[ilist1!=i1&ilist1!=i2&ilist1!=i3]
ic=0
for(i in ilist2){
ic=ic+1
X[ic,]=Rplus[i,]}
for (i4 in ilist2) {
X[6,]=Rplus[i4,]
j=j+1
Info=t(X)%*%P%*%X
a=sum(diag(ginv(Info)))
d=det(Info)
Answers[j,1]=ilist2[1]
Answers[j,2]=ilist2[2]
Answers[j,3]=ilist2[3]
Answers[j,4]=ilist2[4]
Answers[j,5]=ilist2[5]
Answers[j,6]=i4
Answers[j,7]=a
Answers[j,8]=d
}}}}

o = order(Answers[,7])
Answers[o,]
```

Appendix B: Solution notes for selected exercises

Chapter 1

1.

(a) Experimental units include the quantities of milk and tea used to make each serving, possibly also the cup itself (since making a cup of tea without a cup would be difficult), and any other physical material that was used in any given serving. It can also be argued that the time interval actually used to make the tea is "experimental material" as well; that's more critical in some split-plot experiments.

(b) There are two treatments in this experiment – the two ways of making tea. Note that they aren't really "things," but recipes. One is the procedure used when milk is the first ingredient, the other is the procedure used when tea is the first ingredient.

(c) One way to do this would be to select four red cards and four black cards from a standard deck of playing cards, assign one of the treatments to each color, and shuffle them. Suppose the necessary material is available; eight quantities of tea and milk, and eight cups are lined up in a row. Then sequentially, starting from one end of the "material row," draw a card, make the cup of tea in the indicated way, and serve it to the lady. Proceed in this manner until all eight cups have been served. Note that if this approach is used, you are guaranteed to make four cups according to each recipe, and that each possible sequence has the same probability.

At first appearance, it might be tempting to randomize this experiment by starting at one end of the row, and simply flip a fair coin at each cup; "heads" means one recipe, "tails" means the other. But this means that you might finish up with one treatment early and have to use a sequence of cups made with the other recipe at the end. An easier way to see this is to think about what would happen if the (unfortunate) lady had to drink 20,000 cups, 10,000 of each type. The probability that the first two cups are of different types is 1/2, but the probability that the *last* two cups are different is very close to zero (because the probably of getting 10,000 heads or tails well before the end of the sequence is very close to one). So this technique wouldn't give equal probability to

all possible sequences, and so shouldn't be used for treatment randomization in this case.

(d) The experiment could be divided into two "subexperiments" of four cups
each, two made each way. One kind of cup would be used in subexperi-
ment (or block) 1, and the other kind of cup would be used in subexperi-
ment 2. This is called an "extended randomized complete block design,"
to be described in Chapter 4.

4. This exercise is quite artificial, but shows that careful thought is sometimes
needed to determine the elements of an experiment. A *treatment* in this case
might be regarded as the meeting of a particular *pair* of teams. The game
is played under certain circumstances (e.g., in a ballpark, at a given time),
which could be regarded as the *unit*. If we adopt a statistical model that
regards the result of a game between a given pair of teams as a binomial
event, with a probability specific to that pair of teams, *replication* could
be included by having each pair of teams compete more than once. If there
are concerns of possible extraneous effects that change from day to day,
this might be addressed by *randomization* of the order in which teams play,
while control of potential effects associated with longer periods of time
(e.g., changing weather trends over the summer) might be exercised by
blocking complete round-robin schedules to contiguous time intervals (e.g.,
months).

Chapter 2

1.

(a) $\mathbf{H}_1 = \begin{pmatrix} \frac{1}{3}\mathbf{J}_{3\times3} & \mathbf{0}_{3\times3} \\ \mathbf{0}_{3\times3} & \frac{1}{3}\mathbf{J}_{3\times3} \end{pmatrix}$

(b) $\mathbf{X}_{2|1} = \mathbf{X}_2 - \frac{1}{3}\mathbf{J}_{6\times3}$

(c) $\mathbf{c}'\boldsymbol{\theta}_2$ is estimable if $\mathbf{c}'$ can be expressed as a linear combination of:

$$\left(\frac{2}{3}, -\frac{1}{3}, -\frac{1}{3}\right)$$

$$\left(-\frac{1}{3}, \frac{2}{3}, -\frac{1}{3}\right)$$

$$\left(-\frac{1}{3}, -\frac{1}{3}, \frac{2}{3}\right)$$

4.

(a) $\boldsymbol{\theta}_2' = (1, 0, 0)$

(b) $\boldsymbol{\theta}_2' = (0, 0, 1)$

(c) σ^2, rank($\mathbf{X}$), and N

Chapter 3

1.

(a) Under this model, the unique rows of the model matrix and the parameter vector may be written as:

$$\mathbf{X} = \begin{pmatrix} 1 & 1 & & & \\ 1 & & 1 & & \\ 1 & & & 1 & \\ 1 & & & & 1 \end{pmatrix}, \quad \boldsymbol{\theta}' = (\alpha, \tau_1, \tau_2, \tau_3, \tau_4).$$

τ_3 and $\tau_3 + \tau_2$ are not estimable because neither $(0, 0, 0, 1, 0)$ nor $(0, 0, 1, 1, 0)$ can be expressed as a linear combination of the rows of $\mathbf{X}$. $\tau_3 - \tau_2$ is estimable because:

$$(0, 0, -1, +1, 0) = (0, -1, +1, 0)\mathbf{X}.$$

(b) Under this model, the unique rows of the model matrix and the parameter vector may be written as:

$$\mathbf{X} = \begin{pmatrix} 1 & & & \\ 1 & 1 & & \\ 1 & & 1 & \\ 1 & & & 1 \end{pmatrix}, \quad \boldsymbol{\theta}' = (\mu_1, \theta_2, \theta_3, \theta_4).$$

θ_3, $\theta_3 - \theta_2$, and $\theta_3 + \theta_2$ are estimable because:

$$(0, 0, 1, 0) = (-1, 0, 1, 0)\mathbf{X}$$
$$(0, -1, 1, 0) = (0, -1, 1, 0)\mathbf{X}$$
$$(0, 1, 1, 0) = (-2, 1, 1, 0)\mathbf{X}.$$

(c) τ_2 is the treatment-specific effect of treatment 2, beyond the experiment-specific effects common to all units. θ_2 is the difference between treatment-specific effects for treatments 1 and 2:

$$\theta_2 = (\alpha + \tau_2) - (\alpha + \tau_1) = \tau_2 - \tau_1.$$

(d) No. Under the first model, the estimable functions $\mathbf{c}'\boldsymbol{\tau}$ are those for which $\mathbf{c}'\mathbf{1} = 0$. Under the second model, $\mathbf{X}$ is a 4-by-4 matrix of full rank, so *any* four-element vector can be represented as a linear combination of its rows (and so, any linear combination of μ_1, θ_2, θ_3, and θ_4 is estimable).

9. In this case, x is part of the overall experimental effect summarized by α_0; "units" can be thought of abstractly as measurement periods or set-ups, and "treatments" are the methods used in each trial. An effects model for

the 50-run experiment can then be written in matrix form as:

$$y = \begin{pmatrix} 1 & 1 & 0 & 0 & 0 & 0 \\ 1 & 0 & 1 & 0 & 0 & 0 \\ 1 & 0 & 0 & 1 & 0 & 0 \\ 1 & 0 & 0 & 0 & 1 & 0 \\ 1 & 0 & 0 & 0 & 0 & 1 \end{pmatrix} \begin{pmatrix} \alpha_0 \\ \beta_1 \\ \beta_2 \\ \beta_3 \\ \beta_4 \\ \beta_5 \end{pmatrix} + \epsilon$$

where each submatrix is a 10×1 vector of ones or zeros, as indicated.

(a) $\beta_i - \beta_j$ are estimable because they are measurement method contrasts with coefficient vectors that can be formed as linear combinations of the rows of the model matrix.

(b) Individual β_i values are not estimable, nor are their absolute values.

(c) Given independent experimental trials, the sample variance of the ith treatment group is an unbiased estimate of σ_i^2, and the variances of any pair of methods can be tested for equality using an F-test.

Chapter 4

2.

(a) As with exercise 1,

$$\mathbf{X}_1 = \begin{pmatrix} 1 & & & \cdots \\ & 1 & & \cdots \\ \cdots & \cdots & \cdots & \cdots \\ & & & \cdots & 1 \end{pmatrix}.$$

If the data are grouped by block, and by treatment within block:

$$\mathbf{X}_2 = \begin{pmatrix} 1 & & \cdots \\ \hline & \mathbf{I} & \\ \hline\hline 1 & \cdots \\ \hline & \mathbf{I} & \\ \hline\hline \cdots & \cdots & \cdots & \cdots \\ \hline\hline & & \cdots & 1 \\ \hline & \mathbf{I} & \end{pmatrix}$$

with each $\mathbf{1}$ and $\mathbf{0}$ containing r elements, and each $\mathbf{I}$ of dimension $t \times t$.

(b) $\mathbf{H}_1 = \frac{1}{t+r} \begin{pmatrix} \mathbf{J} & & \cdots & \\ & \mathbf{J} & \cdots & \\ \cdots & \cdots & \cdots & \cdots \\ & & \cdots & \mathbf{J} \end{pmatrix}$, i.e., as with a CRD with block size of

$t+r$. Then

$$\mathbf{H}_1\mathbf{X}_2 = \frac{1}{t+r} \begin{pmatrix} (r+1)\mathbf{1} & \mathbf{1} & \cdots & \mathbf{1} \\ \mathbf{1} & (r+1)\mathbf{1} & \cdots & \mathbf{1} \\ \cdots & \cdots & \cdots & \cdots \\ \mathbf{1} & \mathbf{1} & \cdots & (r+1)\mathbf{1} \end{pmatrix}$$

$$\mathbf{X}_2'\mathbf{H}_1\mathbf{X}_2 = \frac{1}{t+r}$$

$$\begin{pmatrix} (r+1)^2 + (b-1) & 2r+b & \cdots & 2r+b \\ 2r+b & (r+1)^2 + (b-1) & \cdots & 2r+b \\ \cdots & \cdots & \cdots & \cdots \\ 2r+b & 2r+b & \cdots & (r+1)^2 + (b-1) \end{pmatrix}$$

$$= \frac{1}{t+r}(r^2\mathbf{I} + (2r+t)\mathbf{J})$$

since $b = t$. Also, $\mathbf{X}_2'\mathbf{X}_2 = (r+t)\mathbf{I}$, so

$$\mathcal{I} = \mathbf{X}_2'\mathbf{X}_2 - \mathbf{X}_2'\mathbf{H}_1\mathbf{X}_2 = (r+t)\mathbf{I} - \frac{1}{t+r}(r^2\mathbf{I} + (2r+t)\mathbf{J}) = \frac{2r+t}{r+t}[t\mathbf{I} - \mathbf{J}]$$

(c) One generalized inverse for $\mathcal{I} = \frac{2r+t}{r+t}[t\mathbf{I} - \mathbf{J}]$ is $\mathcal{I}^- = \frac{r+t}{2r+t}\frac{1}{t}\mathbf{I}$ (check this with $\mathcal{I}\mathcal{I}^-\mathcal{I} = \mathcal{I}$). So, for $\mathbf{c}'\mathbf{1} = 0$,

$$Var(\widehat{\mathbf{c}'\boldsymbol{\tau}}) = \frac{r+t}{2r+t}\frac{1}{t}\mathbf{c}'\mathbf{c}\sigma^2$$

or in this case, $2\frac{r+t}{(2r+t)t}\sigma^2$.

(d) For a CRD, with equal sample sizes for each treatment, $\mathbf{X}_2'\mathbf{H}_1 = \frac{t}{N}\mathbf{J}$, not equivalent to the corresponding matrix product for this design. So the designs are not Condition E equivalent.

8.

(a) $\lambda = \boldsymbol{\tau}'\mathcal{I}\boldsymbol{\tau}/\sigma^2 = \sum_j b(\tau_i - \bar{\tau})^2/\sigma^2 = b(0.2^2 + 0.2^2 + 0.4^2)/2.4 = b/10$. Minimum value of b for which:

$$P\{W > F_{0.95}(2, 2(b-1))\} > 0.8, \quad \text{where} \quad W \sim F(2, 2(b-1), b/10)$$

is 98, as evaluated in R by:

```
critval <- qf(.95,2,2*(b-1))
1-pf(critval,2,2*(b-1),b/10)
```

(b) $EL^2 = 4t_{0.975}^2(18)\sigma^2\frac{\mathbf{c'c}}{10}$, where $\mathbf{c'c} = 1.5$, is 6.356.

Chapter 5

5.

   (a)

source	d.f.	
replicates	1	
automobiles	6	(different autos in each rep)
positions	3	(same positions in each rep)
tire type	3	
residual	18	
corrected total	31	

   (b) $\lambda = r\sum_{i=1}^{4}(\tau_i - \bar{\tau})^2/\sigma^2 = 8 \times 2/4 = 4$.

6.

   (a)

source	d.f.
rows	2
columns	2
treatments	2
residual	2
corrected total	8

   (b) As shown in Section 5, $\mathbf{H}_1\mathbf{X}_2$ takes the same form for LSD, CBD, and CRD with $r = t$. Hence for this design

$$\mathbf{H}_1\mathbf{X}_2 = \frac{1}{3}\mathbf{J}_{9\times 3}.$$

   In addition, $\mathbf{X}_2$ is the same matrix, within row rearrangements, for this design as for a CRD with $r = t = 3$, hence the reduced normal equations are the same for the two designs:

$$\hat{\tau}_i - \frac{1}{3}\sum_{j=1}^{3}\hat{\tau}_j = \bar{y}_{..i} - \bar{y}_{...}, \quad i = 1, 2, 3.$$

   (c) For $\mathbf{c'1} = 0$, $Var(\mathbf{c'}\boldsymbol{\tau}) = \frac{1}{3}\sum_{i=1}^{3}c_i^2\sigma^2$; $Var(\tau_1 \overset{\frown}{-} \tau_2) = \frac{2}{3}\sigma^2$.

Chapter 6

2.

(a) Transforming the data as required for the Levene test, we have:

	Treatment		
1	2	3	4
0.000	1.365	1.482	0.000
0.115	0.680	0.960	1.655
0.759	0.000	0.000	0.197
0.172	0.146	0.653	0.477
0.795	1.872	0.060	0.494

Applying one-way ANOVA to these transformed data indicates a sig-nificant difference among groups ($F = 0.4251$ with 3 and 16 degrees of freedom), suggests no evidence that variance is not consistent across treatments.

(b) Using the glht package in R,

```
y <- matrix( c(9.934,9.819,10.693,10.106,9.139,
               8.675,10.720,10.040,9.894,11.912,
               10.509,8.067,9.027,9.680,8.967,
               8.829,10.484,8.632,8.352,9.323),
             ncol=1)
dose <- matrix(c(1,1,1,1,1,2,2,2,2,2,
                 3,3,3,3,3,4,4,4,4,4),ncol=1)
Dose <- factor(dose)
amod <- aov(y~Dose)
fit <- glht(amod,linfct=mcp(Dose="Dunnett"))
confint(fit,level=0.95)
```

leads to simultaneous intervals:

```
            Estimate  lwr      upr
2 - 1 == 0   0.3100  -1.1682  1.7882
3 - 1 == 0  -0.6882  -2.1664  0.7900
4 - 1 == 0  -0.8142  -2.2924  0.6640
```

i.e., none of doses 2, 3, or 4 lead to responses with demonstrably different means than the control dose.

8. For this balanced and orthogonal design, the diagonal elements of **H** are all the same (0.625), so the studentized residuals are each a (common) multiple of the respective ordinary residuals, ranging from a minimum of -1.6212 to a maximum of 2.1098. The latter value corresponds to the response value 24.1 (third row-block, first column-block, treatment 4), which might be rechecked for accuracy. However, given the fact that 16 residuals are

computed, it would not be especially unusual to see at least one of this size.

Chapter 7

2.

(a) The study described isn't a "true" experiment, because the investigator cannot really *apply* a treatment to a unit. The physical material used in the construction of any detector might be regarded as the *unit*, while the arrangement of that material into a type of detector might be regarded as the *treatment*.

(b) The residual sum of squares for a model containing only block effects is 0.5333, while that for a model containing both blocks and treatments is 0.0684. So for these data, an ANOVA decomposition is:

source	d.f.	sum of squares
blocks (chamber sessions)	3	$0.3475 = 0.8808 - 0.5333$
treatments after blocks	3	$0.4649 = 0.5333 - 0.0684$
residuals	5	0.0684
corrected total	11	0.8809

(c) From the ANOVA decomposition in part (b),

$$F = [0.4649/3]/[0.0684/5] = 11.33.$$

The critical value $F_{0.95}(3,5) = 5.41$.

(d) From coefficient estimates of the full model:

$\widehat{\tau_1 - \tau_2}$	$\widehat{\tau_1 - \tau_3}$	$\widehat{\tau_1 - \tau_4}$	$\widehat{\tau_2 - \tau_3}$	$\widehat{\tau_2 - \tau_4}$	$\widehat{\tau_3 - \tau_4}$
0.455	0.156	0.504	-0.299	0.049	0.348

Each has standard error $\sqrt{(c'c \times MSE \times k)/(\lambda \times t)} = \sqrt{(2 \times 0.01368 \times 3)/(2 \times 4)} = 0.1013$.

4.

(a) For the original BIBD, $b = 40$, $k = 2$, $t = 5$, $r = 16$, and $\lambda = 4$, so the design information matrix is:

$$\mathcal{I} = \frac{\lambda t}{k}\left[\mathbf{I} - \frac{1}{t}\mathbf{J}\right] = \frac{4 \times 5}{2}\left[\mathbf{I} - \frac{1}{5}\mathbf{J}\right] = 10\left[\mathbf{I} - \frac{1}{5}\mathbf{J}\right].$$

For the extended complete block design, $k = 7$, so:

$$\mathcal{I} = b\frac{k(k-3) + 2t}{k(t-1)}\left[\mathbf{I} - \frac{1}{t}\mathbf{J}\right] = 40 \times \frac{7 \times 4 + 2 \times 5}{7 \times 4}\left[\mathbf{I} - \frac{1}{5}\mathbf{J}\right]$$

$$= 54.286\left[\mathbf{I} - \frac{1}{5}\mathbf{J}\right].$$

(b) For the original BIBD, the degrees of freedom associated with MSE is $N - b - t + 1 = 80 - 40 - 5 + 1 = 36$. Hence the critical value for the F-test is $F_{0.95}(4, 36) = 2.6335$. For the hypothetical parameter values, the noncentrality parameter is

$$\tau' \mathcal{I} \tau / \sigma^2 = 800/100 = 8.$$

The power is determined by the $F'(4, 36, 8)$ distribution, and is 0.5419. For the extended complete block design, the degrees of freedom associated with MSE is $N - b - t + 1 = 280 - 40 - 5 + 1 = 236$, the critical value is $F_{0.95}(4, 236) = 2.4099$, the noncentrality parameter is 43.43, and the power, based on the $F'(4, 236, 43.43)$ distribution, is nearly 1.

Chapter 8

2.

(a) If we focus on the specific plot of ground selected for each shelter as the most important component of a unit, then it is reasonable to block units by selecting sets of neighboring or spatially contiguous plots. However, this does not address how the "sets" were selected. If they were chosen randomly from the area over which inferences are to be made (or more likely, from an area deemed typical of those over which inferences are to be made), a random-blocks assumption may be appropriate. However, if the block groups are located for convenience along, say, different road-ways, this suggests that there could possibly be systematic differences among them.

(b) This Latin square experiment involves two "crossed" systems of blocks. One set of blocks is the seven "visitor groups" into which the Web site visitors were *randomly* divided; this suggests that *this* collection of blocks may be reasonably regarded as having a random effect on the data. The other system of blocks was associated with the seven links appearing on each Web page. These are clearly not random draws from any meaningful population, since the entire study is predicated on the specific form of the Web site being investigated, and so these blocks should likely not be regarded has having a random effect on the responses.

(c) The 40 patients recruited into the study serve as blocks, and were randomly assigned to treatment protocols *once they were recruited*. But information is not given on how these individuals were recruited in the first place. Were they selected randomly from a larger group of patients typical of those of interest to the investigators, or were they selected by a process (based perhaps on convenience) that might have resulted in, say, the selection of two distinct groups of patients that are systematically different in one or more characteristics that might affect the response? In the first case, an assumption of random blocks might be appropriate, but it likely would not be in the second.

5.

(a) Fitting a full fixed effect model to rows, columns, and treatments yields
    least-squares estimates of differences:

$$\widehat{\tau_1 - \tau_2} \quad 1.1333$$

$$\widehat{\tau_1 - \tau_3} \quad 1.5666$$

$$\widehat{\tau_1 - \tau_4} \quad 5.4333$$

$$\widehat{\tau_1 - \tau_5} \quad -2.4667$$

$$\widehat{\tau_2 - \tau_3} \quad 0.4333$$

$$\widehat{\tau_2 - \tau_4} \quad 4.3000$$

$$\widehat{\tau_2 - \tau_5} \quad -3.6000$$

$$\widehat{\tau_3 - \tau_4} \quad 3.8667$$

$$\widehat{\tau_3 - \tau_5} \quad -4.0333$$

$$\widehat{\tau_4 - \tau_5} \quad -7.9000$$

The estimated standard deviation of each difference based on MSE is
0.78525; the 97.5 quantile of $t(23)$ is 2.068658, yielding a margin of error
of 1.6244.

(b) Note that the expectation of each block total contains the sum of pa-
    rameters associated with column blocks; since these are the same in
    each block, they are confounded with the ovarall mean/intercept for the
    intra-block model, and can be ignored. Fitting the vector of 10 row-block
    means to a model with

$$\mathbf{X} = \begin{pmatrix} 1 & 1 & 1 & 1 & 1 & 0 \\ 1 & 0 & 1 & 1 & 1 & 1 \\ 1 & 1 & 0 & 1 & 1 & 1 \\ 1 & 1 & 1 & 0 & 1 & 1 \\ 1 & 1 & 1 & 1 & 0 & 1 \\ 1 & 1 & 1 & 1 & 1 & 0 \\ 1 & 0 & 1 & 1 & 1 & 1 \\ 1 & 1 & 0 & 1 & 1 & 1 \\ 1 & 1 & 1 & 0 & 1 & 1 \\ 1 & 1 & 1 & 1 & 0 & 1 \end{pmatrix}$$

leads to estimates:

$$\widehat{\tau_1 - \tau_2} \qquad -5.00$$
$$\widehat{\tau_1 - \tau_3} \qquad 32.50$$
$$\widehat{\tau_1 - \tau_4} \qquad 28.50$$
$$\widehat{\tau_1 - \tau_5} \qquad 15.00$$
$$\widehat{\tau_2 - \tau_3} \qquad 37.50$$
$$\widehat{\tau_2 - \tau_4} \qquad 33.50$$
$$\widehat{\tau_2 - \tau_5} \qquad 20.00$$
$$\widehat{\tau_3 - \tau_4} \qquad -4.00$$
$$\widehat{\tau_3 - \tau_5} \qquad -17.50$$
$$\widehat{\tau_4 - \tau_5} \qquad -13.50$$

The estimated standard deviation of each difference based on MSE is 11.5920; the 97.5 quantile of $t(5)$ is 2.570582, yielding a margin of error of 29.80. In this case, even if a random block assumption is reasonable, the intra-block analysis is of very little practical value.

Chapter 9

2.

(a)

rad	2
chem	3
rad×chem	6
residual	24

(b)

blocks	2
rad	2
chem	3
rad×chem	6
residual	22

(c) No. The value of σ^2 is also needed to calculate power.

(d) Yes. The expected squared length of confidence intervals based on the CRD is proportional to

$$t_{0.975}^2(24) \times \sigma^2 = 4.2597\sigma^2,$$

while that based on the CBD is proportional to

$$t_{0.975}^2(22) \times 0.95 \times \sigma^2 = 4.0859\sigma^2.$$

So the CBD supports more precise estimation in this sense.

6.

(a) $\sum_j 8(\bar{y}_{.j.} - \bar{y}_{...})^2 = 8 \times 10 = 80$.

(b) 2.

(c) $\sum_{ijl}(y_{ijl} - \bar{y}_{...})^2 - \sum_{ij} 4(\bar{y}_{ij.} - \bar{y}_{...})^2 = 360 - 4 \times 60 = 120$.

(d) 18.

Chapter 10

1. The remaining sums of squares can be calculated by performing an analysis of variance on the 24 treatment averages. Since each treatment is actually associated with *three* observations (since three "batches" are produced for each temperature), these sums of squares must be multiplied by 3 to put them on a "per observation" basis. For the described arrangement, batches (within temperatures) is the correct denominator component for testing the temperature main effect; the remaining factorial effects are compared to residual variation in the split-plot (within batch) portion of the analysis:

source	d.f.	SS (trt means)	SS (all data)	MS	F
Temp	3	0.19171	0.57513	0.19171	5.112
Batch	8		0.30000	0.03750	
BCon	2	0.53972	1.61916	0.80958	10.794
BFib	1	0.92434	2.77302	2.77302	36.974
Temp×BCon	6	0.08067	0.24201	0.04035	0.538
Temp×BFib	3	0.05958	0.17874	0.05958	0.794
BCon×BFib	2	0.31343	0.94029	0.47013	6.268
Temp×BCon×BFib	6	0.05771	0.17313	0.02886	0.385
Residual	40		3.00000	0.07500	

7.

(a)

source	d.f.
-whole plots-	
A	$l_1 - 1$
W.P. Residual	$r - l_1$
C. Total (Latin square Rep's)	$r - 1$
-split plots-	
Blocks (Latin square Rep's)	$r - 1$
B	$l_2 - 1$
A×B	$(l_1 - 1)(l_2 - 1)$
S.P. Residual	$r(l_2 - 1)^2 - l_1(l_2 - 1)$
C. Total	$rl_2^2 - 1$
Latin Square rows	$r(l_2 - 1)$
Latin Square columns	$r(l_2 - 1)$

(b) If all variation due to Latin Square reps, row-blocks, and column-blocks is ignored in the ANOVA decomposition, the (single) "residual" mean square is likely to be inflated; at least the tests for the B main effect and the A × B interaction would likely be conservative.

Chapter 11

1.

(a) Using R, the factorial effect estimates and MSE can be computed as:

```
data<-matrix(c(
-1,-1,-1,-1,10.7,11.0,
-1,-1,-1, 1, 8.7, 8.0,

        ...

 1, 1, 1,-1,11.2,11.4,
 1, 1, 1, 1,11.8, 8.9),
nrow=16,ncol=6,byrow=T)

X16<-cbind(data[,1],data[,2],data[,3],data[,4],
           data[,1]*data[,2],data[,1]*data[,3],data[,1]*data[,4],
           data[,2]*data[,3],data[,2]*data[,4],data[,3]*data[,4],
           data[,1]*data[,2]*data[,3],
           data[,1]*data[,2]*data[,4],
           data[,1]*data[,3]*data[,4],
           data[,2]*data[,3]*data[,4],
           data[,1]*data[,2]*data[,3]*data[,4])

X32<-rbind(X16,X16)

y32<-matrix(c(data[,5],data[,6]),nrow=32)

theta32<-t(X32)%*%y32/32

                   [,1]
 [1,]   5.187500e-01   A
 [2,]   1.500000e-01   B
 [3,]   2.125000e-01   C
 [4,]  -7.312500e-01   D
 [5,]   7.500000e-02   AB
 [6,]  -1.750000e-01   AC
 [7,]   1.187500e-01   AD
 [8,]   1.875000e-02   BC
 [9,]   2.750000e-01   BD
[10,]   2.220446e-16   CD
[11,]  -2.687500e-01   ABC
[12,]  -5.000000e-02   ABD
[13,]  -1.250000e-02   ACD
[14,]  -6.875000e-02   BCD
[15,]  -6.250000e-03   ABCD
```

```
dif<-matrix((data[,5]-data[,6])/sqrt(2),nrow=16)
MSE<-t(dif)%*%dif/16   ...   0.7000
```

(b) Continuing using the R from part (a), F-statistics and critical values can be calculated as:

```
32*theta32[15,]**2/MSE
[1,] 0.001785714
qf(.95,1,16)
[1] 4.493998

32*(theta32[11,]**2+theta32[12,]**2+theta32[13,]**2+theta32[14,]**2)/MSE
[1] 3.639286
qf(.95,4,16)
[1] 3.006917

32*(theta32[5,]**2+theta32[6,]**2+theta32[7,]**2+
    theta32[8,]**2+theta32[9,]**2+theta32[10,]**2)/MSE
[1,] 5.775
qf(.95,6,16)
[1] 2.741311
```

(c) The noncentrality parameter is $\lambda = 32[(\frac{1}{3}\sigma)^2 + (\frac{1}{3}\sigma)^2]/\sigma^2 = 7.1111$, the critical value of the test is $F_{0.95}(4, 16) = 3.0069$, and the power is the probability that a random variable with distribution $F'(4, 16, 7.1111)$ is larger than this value; that probability is 0.4228.

4.

(a) $\frac{1}{N}\sigma^2 = \frac{1}{128}\sigma^2$.

(b) $E(\widehat{y_{111222}}) = \hat{\mu} - \hat{\alpha} - \hat{\beta} - \hat{\gamma} + \hat{\delta} + \hat{\zeta} + \hat{\eta} + (\widehat{\alpha\beta}) + (\widehat{\alpha\gamma}) - (\widehat{\alpha\delta}) - (\widehat{\alpha\zeta}) - (\widehat{\alpha\eta})$. $Var(E(\widehat{y_{111222}})) = \frac{12}{128}\sigma^2 = 0.09375\sigma^2$. The quantity being estimated is not a contrast in treatment means.

(c) $E(\widehat{y_{111222} - y_{111111}}) = 2\hat{\delta} + 2\hat{\zeta} + 2\hat{\eta} - 2(\widehat{\alpha\delta}) - 2(\widehat{\alpha\zeta}) - 2(\widehat{\alpha\eta})$. $Var(E(\widehat{y_{111222} - y_{111111}})) = \frac{6}{128}4\sigma^2 = 0.1875\sigma^2$.

Chapter 12

1. $t = 8$, $k = 6$, $b = 28$; $r = \frac{bk}{t} = 21$, $\lambda = r\frac{k-1}{t-1} = 15$. Such a design can be constructed by applying each of the 21 distinct combinations of six from the eight treatments to the units within one block:

(1,1,1)	(1,1,2)	(1,2,1)	(1,2,2)	(2,1,1)	(2,1,2)
(1,1,1)	(1,1,2)	(1,2,1)	(1,2,2)	(2,1,1)	(2,2,1)
(1,1,1)	(1,1,2)	(1,2,1)	(1,2,2)	(2,1,1)	(2,2,2)
(1,1,1)	(1,1,2)	(1,2,1)	(1,2,2)	(2,1,2)	(2,2,1)
...	...	...	...	...	...
(2,2,2)	(2,2,2)	(2,2,2)	(2,2,2)	(2,2,2)	(2,2,2)

$$\alpha = \frac{1}{8}[E(y_{211}) + E(y_{212}) + E(y_{221}) + E(y_{222}) - E(y_{111}) - E(y_{112})$$
$$-E(y_{121}) - E(y_{122})]$$

or in unstructured treatment notation, $c'\tau$ where:

$$c' = \left(+\frac{1}{8}, +\frac{1}{8}, +\frac{1}{8}, +\frac{1}{8}, -\frac{1}{8}, -\frac{1}{8}, -\frac{1}{8}, -\frac{1}{8}\right).$$

So $Var(\hat{\alpha}) = \frac{k}{\lambda t}c'c\sigma^2 = \frac{6}{15\times 8}\frac{1}{8}\sigma^2 = 0.00625\sigma^2$.

7.

(a) $I = \pm ACD = \pm BDE \ (= \pm ABCE)$.

(b)

A	B	C	D	E
+	+	+	+	−
+	−	+	+	+
+	+	−	−	+
+	−	−	−	−
−	+	+	−	+
−	−	+	−	−
−	+	−	+	−
−	−	−	+	+

A	B	C	D	E
+	+	+	−	−
+	−	+	−	+
+	+	−	+	+
+	−	−	+	−
−	+	+	+	+
−	−	+	+	−
−	+	−	−	−
−	−	−	−	+

A	B	C	D	E
+	+	+	−	+
+	−	+	−	−
+	+	−	+	−
+	−	−	+	+
−	+	+	+	−
−	−	+	+	+
−	+	−	−	+
−	−	−	−	−

Chapter 13

1. The highest-order effect used in this generating relation involves four factors, so the highest-resolution half-fraction that can be obtained in two doublings is a 2^{6-1}_{III}. The 2^{6-3}_{II} specified by:

$$I \ = +AB$$
$$= +CDE(= +ABCDE)$$
$$= -ADF(= -BDF = -ACEF = -BCEF)$$

can be doubled once to obtain the 2^{6-2}_{III}:

$$I = +ABCDE$$
$$= -ADF(= -BCEF)$$

and doubled a second time to obtain the 2^{6-1}_{V}:

$$I = +ABCDE.$$

However, the main effects for the first two factors are not individually estimable until after the first doubling.

4. Denote the $N \times f$ design matrix for the initial Plackett-Burman plan by $\mathbf{D}$, and the $N \times (1 + f)$ and $N \times \binom{f}{2}$ model matrices associated with the intercept and main effects, and two-factor interactions, by:

$$\mathbf{X}_1 = (\mathbf{1}|\mathbf{D}) \quad \text{and} \quad \mathbf{X}_2$$

respectively. Then the corresponding $2N$-row model matrices for the fold-over design are:

$$\mathbf{X}_1^{fo} = \left(\begin{array}{c|c} \mathbf{1} & \mathbf{D} \\ \hline \mathbf{1} & -\mathbf{D} \end{array}\right) \quad \text{and} \quad \mathbf{X}_2^{fo} = \left(\begin{array}{c} \mathbf{X}_2 \\ \hline \mathbf{X}_2 \end{array}\right).$$

The alias matrix for the main effects model is then:

$$\left(\mathbf{X}_1^{fo'}\mathbf{X}_1^{fo}\right)^{-1}\mathbf{X}_1^{fo'}\mathbf{X}_2^{fo} = \frac{1}{2N}\mathbf{I}\left(\begin{array}{c} \mathbf{2}\mathbf{1}'\mathbf{X}_2 \\ \hline \mathbf{0} \end{array}\right) = \frac{1}{N}\left(\begin{array}{c} \mathbf{1}'\mathbf{X}_2 \\ \hline \mathbf{0} \end{array}\right).$$

So, while the intercept may be aliased with two-factor interactions, main effects are not.

Chapter 14

3.

(a) Critical value for F is $F_{.90}(1, 13) = 3.1362$, the relevant noncentrality parameter is $\lambda = \frac{24 \times 1^2}{2^2} = 6$, so the power of the test is $Prob(W > 3.1362)$ where $W \sim F'(1, 13, 6)$, or 0.7499.

(b) $4 \times [0.7499 + 9 \times 0.10] = 6.6$.

4.

(a) Because the experiment is unreplicated, the 11 group main effect estimates will be assessed via a normal or half-normal plot. But since the group main effects are normally distributed with mean 0 and variance approximately $4\sigma_E^2$, the plot pattern will likely be linear and no factor groups will be identified for second-stage follow-up.

(b) With replication, σ^2 can be estimated with 12 degrees of freedom in an analysis of variance. Since most group main effects (with variance $4\sigma_E^2$) will likely be substantially larger than σ^2, most factor groups will likely be identified for second-stage follow-up.

Chapter 15

2. The standardized and studentized residuals for the 11 data values as labeled in Table 15.1, based on a first-order model, are:

	residuals	
run	standardized	studentized
1	−0.322	−0.806
2	−0.322	−0.806
3	−0.395	−0.990
4	−0.395	−0.990
5	−0.322	−0.806
6	−0.395	−0.990
7	−0.395	−0.990
8	−0.322	−0.806
9	0.639	0.671
10	0.918	0.963
11	1.308	1.372

Although the response at run 7 appears most out of line with the rest of the data, run 11 results in the largest scaled residual. Its magnitude is not so large as to be strong evidence that the data value should be ignored. However, there is clearly inadequacy in the first-order model being used here; for example, all residuals at the center point runs are positive, while all residuals at the factorial points are negative. While the model we are using in this diagnostic is questionable, the data set is too small to support estimation of a complete quadratic model. Hence it is difficult to argue, on the basis of these data alone, that any of the response values should be regarded as outliers.

7.

(a) Design A: $N - 6 = 36 - 6 = 30$. Design B: $N - 6 = 20 - 6 = 14$. Design C: $N - 6 = 12 - 6 = 6$.

(b) Design A: $\mathcal{I} = 32\mathbf{I}$, $\lambda = \frac{32 \times (1^2 + 1^2 + 1^2 + 0^2 + 0^2)}{3^2} = 10.67$. Design B: $\mathcal{I} = 16\mathbf{I}$, $\lambda = \frac{16 \times (1^2 + 1^2 + 1^2 + 0^2 + 0^2)}{3^2} = 5.33$. Design C: $\mathcal{I} = 8\mathbf{I}$, $\lambda = \frac{8 \times (1^2 + 1^2 + 1^2 + 0^2 + 0^2)}{3^2} = 2.67$.

(c) Design A: $Prob[W > F_{.95}(5, 30)] = 0.619$ for $W \sim F'(5, 30, 10.67)$. Design B: $Prob[W > F_{.95}(5, 14)] = 0.272$ for $W \sim F'(5, 14, 5.33)$. Design C: $Prob[W > F_{.95}(5, 6)] = 0.115$ for $W \sim F'(5, 6, 2.67)$.

Chapter 16

3.

(a) For this design, $N = 11$, $a_{PQ} = (4 + 2 \times 1.5^2)/11 = 0.7727$, and the information matrix is:

$$\mathcal{I} = \begin{pmatrix} 8.5\mathbf{I}_{2\times2} & 0 & 0 \\ 0 & 10.125\mathbf{I}_{2\times2} - 2.568\mathbf{J}_{2\times2} & 0 \\ 0 & 0 & 4\mathbf{I}_{1\times1} \end{pmatrix}.$$

With $\boldsymbol{\beta}' = (2, 2, -1, -1, 0)$, the noncentrality parameter for the test of model effectiveness is:

$$\boldsymbol{\beta}'\mathcal{I}\boldsymbol{\beta}/\sigma^2 = 8.664.$$

(b) There are nine distinct experimental runs (e.g., ignoring replicates) in the experiment and the model contains six parameters. Further, there are two degrees of freedom for "pure error" based on the three replicate center point runs. So the test for lack of fit is based on $F(3, 2)$.

8.

design	$Var(\hat{\beta}_1)$	$Var(\hat{\beta}_{11})$	$Var(\hat{\beta}_{12})$
1	$0.0833\sigma^2$	$0.2500\sigma^2$	$0.1250\sigma^2$
2	$0.0714\sigma^2$	$0.3684\sigma^2$	$0.0833\sigma^2$
3	$0.1000\sigma^2$	$0.2564\sigma^2$	$0.2500\sigma^2$
4	$0.1667\sigma^2$	$0.3333\sigma^2$	$0.2500\sigma^2$

Chapter 17

1. Let $D = \{(x_{11}, x_{21}), (x_{12}, x_{22}), (x_{13}, x_{23}), (x_{14}, x_{24})\}$, where each $-1 \le x_{ij} \le +1$. As noted in the problem, the information matrix

$$\mathcal{I} = \mathbf{X}_2' \left(\mathbf{I} - \frac{1}{4}\mathbf{J}\right) \mathbf{X}_2$$

can be written as

$$\mathcal{I} = \begin{pmatrix} \sum_i (x_{1i} - \bar{x}_{1\cdot})^2 & \sum_i (x_{1i} - \bar{x}_{1\cdot})(x_{2i} - \bar{x}_{2\cdot}) \\ \sum_i (x_{1i} - \bar{x}_{1\cdot})(x_{2i} - \bar{x}_{2\cdot}) & \sum_i (x_{2i} - \bar{x}_{2\cdot})^2 \end{pmatrix} = 3 \begin{pmatrix} S_1^2 & S_{12} \\ S_{12} & S_2^2 \end{pmatrix}$$

where S_1^2, S_2^2, and S_{12} denote the functional form of sample "variances" and the "covariance" of selected x_1 and x_2 values, respectively. Hence

$$|\mathcal{I}| = 9(S_1^2 S_2^2 - S_{12}^2)$$

is maximized if D is selected so that:

- S_1^2 is as large as is possible,
- S_2^2 is as large as is possible,
- S_{12} is as near zero as possible.

This is achieved by $\{(-1,-1),(-1,+1),(+1,-1),(+1,+1)\}$ because

- S_1^2 is 1, the largest "variance" that can be achieved for any distribution of range 2,
- S_2^2 is 1, and
- S_{12} is 0.

5.

(a) $\phi_M(D) = ||(\mathbf{X}_2'\mathbf{X}_2)^{-1}\mathbf{X}_2'\mathbf{X}_3|| = \frac{1}{N^2}||\mathbf{X}_2'\mathbf{X}_3|| = \frac{1}{N^2}\text{trace}(\mathbf{X}_3'\mathbf{X}_2\mathbf{X}_2'\mathbf{X}_3)$.

(b) If N is large enough to accommodate a Resolution IV (or more) fraction, this is optimal because $\mathbf{X}_2'\mathbf{X}_3 = \mathbf{0}$ for such designs. If N is smaller, the non-zero elements of $\mathbf{X}_2'\mathbf{X}_3$ all have absolute value N, and the number of them is 3 times the number of words of length 3 in the generating relation; therefore the optimal designs are minimum aberration designs of Resolution III.

References

Addelman, S. (1961). "Irregular Fractions of the 2^n Factorial Experiments," *Technometrics* **3**, 479–496.

Atkinson, A.C., and A.N. Donev (1992). *Optimal Experimental Designs*, Clarendon Press, Oxford.

Barton, R.R. (1999). *Graphical Methods for the Design of Experiments* (Lecture Notes in Statistics), Springer, New York.

Beyer, W.H., editor (1968). *CRC Handbook of Tables for Probability and Statistics*, 2nd edition. The Chemical Rubber Company, Cleveland, Ohio.

Bingham, D., and R.R. Sitter (2001). "Design Issues in Fractional Factorial Split-Plot Experiments," *Journal of Quality Techology* **33**, 2–15.

Box, G.E.P. (1952). "Multifactor Designs of First Order," *Biometrika* **39**, 49–57.

Box, G.E.P., and D.W. Behnken (1960). "Some New 3 Level Designs for the Study of Quantitative Variables," *Technometrics* **2**, 455–475.

Box, G.E.P., and D.R. Cox (1964). "An Analysis of Transformations" (with discussion), *Journal of the Royal Statistical Society, Series B* **26**, 211–246.

Box, G.E.P., and J.S. Hunter (1961). "The 2^{k-p} Fractional Factorial Designs," *Technometrics* **3**, 311–351.

Box, G.E.P., J.S. Hunter, and W.G. Hunter (2005). *Statistics for Experimenters: Design, Innovation, and Discovey*, 2nd edition, Wiley Interscience, Hoboken, NJ.

Box, G.E.P., and D. Meyer (1986). "An Analysis for Unreplicated Fractional Factorials," *Technometrics* **28**, 11–18.

Box, G.E.P., and K.B. Wilson (1951). "On the Experimental Attainment of Optimum Conditions," *Journal of the Royal Statistical Society, Series B* **13**, 1–45.

Chen, Q.-h., H. Ruan, H.-f. Zhang, H. Ni, and G.-q. He (2007). "Enhanced Production of Elastase by *Bacillus licheniformis* ZJUEL31410: Optimization of Cultivation Conditions Using Response Surface Methodology," *Journal of Zhejiang University Science* **8**, 845–852.

Christensen, R. (2002). *Plane Answers to Complex Questions: The Theory of Linear Models*, 3rd edition. Springer, New York.

Clatworthy, W.H. (1973). "Tables of Two-Associate-Class Partially Balanced Designs," *NBS Applied Mathematics Series*, 63.

Cobb, G.W. (1999). *Introduction to Design and Analysis of Experiments*, Key College Publishing, Emeryville, CA.

Colburn, C.J., and J.H. Dinitz (eds.) (1996). *The CRC Handbook of Combinatorial Designs*, CRC Press, Boca Raton, FL.

Coleman, D.E., and D.C. Montgomery (1993). "A Systematic Approach to Planning for a Designed Industrial Experiment," *Technometrics* **35**, 1–12.

Conover, W.J., M.E. Johnson, and M.M. Johnson (1981). "A Comparative Study of Tests for Homogeneity of Variances, with Application to the Outer Shelf Bidding Data," *Technometrics* **23**, 351–361.

Cox, D.R., and N. Reid (2000). *The Theory of the Design of Experiments*, Chapman and Hall, London.

Daniel, C. (1959). "Use of Half-Normal Plots in Interpreting Factorial Two Level Experiments," *Technometrics* **1**, 311–341.

Dayal, P., V. Pillay, J. Babu, and M. Singh (2005). "Box-Behnken Experimental Design in the Development of a Nasal Drug Delivery System of Model Drug Hydroxyurea: Characterization of Viscosity, In Vitro Drug Release, Droplet Size, and Dynamic Surface Tension," *AAPS PharmSciTech* **6**, E573–E586.

Dean, A.M., and D. Voss (1999). *Design and Analysis of Experiments*, Springer, New York.

Dedidenko, E. (2002). *Mixed Models: Theory and Applications*, John Wiley and Sons, Hoboken, N.J.

Denton, C.A., J.L. Anthony, R. Parker, and J.E. Hasbrouck (2004). "Effects of Two Tutoring Programs on the English Reading Development of Spanish-English Bilingual Students," *The Elementary School Journal* **104**, 289–305.

Dorfman, R. (1943). "The Detection of Defective members of Large Populations," *Annals of Mathematical Statistics* **14**, 436–440.

Dunnett, C.W. (1964). "New Tables for Multiple Comparisons with a Control," *Biometrics* **20**, 482–491.

Edwards, C.H. (1994). "Ladders, Moats, and Lagrange Multipliers," *The Mathematica Journal* 4, http://www.mathematica-journal.com/issue/v4i1/

Edwards, D., and J.J. Berry (1987). "The Efficiency of Simulation-Based Multiple Comparisons," *Biometrics* **43**, 913–928.

Fay, P.A., J.D. Carlisle, A.K. Knapp, J.M. Blaire, and S.L. Collins (2000). "Altering Rainfall Timing and Quantity in a Mesic Grassland Ecosystem: Design and Performance of Rainfall Manipulation Shelters," *Ecosystems* **3**: 308–319.

Federer, W.T. (1955). *Experimental Design – Theory and Application*, Macmillan, New York.

Federer, W.T. (1961). "Augmented Designs with One-Way Elimination of Heterogeneity," *Biometrics* **17**, 447–473.

Fedorov, V.V. (1972). *Theory of Optimal Experiments*, Academic Press, New York.

Fisher, R.A. (1926). "The Arrangement of Field Experiments," *Journal of the Ministry of Agriculture (England)* **33**, 503–513.

Fisher, R.A. (1971). *The Design of Experiments*, 9th ed., Hafner Press, London. (The first edition of this classic reference was published in 1935.)

Fries, A., and W.G. Hunter (1980). "Minimum Aberration 2^{k-p} Designs," *Technometrics* **22**, 601–608.

Frigge, M., D.C. Hoaglin, and B. Iglewicz (1989). "Some Implementations of the Boxplot," *The American Statistician* **43**, 50–54.

Gentle, J.E. (2007). *Matrix Algebra: Theory, Computations, and Applications in Statistics*. Springer, New York.

Graybill, F.A. (1983). *Matrices with Applications in Statistics*, Wadsworth Group, Belmont, CA.

Hamada, M., and C.F.J. Wu (1992). "Analysis of Designed Experiments with Complex Aliasing," *Journal of Quality Technology* **24**, 130–137.

Harville, D.A. (2008). *Matrix Algebra from a Statistician's Perspective*, Springer, New York.

Hedayat, A., J. Stufken, and N.J.A. Sloane (1999). *Orthogonal Arrays: Theory and Applications*, Springer, New York.

Hendrix, C.D. (1979). "What Every Technologist Should Know about Experimental Design," *Chemtech* **9**, 1967–1974.

Hinkelmann, K., and O. Kempthorne (2005) *Design and Analysis of Experiments: Advanced Experimental Design*, John Wiley & Sons, Hoboken, N.J.

Hocking, R.R. (2003). *Methods and Applications of Linear Models: Regression and the Analysis of Variance*, 2nd edition, Wiley Inter-Science, New York.

Hunter, J.S. (1989). "Let's All Beware the Latin Square," *Quality Engineering* **1**, 453–465.

Ivanova, T., L. Malone, and M. Mollaghasemi (1999). "Comparison of a Two-Stage Group-Screening Design to a Standard 2^{k-p} Design for a Whole-Line Semiconductor Manufacturing Simulation Model," In: Farrington, P.A., Nembhard, H.B., Sturrock, D.T., and Evans, G.W. (eds.), *Proceedings of the 1999 Winter Simulation Conference*.

Jennrich, R.I., and R.H. Moore (1975). "Maximum Likelihood Estimation by Means of Nonlinear Least Squares," in *Proceedings of the Statistical Computing Section*, 57–65. American Statistical Association, Washington, DC.

Jepsen, P.K., E. Riise, K. Biedermann, P.C.R. Kristensen, and C. Emborg (1987). "Two-Level Factorial Screening for Influence of Temperature, pH, and Aeration on Production of *Serratia marcescens* Nuclease," *Applied and Environmental Microbiology* **53**, 2593–2596.

John, J.A., and E.R. Williams (1995). *Cyclic and Computer Generated Designs*, 2nd edition, Chapman and Hall, London.

John, P.W.M. (1963). "Extended Complete Block Designs," *Australian Journal of Statistics* **5**, 147–152.

John, P.W.M. (1966). "Augmenting 2n-1 Designs," *Technometrics* **8**, 469–480.

John, P.W.M. (1961). "An Application of a Balanced Incomplete Block Design," *Technometrics* **3**, 51–54.

John, P.W.M. (1998). *Statistical Design and Analysis of Experiments*, Society for Applied and Industrial Mathematics.

Kackar, R.N. (1985). "Off-Line Quality Control, Parameter Design, and the Taguchi Method," *Journal of Quality Technology* **17**, 176–188.

Kiefer, J., and J. Wolfowitz (1960). "The Equivalence of Two Extremum Problems," *Canadian Journal of Mathematics* **12**, 363–366.

Kocaoz, S., V.A. Samaranayake, and A. Nanni (2005). "Tensile Characterization of Glass FRP Bars," *Composites: Part B* **36**, 127–134.

Kraiczi, H., J. Hedner, Y. Peker, and L. Grote (2000). "Comparison of Atenolol, Amlodipine, Enalapril, Hydrochlorothiazide, and Losartan for Antihypertensive Treatment in Patients with Obstructive Sleep Apnea," *American Journal of Respiratory and Critical Care Medicine* **161**, 1423–1428.

Leal-Sanchez, M., R. Jimenez-Diaz, A. Maldonado-Barragan, A. Garrido-Fernandez, and J. Ruiz-Barba (2002). "Optimization of Bacteriocin Production by Batch Fermentation of *Lactobacillus plantarum LPCO10*," *Applied and Environmental Microbiology* **68**, 4465–4471.

Lenth, R.V. (1989). "Quick and Easy Analysis of Unreplicated Factorials," *Technometrics* **31**, 469–473.

Levene, H. (1960). "Robust Tests for Equality of Variances," in *Contributions to Probability and Statistics*, Olkin, I. (ed.) Stanford University Press, Palo Alto, CA.

Lewis, S.M., and A.M. Dean (2001). "Detection of Interaction in Experiments on Large Numbers of Factors," *Journal of the Royal Statistical Society B* **63**, 633–672.

Loeppky, J.L., and R.R. Sitter (2002). "Analyzing Unreplicated Blocked or Split-Plot Fractional Factorial Designs," *Journal of Quality Technology* **34**, 229–243.

Matsuu, M., K. Shichijo, Y. Ikeda, M. Ito, S. Naito, K. Okaichi, M. Nakashima, T. Nakayama, and I. Sekine (2005). "Sympathetic Hyperfunction Causes Increased Sensitivity of the Autonomic Nervous System to Whole-Body X Irradiation," *Radiation Research* **163**, 137–143.

Mauro, C.A., and D.E. Smith (1982). "The Performance of Two-Stage Group Screening in Factor Screening Experiments," *Technometrics* **24**, 325–330.

Mee, R.W., and M. Peralta (2000). "Semifolding $2k\text{-}p$ Designs," *Technometrics* **42**, 122–134.

Mee, R. (2009). *A Comprehensive Guide to Factorial Two-Level Experimentation*, Springer, New York.

Miller, R.G. Jr. (1981). *Simultaneous Statistical Inference*, 2nd edition, Springer, New York.

Milliken, G.A., and D.E. Johnson (2002). *Analysis of Messy Data, Volume III: Analysis of Covariance*, Chapman and Hall/CRC, Boca Raton, FL.

Mitchell, T.J. (1972). "An Algorithm for the Construction of 'D-Optimum' Experimental Designs," *Technometrics* **16**, 203–210.

Montgomery, D.C., and G.C. Runger (1996). "Foldovers of $2k\text{-}p$ Resolution IV Experimental Designs," *Journal of Quality Technology* **28**, 446–450.

Morris, M.D. (2000). "A Class of Three-Level Experimental Designs for Response Surface Modeling," *Technometrics* **42**, 111–121.

Morris, M.D. (2006). "An Overview of Group Factor Screening," In: Dean, A.M., and Lewis, S. (eds.), *Screening: Methods of Experimentation in Industry, Drug Discovery, and Genetics*, Springer, ISBN 0387280138.

Murphy, J., C. Hofacker, and R. Mizerski (2006). "Primacy and Recency Effects on Clicking Behavior," *Journal of Computer-Mediated Communications* **11**, article 7.

Myers, R.H., and D.C. Montgomery (2002). *Response Surface Methodology: Process and Product Optimization Using Designed Experiments*, 2nd edition, John Wiley and Sons, New York.

Patel, M.S. (1962) "Group-Screening with More than Two Stages," *Technometrics* **4** 209–217.

Plackett, R.L., and J.P. Burman (1946). "The Design of Optimum Multifactorial Experiments," *Biometrika* **33**, 305–325.

Preece, D.A. (1990). "R. A. Fisher and Experimental Design: A Review," *Biometrics* **46**, 925–935.

Prestwich, S. (2003). "A Local Search Algorithm for Balanced Incomplete Block Designs," in Recent Advances in Constraints: Joint ERCIM/CologNet International Workshop on Constraint Solving and Constraint Logic Programming, Cork, Ireland, June 19–21, 2002, Springer.

Ravishanker, N., and D. Dey (2002). *A First Course in Linear Model Theory*, Chapman and Hall/CRC, Boca Raton, FL.

Rechtschaffner, R.L. (1967). "Saturated Fractions of 2^n and 3^n Factorial Designs," *Technometrics* **9**, 569–575.

Rong, H., R.V. Leon, and G.S. Bhat (2005). "Statistical Analysis of the Effect of Processing Conditions on the Strength of Thermal Point-Bonded Cotton-Based Nonwovens," *Textile Research Journal* **75**, 35–38.

Scheffé, H. (1953). "A Method for Judging All Contrasts in the Analysis of Variance," *Biometrika* **40**, 87–104.

Searle, S.R. (1971). *Linear Models*. Wiley, New York.

Silvey, S.D. (1980). *Optimum Design*, Chapman & Hall, London.

Soudki, K.A., E.F. El-Salakawy, and N.B. Elkum (2001). "Full Factorial Optimization of Concrete Mix Design for Hot Climates," *Journal of Materials in Civil Engineering* **13**, 427–433.

Trail, S.M., and D.L. Weeks (1973). "Extended Complete Block Design Generated by BIBD," *Biometrics* **29**, 565–578.

Tukey, J. (1949). "One Degree of Freedom for Non-Additivity," *Biometrics* **5**, 232–242.

Tukey, J. (1953). "The Problem of Multiple Comparisons," Unpublished Notes, Princeton University.

Tukey, J.W. (1977). *Exploratory Data Analysis*, Addison-Wesley, Reading, MA.

Watson, G.S. (1961). "A Study of the Group Screening Method," *Technometrics* **3**, 371–388.

Werner, S.F., D.L. Nolte, and F.D. Provenza (2005). "Proximal Cues of Pocket Gopher Burrow Plugging Behavior: Influence of Light, Burrow Openings, and Temperature," *Physiology and Behavior* **85**, 340–345.

Wildt, A.R., and O.T. Ahtola (1978). *Analysis of Covariance: Quantitative Applications in the Social Sciences*, Sage Publications, London.

Wu, C.F.J., and M. Hamada (2009). *Experiments: Planning, Analysis, and Optimization*, Wiley and Sons, New York.

Yates, F. (1940). "The Recovery of Inter-Block Information in Balanced Incomplete Block Designs," *The Annals of Eugenics* **10**, 317–325.

Youden, W.J. (1940). "Experimental Designs to Increase Accuracy of Green-House Studies," *Contributions of the Boyce Thompson Institute* **11**, 219–228.

Youden, W.J. (1972). "Enduring Values," *Technometrics* **14**, 1–11.

Index

Printed in the United States
by Baker & Taylor Publisher Services

Printed in the United States
by Baker & Taylor Publisher Services